TURBULENCE IN OPEN CHANNELS AND RIVER FLOWS

This book is a monograph on the open-channel turbulence. The former part covers the basic characteristics, including derivation of governing equations. The latter part focuses on various topics related to environmental management and disaster problems in river turbulence.

The turbulent flow is a momentary phenomenon, but there are also organized motions, not just random motions. How to feed back this fundamental knowledge to the practice of hydraulic engineering including mass and sediment transport is a big issue. This book explains in detail the basic points and various specific problems of river turbulence in order for abecedarians to easily learn for themselves. By including application examples, the understanding of readers is deepened. In addition, it also outlines the methods of hydraulics experiment and turbulence measurements. So, it will be very useful for readers themselves to measure turbulence with their own hands in the future.

It is intended for a wide range of readers, from students studying hydraulics to graduate students who want to know more about the essence of phenomena, and engineers using it as a reference for hydraulic works and practice.

Michio Sanjo was awarded a PhD in 2003 from Kyoto University under Prof. Iehisa Nezu. He has been working on hydraulics as an associate professor at Kyoto University for several years. He specializes in experimental investigations of turbulence phenomena in open-channel flows, such as the secondary currents, horizontal vortex, free-surface divergence, gas transfer and influence of bottom roughness. Further, he conducted field work with the development of the autonomous robot boat for automatic velocity measurement in river.

TURBULENCE IN OPEN CHANNELS AND RIVER FLOWS

Michio Sanjou

CRC Press
Taylor & Francis Group
Boca Raton London New York

CRC Press is an imprint of the
Taylor & Francis Group, an **informa** business

First edition published 2023
by CRC Press
6000 Broken Sound Parkway NW, Suite 300, Boca Raton, FL 33487-2742

and by CRC Press
4 Park Square, Milton Park, Abingdon, Oxon, OX14 4RN

CRC Press is an imprint of Taylor & Francis Group, LLC

Library of Congress Cataloging-in-Publication Data
Names: Sanjou, Michio, author.
Title: Turbulence in open channel and river flows / Michio Sanjou.
Description: First edition. | Boca Raton, FL : CRC Press, 2022. | Includes bibliographical references and index. |
Summary: "This book is a monograph on the open-channel turbulence. Former part covers the basic characteristics including derivation of governing equations. The latter part focuses on various topics related to environmental management and disaster problem in river turbulence"-- Provided by publisher.
Identifiers: LCCN 2022001845 (print) | LCCN 2022001846 (ebook) | ISBN 9780367630966 (hbk) | ISBN 9780367631086 (pbk) | ISBN 9781003112198 (ebk)
Subjects: LCSH: Open-channel flow. | Streamflow. | Turbulence.
Classification: LCC TC175 .S326 2022 (print) | LCC TC175 (ebook) | DDC 627/.12--dc23/eng/20220405
LC record available at https://lccn.loc.gov/2022001845
LC ebook record available at https://lccn.loc.gov/2022001846

ISBN: 978-0-367-63096-6 (hbk)
ISBN: 978-0-367-63108-6 (pbk)
ISBN: 978-1-003-11219-8 (ebk)

DOI: 10.1201/9781003112198

Typeset in Sabon
by MPS Limited, Dehradun

Contents

Preface vii

1 **Introduction** 1

2 **Statistical Methods** 5

3 **2-D Open-Channel Turbulence** 29

4 **Horizontal Open-Channel Turbulence** 93

5 **3-D Turbulence Structure** 111

6 **Turbulence Transport** 129

7 **Applications to River Flow** 141

8 **Measurements in Open-Channel Turbulence** 223

Appendix 1: Reynolds Decomposition 255
Appendix 2: Calculation Procedure of Blasius Solution 261
Appendix 3: Mathematical Derivation of Equation in the K-H Instability 269
References 275
Index 289

Preface

Flow resistance, energy head loss or mass transport in the open channel seen in rivers, sewers or flood bypasses, cannot be talked about without turbulence. This book is a monograph on the open-channel turbulent flow, which reflects the questions and requests the author received from many undergraduate and graduate students. The purpose of the reader who picks up this book may vary. It is intended for a wide range of readers, from students studying hydraulics to graduate students who want to know more about the essence of phenomena, and engineers using it as a reference for hydraulic works and practice.

In the former part, the basic characteristics of the open-channel turbulence are explained and, in the latter part, various topics found in rivers are introduced, including issues related to environmental management and disaster problems. However, it is difficult to understand the turbulence phenomenon from the theoretical framework alone. Therefore, the final chapter outlines the methods of hydraulics experiments and turbulence measurements.

It is essential to know the basic flow equations when reading specialized literature. In Chapter 2, the fundamental characteristics of turbulence, average processing and deviations of Navier Stoke equations and RANS equations are necessary in turbulence hydraulics, and are explained using figures. Further, the concept of viscosity, Euler and Lagrange methods and shear and normal stresses are described in detail.

The two-dimensional open-channel flow is a kind of turbulent boundary layer. The flow velocity decreases at the bottom due to friction, and the gradient in the depth direction of streamwise means velocity increases. As a result, the generation of turbulence becomes active near the bottom surface. Chapter 3 explains the relationship of the development of boundary layer thickness and bottom shear stress, as well as the boundary layer theory. In addition, the mechanism of turbulence generation is explained by stability analysis. It derives the two-dimensional momentum equations in open-channel turbulent flow, including the modeling of Reynolds stress. From these, the flow velocity profile formula is obtained, and the mean velocity profile in each layer in the depth direction is considered. Also, this chapter discusses the distribution of turbulence statistics such as turbulent kinetic energy and their production, dissipation and transport. Spectral and frequency analysis will also be taken up to explain cascade process and Kolmogorov's -5/3 law. Finally, it describes the basic concept of bursting, which is the organized structure of turbulent flow, and illustrates low-speed streak and hairpin vortex.

Chapter 4 deals with the open-channel flows, in which horizontal phenomena predominate. The compound open-channel flow, the circulation in dead water zones and the wakes behind piers are examples. First, the basic theory of the horizontal velocity shear was taken up, and the relationship between the frequency of small fluctuation and development rate are shown. Next, it focuses on a large-scale horizontal vortex and derives the K-H instability theory that can support its occurrence. Also, it introduces specific examples of the flows in rivers where horizontal vortex is generated. In addition, it explains the surface velocity divergence related to heat and gas transports between the water layer and the air layer. Finally, a shallow water momentum equation is derived mathematically that describes the fluid motion where the horizontal scale is predominant compared to the vertical direction.

Chapter 5 deals with the turbulence of open-channel flow with strong three dimensionality. The characteristics of second-kind secondary current, which is said to be caused by the anisotropy of turbulence, is also described, in addition to the first-kind secondary currents due to the centrifugal force. Regarding the former, it also introduces another idea that is attributed to the transverse gradient of the streamwise velocity. Further, the characteristics of VLSM, which is a three-dimensional structure, is described. The VLSM is the result of interactions between the streaks and the coherent vortices. In addition, it explains the secondary currents and horizontal vortex generated in the curved flow or meandering flow.

Chapter 6 explains turbulence transport, turbulence diffusion and derivation of basic equations to particle analysis and dispersion. In particular, it introduces Taylor's theory, which is one particle analysis, and Richardson's theory, which is two-particle analysis, and explains the concept of temporal and spatial scales of turbulent diffusion.

Chapter 7 describes the applications to river turbulence. it explains each feature by focusing on some flows with reference to previous studies. Since any topic will be further researched in the future, I would like the readers themselves to check the final research and try to understand the content and fill in the missing parts.

Readers who specialize in computer simulation may also use experimental data for accuracy verification. In that case, it is also necessary to access or evaluate the quality of the experimental data. Chapter 8 summarizes the principles and features of velocimetry used in the flume experiments and field survey. The readers who specialize in hydraulic experiments will use this as a reference when selecting the most suitable device for the flow that they are targeting. In addition, it introduces the turbulence sampling method, detection technique for coherent turbulence and the specific experimental setup, so this chapter will help when constructing a new channel system. I would like you to deepen your understanding by actually moving your hands and measuring and observing turbulence, not just in the framework of theory.

I hope that this book will help students, hydraulic engineers and scientists who will lead the future.

The author wishes to thank Prof. Iehisa Nezu, who enthusiastically guided my study and raised me as a researcher. He taught me the charm of the study on the open-channel turbulence, and I was greatly inspired by his great achievements. I also thank Prof. Akihiro Tominaga for providing interesting topics about river turbulence, Prof. Koki Onitsuka for advising various methods of experiments and Dr. Takaaki Okamoto for supporting my research works. Further, I would also like to expresses my appreciation to Prof. Tetsuya Sumi for financial support and advice for practical topics in hydraulics, and to Prof. Yuji Sugihara for valuable discussion on turbulent transport related to environment issues as well as financial support.

Also, I am grateful to my family for their encouragement during this project over the years.

Michio Sanjou
Katano Osaka

Chapter 1

Introduction

Open-channel flows are unidirectional streams with free surface found in rivers, canals, spillways and raceway tanks, etc. In hydraulics, a one-dimensional analysis for calculating the streamwise variations of bulk-mean velocity and water depth has been established, and is widely used in practice. In addition, two-dimensional analysis by shallow water momentum equations that average the hydraulic values with the depth direction are often used to predict the meandering flow, inundation over the bank and space-time changes of channel morphology. Such macroscopic handlings are excellent in practice, but the effects of turbulence are pushed into model parameters, so microscopic consideration is limited. If we do not understand the essence of turbulence hidden in modeling and perform the analysis, we may not be able to interpret the output results of the experiments and calculations correctly.

Therefore, it is necessary to learn the basics of turbulent dynamics, especially in the open-channel flows. In fact, river flows have various turbulent phenomena. It is no exaggeration to say that the practical topics related to river management and design are largely dominated by turbulence as follows:

1. Variation mechanism in water flow resistance due to aquatic vegetation
2. Influences of turbulence on flow capacity in compound open-channel
3. Effect of bottom roughness on flow resistance
4. Interaction between sand wave formation and turbulence
5. Mass/momentum exchanges between mainstream and floodplain
6. Mechanism of material retention in dead water zone
7. Transport of sediment and driftwood in river flow
8. Physical modeling of turbulent density current
9. Numerical prediction of debris flows considering turbulence
10. Evaluation of fluid force due to high-speed flow in flood bypass of dam
11. Three-dimensional flow structure in meandering and bending flows

DOI: 10.1201/9781003112198-1

12. Secondary current generation and transport mechanism
13. Gas and heat transfers through the free surface
14. Energy dissipation due to turbulence generation

Turbulence is also very important in considering the maintenance of aquatic ecosystems, and the distribution of mean currents alone cannot sufficiently discuss the formation of boundary layers, the transport of dissolved substances, etc. From the standpoint of hydraulics, the objectives of river management and design can be classified into three types: flood control, water utilization and environment protection.

The flow resistance in the previous topics is one of the important factors in hydraulic control measures, such as flood prevention and flood prediction, because it is involved not only in the bottom deformation but also increase of flow depth. In particular, submerged vegetation and the large roughness promote the generation of turbulence, which can significantly change the flow property. Also, in flood currents mixed with gravel and driftwood, the interaction between the solid and liquid phases affects the turbulent motions. In general, hydraulic control measures often deal with large discharge and high-speed flows accompanied by large turbulence.

Since the regular calm condition has a longer period than the temporally flood event, in environmental management, it is important to understand the slow-speed hydrodynamics. Reliable predictions are needed about how suspended sediments, nutrients, gas and pollutants taken into the water stream will be transported and locally deposited. The suspended sediments created on the upstream side is transported downstream by the turbulence current and deposited in a slow-speed basin. How the bottom configuration changes due to the construction of artificial structure such as dam and weir greatly affect the survival of endemic species. Also, dissolved gas and heat are exchanged on the free surface, but it has become clear that they are controlled by the bottom-oriented turbulence. Furthermore, the transport of plant species causes forestation in trapped areas and affects the ecosystem. Turbulent motion is also greatly involved in how to accurately predict and analyze such a mass transport.

We cannot live without using the water resources of rivers and lakes. Most of them are used for agriculture, irrigation, water supply, industrial water and power generation. A river traffic is also one of the important transport infrastructures. These systems include many related to open-channel flow such as intake facility, headrace, waterway, lock gate and weir. Many problems have not yet been solved, such as sediment accumulation and air entrainment around the intake and the generation of blue-green algae caused by sedimentation in agricultural canals.

These problems have different scales and characteristics, but all involve turbulence. The Reynolds number is the ratio of inertial force to viscous force of fluid motion. When the Reynolds number is small, the influence of the viscous force is greater than the inertial force. The viscous force is related to the shear friction and, for example, even if a sudden high-speed flow is locally generated due to disturbance, it is dragged by the surrounding slow flow. As a result, the flow becomes regular and stable, even when it receives an external stimulus. The flow of recovery to the original state is called laminar flow. In the famous Reynolds experiment, the dye injected into the pipe flow travels downstream without spreading to the surroundings. On the other hand, when the Reynolds number is large, the inertial force is stronger than the viscous force, and the fluid becomes easier to move freely. In the Reynolds experiment, the dye fluctuates greatly during a streamwise travel and eventually mixes with the surrounding fluid having vortices. This is called turbulence flow. The turbulence is observed in various phenomena regardless of scale, from familiar places to huge natural fields such as atmosphere, river and ocean.

It is known that vortices of various sizes are mixed, and large vortices scale down to small vortices. Which scale vortex contributes to the matter transport we are thinking about? What is the energy transport rate (turbulent energy dissipation rate) from a large vortex to a small vortex or heat? These answers will also help solve some of the previous topics.

The open-channel flow is a kind of boundary layer, and the boundary layer occupies the entire area from the bottom to the water surface in the fully developed open channel. The characteristic of the open channel is that it has a free surface that is a non-rigid boundary. This is quite different from the closed pipe flow and atmospheric flow. The bottom-oriented turbulence dissipates energy while being transported to the free surface. Furthermore, the transported turbulence is dissipated on the free surface and, at the same time, it is responsible for the gas and heat exchange between the air and water. The turbulent structure of a steady vertical two-dimensional flow with a flat bottom has been elucidated to some extent. The immediate task is to apply these findings to the problems of actual rivers and waterways with vegetation and movable beds, etc. It is also important to extend them to three dimensions. There is a three-dimensional interaction between longitudinal eddy related to secondary currents and the horizontal vortex in the compound open-channel flows or vegetation flows. The detailed investigation of them is required for the accurate evaluation of the flow resistance and the prediction of the turbulent transport.

The free-surface of the water is the only thing that can be seen directly in rivers and concrete canals. The bottom-oriented turbulence arrears in the form of divergence or boil, updating the fluid elements near the free surface (surface renewal). In other words, the free-surface layer and the bottom layer are exchanged due to the turbulence. Many researchers have been studying the relationship between the bottom and free-surface turbulences. In the future, as the research progresses, it may be possible to predict the internal flow velocity profile, the bed formation and the scale of the turbulence, etc. from only the information of the turbulence on the free surface.

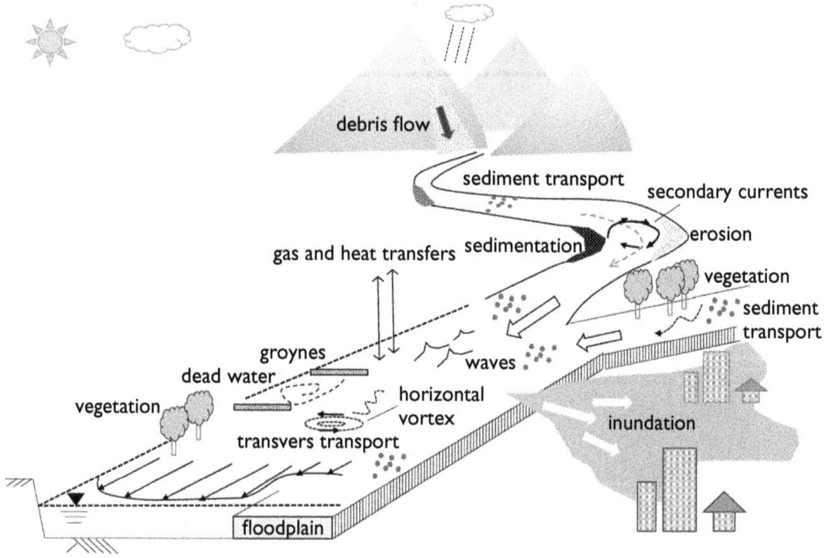

Figure 1.1 Examples of hydraulic phenomena in rivers related to turbulent dynamics.

The turbulent flows are a momentary phenomenon, but there are also organized motions, not just random fluctuations. How to feed back this knowledge to the practice of hydraulic engineering will be a big issue in the future.

Chapter 2 describes the statistical analysis and basic equations for study about open-channel turbulence. The basic matters about two-dimensional fields are summarized in Chapters 3 and 4. Chapter 5 mainly deals with three-dimensional structures, and especially explains the generation and distribution of secondary currents with the results of previous studies. You can refer to the basics of turbulent diffusion and dispersion in Chapter 6. Chapter 7 picks up typical examples of turbulent flow phenomena in rivers, as shown in FIG. 1.1, as applications. Chapter 8 describes how to measure and analyze data. We hope that readers who specialize in numerical calculations will also be able to experience the turbulence phenomena through a tabletop experiment, at least once referring to this chapter.

Chapter 2

Statistical Methods

2.1 INTRODUCTORY REMARKS

Turbulent flow is closely related to Reynolds number, which is a ratio of inertial force to viscous force. In small Reynolds number condition, viscous force is comparatively significant. The viscous force is well known as shear friction. Even if a local fluid parcel is accelerated by outer disturbance, surrounding low-speed fluid restricts such an irregular motion, and then it returns to a stable and regular situation. This is a laminar flow. In contrast, in large Reynolds number conditions, the inertial force is dominant compared to the viscous one. Hence, only a small disturbance generates large fluctuations in whole fluid field. This is a turbulent flow. This chapter explains basic features of turbulence, statistics methods and governing equations.

2.2 BASIC FEATURES OF TURBULENCE

2.2.1 What Is Turbulence?

Turbulence plays important roles for promotion of mass transport and variation of friction loss in the solid–liquid boundary. Several basic characteristics in turbulence are known, although the turbulence flow is much more complicated compared to the laminar flow.

2.2.1.1 Irregularity

We know the rising smoke from the chimney fluctuates randomly in the horizontal direction, as shown in FIG. 2.1. It is difficult to solve deterministically because of high irregularity in turbulent flow. Hence, the analysis of turbulence is often conducted based on statistical methods. For example, one of the popular methods is Reynolds decomposition, which extracts the time-mean and the fluctuating components (see section 2.6). The turbulence strength is evaluated from the root mean square of the fluctuating component.

DOI: 10.1201/9781003112198-2

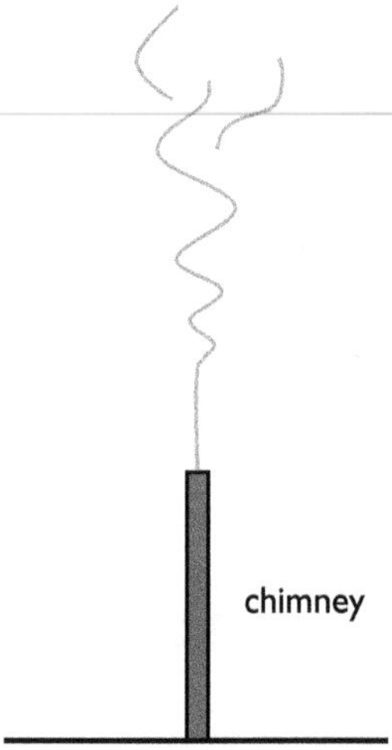

Figure 2.1 Turbulence observed in rising smoke from chimney.

2.2.1.2 *Diffusibility*

Turbulence promotes significant diffusion of heat, concentration, momentum and so on. Let's consider that you dissolve milk into coffee. When you stir by using a spoon, the coffee and the milk are mixed in a shorter time compared to a natural mixture without stirring. This is because the turbulence and vortex generated by the stirring transport very effectively in the coffee and the milk in a cup. FIG. 2.2 shows a sketch of time variation in milk concentration with an initial value C at the cup center. FIG. 2.2 (a) and (b) correspond to the case of only molecular diffusion without turbulence and the case of the turbulence diffusion generated by spoon stirring. At $t = T$ after starting diffusion, turbulent diffusion transports milk farther in the radial direction compared to molecular diffusion. That is to say, the turbulence increases a diffusion velocity and reduces a diffusion time.

2.2.1.3 *Three dimensionality*

Three-dimensional vortex is generally observed in fully developed turbulence. The vortex has a tube-like structure, as shown in FIG. 2.3. This is

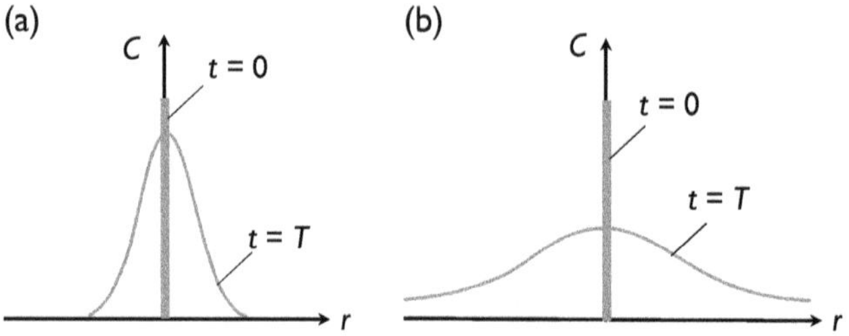

Figure 2.2 Sketch of concentration diffusion: (a) molecular diffusion without stirring and (b) turbulence diffusion with stirring.

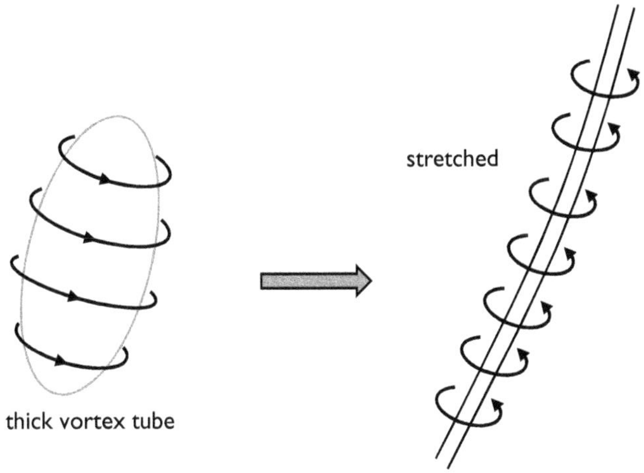

Figure 2.3 Vortex stretching.

stretched and compressed, interacting with surrounding currents. Being stretched more significantly, the circulation becomes larger, and vice versa.

2.2.1.4 Dissipation of kinetic energy

A local gradient of time-mean velocity induces the shear stress, deforming a fluid parcel. It results in the generation of vortex. Then the fluid parcel works against it and the consumed kinetic energy is dissipated to heat.

2.2.1.5 Multi-scales

A large vortex in the turbulence divides stepwise into smaller vortices and, finally, a minimum vortex appears, as shown in FIG. 2.4. Further, a minimum vortex is converted to heat. This is called the cascade process. Hence, several sizes of vortices are mixed well.

2.2.2 Generation of Turbulence

Turbulent flow is decomposed to the time-mean velocity component and the fluctuating component. Appearance of the spatial gradient of mean

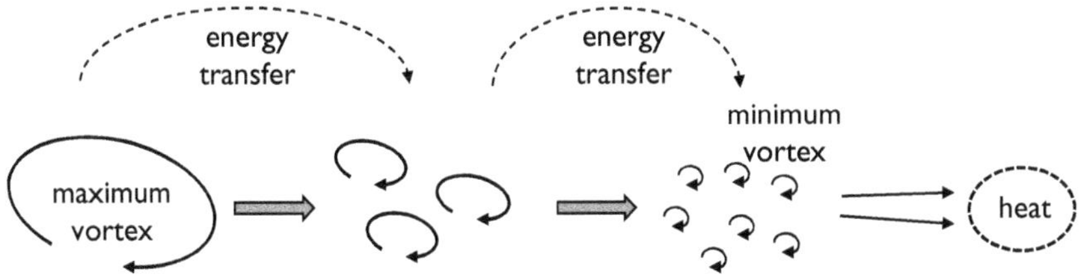

Figure 2.4 Vortex division in multi-scale turbulence.

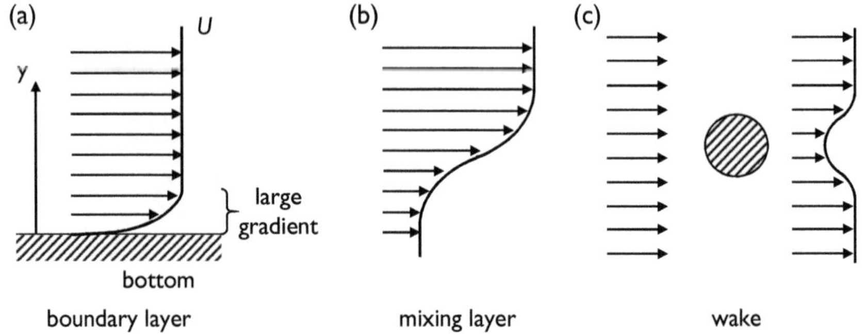

Figure 2.5 Flow types accompanied with velocity shear.

velocity called velocity shear, the energy in the time-mean field, is transferred to the turbulence component. Therefore, large-scale turbulence is often observed around the velocity shear. FIG. 2.5 shows some examples of flow type accompanied by the velocity shear, i.e., a) boundary layer near the solid wall, b) mixing layer of high- and low-speed currents, c) wake behind cylinder. They are also observed in natural rivers. The boundary layer is formed near the bottom and side walls. The conjunction of two rivers with different velocities promotes a mixing layer downstream. The wake is formed behind the bridge peer with flow separation and shedding vortices. A strong turbulence distributes locally in these regions.

It is found by the linear stability theory that the velocity shear introduces a flow instability. This topic is explained in section 3.3.

2.2.3 Diffusion in Turbulence

2.2.3.1 Molecular diffusion

Let us consider thermal diffusion from a stove in a closed calm room, as shown in FIG. 2.6. The one-dimensional diffusion equation of heat is given by

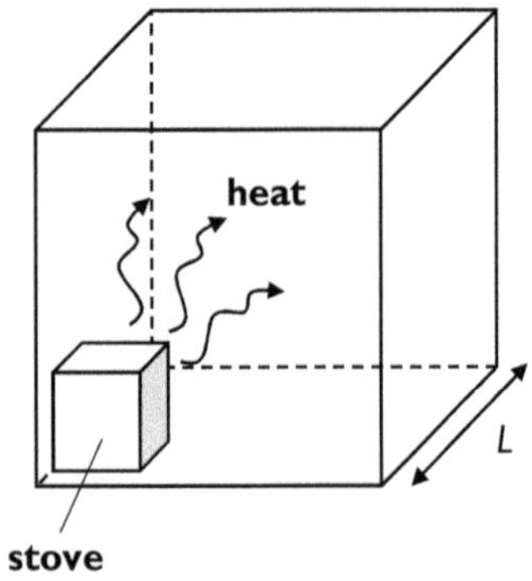

Figure 2.6 Thermal diffusion in a room.

$$\frac{\partial \theta}{\partial t} = \gamma \frac{\partial^2 \theta}{\partial x^2} \tag{2.2.1}$$

in which θ is a temperature and γ is a thermal diffusivity. T_m is defined as the time scale of molecular diffusion, which means an approximate time until heat is diffused throughout the room. L is defined as a characteristic length scale that corresponds to the depth length of the room.

After-dimension analysis of Eq. (2.2.1) leads to the following relation:

$$T_m \sim \frac{L^2}{\gamma} \tag{2.2.2}$$

For example, when $\gamma = 0.20\ \text{cm}^2/\text{s}$ and $L = 5$ m are given, $T_m \sim 10^6$ s is obtained. Namely, it takes more than 100 hours! A long time is needed to warm a whole room for only molecular diffusion without turbulent air.

2.2.3.2 Turbulent diffusion

A second situation places a circulator to generate turbulent air in the same room. Here, we introduce the eddy diffusion coefficient K, including the turbulence effects instead of the thermal diffusivity used. Therefore, Eq. (2.2.1) is replaced by the following form:

$$\frac{\partial \theta}{\partial t} = K \frac{\partial^2 \theta}{\partial x^2} \tag{2.2.3}$$

Defining a characteristic time scale in the turbulence diffusion as T_t, Eq. (2.2.3) leads to

$$T_t \sim \frac{L^2}{K} \tag{2.2.4}$$

Furthermore, defining a characteristic velocity of the turbulent diffusion means a propagation speed of the thermal front edge as u,

$$T_t \sim \frac{L}{u} \tag{2.2.5}$$

is obtained. Eqs. (2.2.4) and (2.2.5) give the eddy diffusion coefficient as

$$K \sim uL \tag{2.2.6}$$

Here, consider the relation among K, γ and the kinetic viscosity $\nu \equiv \frac{\mu}{\rho}$, in which μ is the viscosity. A relation $\gamma \sim \nu$ is already known for the gases such as air. Hence

$$\frac{K}{\gamma} \simeq \frac{K}{\nu} \sim \frac{uL}{\nu} = \mathrm{Re} \leftrightarrow K \sim \mathrm{Re}\gamma \tag{2.2.7}$$

is obtained. This suggests that the eddy diffusion coefficient is proportional to the Reynolds number. Namely, larger turbulence included in a flow field promotes more significantly the diffusion. By the way, the ratio of thermal diffusion coefficient and the kinetic viscosity, $\mathrm{Pr} = \frac{\nu}{\gamma}$, is called the Prandtl number, which is about 0.7 for the air and about 7 for the water.

2.2.4 Minimum Vortex in Multi-Scale Turbulence

The nonlinear convection term in the Navier Stokes equation induces the vortex division, i.e., generations of smaller vortices. Then, the turbulent kinetic energy is transferred from the larger vortex to the smaller vortex, as shown in FIG. 2.4. Here, the energy supply rate is called the turbulent energy dissipation rate. Generally, it is expressed by ε. However, an infinitely small vortex rotating freely in the fluid cannot exist due to the viscosity. Hence, the minimum vortex exists whose size depends on a flow condition. Here, we assume that the small-scale motion is governed by two parameters, i.e., the turbulent energy dissipation rate, ε, and the kinetic viscosity, ν. Conducting the dimensional analysis, the following minimum vortex scales can be defined. They are called Kolmogorov scales.

Kolmogorov length scale: $\eta \equiv (\nu^3/\varepsilon)^{1/4}$

Kolmogorov time scale: $\tau \equiv (\nu/\varepsilon)^{1/2}$

Kolmogorov velocity scale: $\upsilon \equiv (\nu\varepsilon)^{1/4}$

The kinetic viscosity ν is a known physical property. Let's think about how to evaluate the turbulent energy dissipation rate, ε. Here, we are going to evaluate it from the large-scale vortex motion. First, a following assumption is introduced, i.e., the energy supply rate is proportional to the time scale of the large scale vortex. A kinetic energy per unit mass of the large vortex is given by u^2. A frequency of the large vortex with length scale l supplies the energy to the smaller vortices is considered u/l (1/s). Hence, the energy supply rate from the large vortex to the small vortex is given by $u^2 \times (u/l) = u^3/l$. Considering the energy supply rate is equal to the energy dissipation rate

$$\varepsilon \sim u^3/l \tag{2.2.8}$$

is obtained. The Kolmogorov scales are normalized in the following forms:

$$\eta/l \sim (ul/\nu)^{-3/4} = \mathrm{Re}^{-3/4} \tag{2.2.9}$$

$$\tau/t \sim \tau u/l = (ul/\nu)^{-1/2} = \mathrm{Re}^{-1/2} \tag{2.2.10}$$

$$v/u \sim (ul/\nu)^{-1/4} = \mathrm{Re}^{-1/4} \tag{2.2.11}$$

Eq. (2.2.9) means the ratio of minimum vortex length to maximum vortex length is a function of the Reynolds number. Similarly, Eqs. (2.2.10) and (2.2.11) mean that relative time scale and velocity scale respond to the Reynolds number. For example, in the case of Re = 1,000, $\eta/l \simeq 5 \times 10^{-3}$ is calculated. Therefore, the minimum vortex is found to be very tiny compared to the maximum vortex. Let's consider another example for a laboratory experiment, in which the water depth is 10 cm and the Reynolds number is 1,000. Assume the maximum vortex size and the water depth equal $l = 10$ cm. Hence, $\eta \simeq 0.5$ mm is obtained, and thus, the non-intrusive accurate velocimetry, such as a laser Doppler anemometer, is required in order to detect the minimum vortex.

2.3 AVERAGING PROCEDURES

2.3.1 Time Average

This term means an average over a long period in time-series of physical value at a certain point. Consider turbulence in a straight open channel as an example, in which a time-series of instantaneous velocity $\tilde{u}$ is measured for the duration T. Then, a time-averaged velocity U is calculated as follows:

$$U(x, y, z) \equiv \lim_{T\to\infty} \frac{1}{T}\int_0^T \tilde{u}(x, y, z, t)\,dt \tag{2.3.1}$$

Note that the time average is applied to only a steady-flow field with a constant time-averaged velocity, as shown in FIG. 2.7.

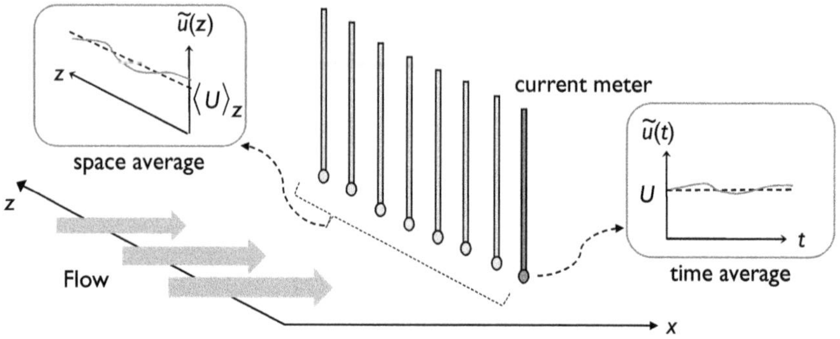

Figure 2.7 Space average and time average.

For the unsteady flow field, such as a flood event, the calculated time average velocity has no significance, as shown in FIG. 2.8, which means a typical flood wave. This is an unsteady turbulence flow with a rising stage and a falling stage. Also, high-frequency fluctuations are superimposed on the low-frequency curve. The time-averaged value calculated by Eq. (2.3.1) appears between the base and peak velocities. This value described with a dot line is constant, and thus such a time average is insignificant to consider the unsteady flood flow. In this case, a time-dependent low-frequency velocity is sometimes used instead of the time-averaged velocity (Nezu et al. 1997). Since the mean velocity varies in time, the turbulent component fluctuating around the mean value can be evaluated reasonably.

Nikora & Goring (2002) examined a statistical stability comparing 2 min and 20 min measurements in the free-surface flow, and concluded that 2 min measurement data provided reasonable results for turbulence statistics when many measured data was obtained.

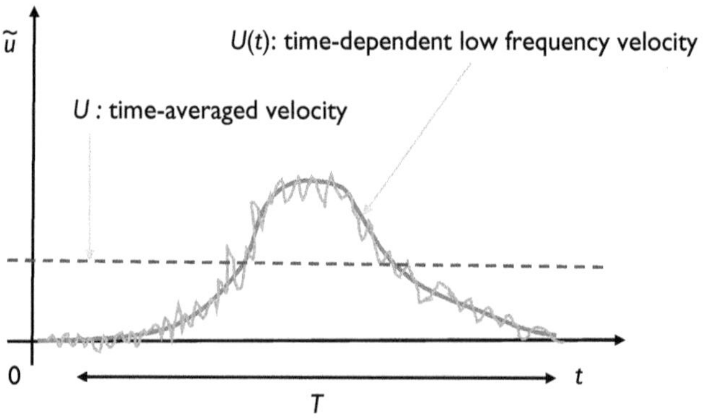

Figure 2.8 Unsteady time-dependent velocity in flood wave.

2.3.2 Space Average

This term means an average of instantaneous physical value with a spatial variation. When velocities are measured simultaneously at several points arranged in the spanwise direction z, as shown in FIG. 2.7, the spatial averaged velocity with the spanwise direction $\langle U \rangle_z$ is defined as follows:

$$\langle U \rangle_z (x, y, t) \equiv \lim_{L \to \infty} \frac{1}{L} \int_0^L \tilde{u}(x, y, z, t) dz \qquad (2.3.2)$$

in which L is a length of the measurement area. However, the space average is applied to only the spatial uniform flow field. Using time series of the space averaged velocity, the time variation of unsteady flow can be drawn.

The time-averaged flow is spatially heterogeneous with roughness elements like aquatic vegetation. The spatial averaging of the time-averaged physical variables can link reasonably the spatial averaged roughness parameters with the local flow characteristics.

Nikora et al. (2007) explained the advantages of such a double-average procedure in space and time for rough bed channel flows.

2.3.3 Ensemble Average

This term means an average of many instantaneously and locally measured data under the same condition. The spatial averaged velocity with the spanwise direction $[U]$ is defined as follows:

$$[U] \equiv \lim_{N \to \infty} \frac{1}{N} \sum_{n=1}^{N} \tilde{u}(x, y, z, t) \qquad (2.3.3)$$

The flow field does not always have to be stead or spatially uniform. When the time average for the steady flow coincides with the ensemble average, this is called ergodicity. In actual experiments, assuming the ergodicity, turbulence is often analyzed, introducing the time average instead of the ensemble average.

Although there are many kinds of dice with different colors and forms, they have a statistically similar characteristic. This implies that each die does not have a personality. If we examine a statistical property of any die randomly extracted, the property of the other dice can be anticipated. As previously mentioned, the ergodicity means that the whole statistic of the phenomenon can be predicted by using the behavior of one sample.

2.4 DIFFERENTIAL FORMS

It is required to observe microscopically the small-scale motions of fluid motion when we try to investigate their detailed mechanism. Focusing on the small fluid parcel in the current, the mass and momentum conservations related to that control volume are mathematically expressed in differential forms. This chapter derived popular differential forms of continuity equation. The explanation of Eulerian and Lagrangian specifications is followed by derivation of the Euler equation.

2.4.1 Einstein Notation and Coordinate System

Let us consider the three-dimensional coordinate with three directions. The turbulent motion is three dimensional, three directions, and thus, is followed by x, y, z. Here, introducing the Einstein notation, complicated mathematical expression of three-dimensional governing equations is simplified, which numbers all direction. That is to say, x, y, z are expressed by x_1, x_2, x_3, respectively. In the same fashion, the instantaneous velocity components $\tilde{u}$, $\tilde{v}$, $\tilde{w}$ are rewritten by $\tilde{u}_1$, $\tilde{u}_2$, $\tilde{u}_3$, respectively. Generally, the axis and instantaneous velocity with i direction are expressed by x_i, $\tilde{u}_i$, respectively, as shown in FIG. 2.9.

Einstein notation makes a rule that all components are summed up when there are two of the same subscripts in one term. For example, the term $\tilde{u}_j \frac{\partial \tilde{u}_i}{\partial x_j}$ has two subscript j, and thus, it is expanded, keeping i. Namely, $\tilde{u}_j \frac{\partial \tilde{u}_i}{\partial x_j} \equiv \tilde{u}_1 \frac{\partial \tilde{u}_i}{\partial x_1} + \tilde{u}_2 \frac{\partial \tilde{u}_i}{\partial x_2} + \tilde{u}_3 \frac{\partial \tilde{u}_i}{\partial x_3}$. Although the component expression on the right-hand side may be familiar with a beginner, many notations are required. In contrast, the Einstein notation makes it short and has an advantage in mathematical expansion of equations.

2.4.2 Continuity Equation

Let us define a small rectangular region in the two-dimensional fluid field, as shown in FIG. 2.10. Inflow and outflow are observed through the four

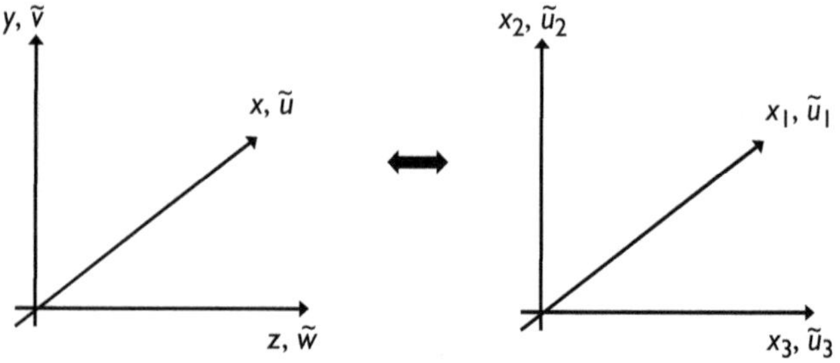

Figure 2.9 Expression of axis and velocity component for Einstein notation.

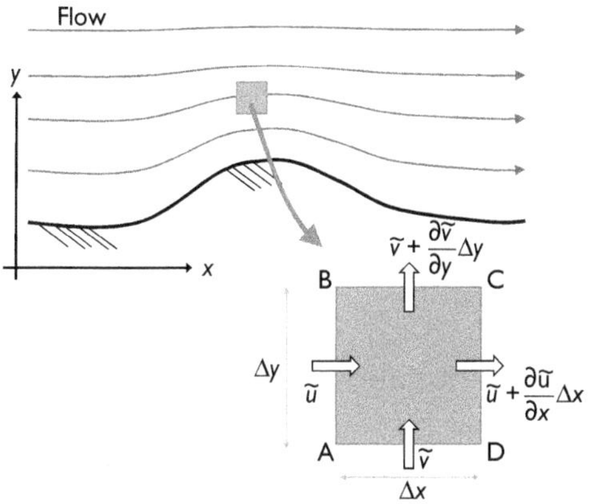

Figure 2.10 Small rectangular fluid parcel in two-dimensional field.

sides. That is to say, a relation, [(1) deposition mass per unit time] = [(2) inflow mass per unit time] – [(3) outflow mass per unit time], holds. Each component is expressed as follows:

$$(1) = \frac{\partial}{\partial t}(\rho \Delta x \Delta y) \times 1$$
$$(2) = \rho(\tilde{u} \times 1(\mathrm{s}) \times \Delta y) + \rho(\tilde{v} \times 1(\mathrm{s}) \times \Delta x)$$
$$(3) = \rho\left\{\left(\tilde{u} + \frac{\partial \tilde{u}}{\partial x}\Delta x\right) \times 1(\mathrm{s}) \times \Delta y\right\} + \rho\left\{\left(\tilde{v} + \frac{\partial \tilde{v}}{\partial y}\Delta y\right) \times 1(\mathrm{s}) \times \Delta x\right\}$$

in which $\tilde{u}$, $\tilde{v}$ are instantaneous velocity components with the x, y directions, respectively.

Therefore, it results in

$$\frac{\partial \rho}{\partial t} = -\rho\frac{\partial \tilde{u}}{\partial x} - \rho\frac{\partial \tilde{v}}{\partial y} \tag{2.4.1}$$

This is called the continuity equation, which means mass conservation. When we deal the incompressible fluid with constant density ρ, this is simplified as follows:

$$\frac{\partial \tilde{u}}{\partial x} + \frac{\partial \tilde{v}}{\partial y} = 0 \tag{2.4.2}$$

In the three-dimensional field, it is expanded to

$$\frac{\partial \tilde{u}}{\partial x} + \frac{\partial \tilde{v}}{\partial y} + \frac{\partial \tilde{w}}{\partial z} = 0 \qquad (2.4.3)$$

The Einstein notation expresses it by

$$\frac{\partial \tilde{u}_i}{\partial x_i} = 0 \qquad (2.4.4)$$

in which $i = 1, 2$ for the two-dimensional field, and $i = 1, 2, 3$ for the three-dimensional one.

2.4.3 Euler Equations

2.4.3.1 Mathematical form of fluid acceleration

The fluid moves according to Newton's second law of motion, i.e., net force equals the product of mass times acceleration, as well as a motion of mass point. The first question is how we express the fluid acceleration. Here, two observation methods for the fluid motion are introduced. One is called the Lagrangian specification (FIG. 2.11), which tracks one particle, assuming that fluid composes of a group of many particles. The fluid particle is accelerated and decelerated by external forces in accordance with Newton's second law. The acceleration is given by a time differential of velocity.

$$a = \frac{\partial \tilde{u}}{\partial t} \qquad (2.4.5)$$

Another is called the Eulerian specification, which evaluates the spatial and time variation of velocity between two test sections with a constant streamwise distance. The fluid passing through section I reaches section II dx downstream of section I after dt, as shown in FIG. 2.12. Hence, the velocity in section II is Taylor – expanded as follows:

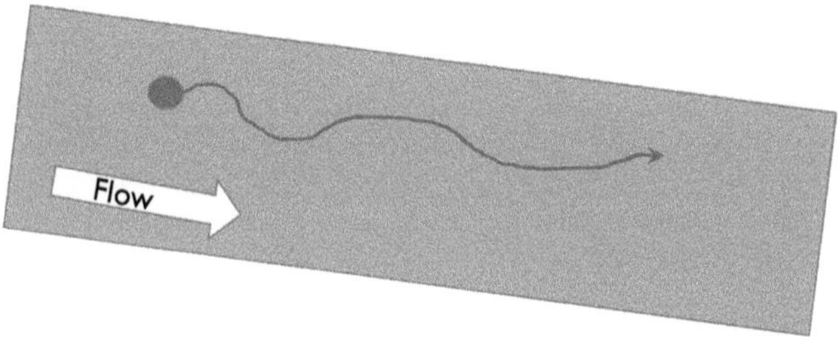

Figure 2.11 Lagrangian specification.

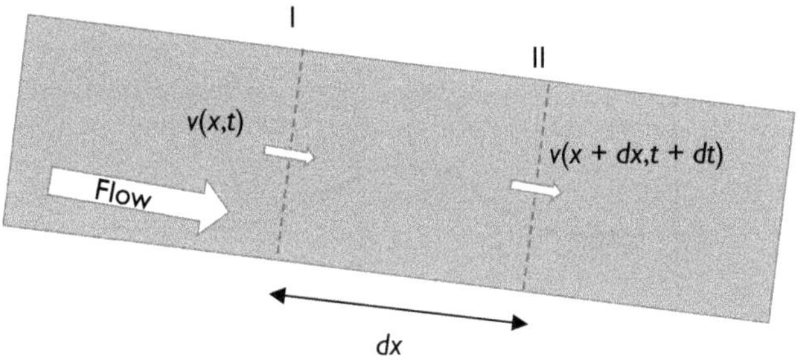

Figure 2.12 Eulerian specification.

$$\tilde{u}(x + dx, t + dt) = \tilde{u}(x, t) + \frac{\partial \tilde{u}(x, t)}{\partial t}dt + \frac{\partial \tilde{u}(x, t)}{\partial x}dx \tag{2.4.6}$$

Thus, the acceleration is written by $a \equiv \lim_{dt \to 0} \frac{\tilde{u}(x + dx, t + dt) - \tilde{u}(x, t)}{dt} = \lim_{dt \to 0} \frac{\partial \tilde{u}(x, t)}{\partial t} + \lim_{dt \to 0} \frac{dx}{dt} \times \frac{\partial \tilde{u}(x, t)}{\partial x}$. Due to $\tilde{u} = \frac{dx}{dt}$, it results in

$$a = \frac{\partial \tilde{u}}{\partial t} + \tilde{u}\frac{\partial \tilde{u}}{\partial x} = \left(\frac{D\tilde{u}}{Dt}\right) \tag{2.4.7}$$

In the same way, given that the fluid at point (x, y, z) reaches another point $(x + dx, y + dy, z + dz)$ in dt., the acceleration in x direction can be expressed by

$$\begin{aligned} a_x &\equiv \lim_{dt \to 0} \frac{\tilde{u}(x + dx, \ y + dy, \ z + dz, \ t + dt) - \tilde{u}(x, t)}{dt} \\ &= \lim_{dt \to 0} \frac{\partial \tilde{u}(x, t)}{\partial t} + \lim_{dt \to 0} \frac{dx}{dt} \times \frac{\partial \tilde{u}(x, t)}{\partial x} + \lim_{dt \to 0} \frac{dy}{dt} \times \frac{\partial \tilde{u}(x, t)}{\partial y} + \lim_{dt \to 0} \frac{dz}{dt} \times \frac{\partial \tilde{u}(x, t)}{\partial z} \\ &\leftrightarrow a_x = \frac{\partial \tilde{u}}{\partial t} + \tilde{u}\frac{\partial \tilde{u}}{\partial x} + \tilde{v}\frac{\partial \tilde{u}}{\partial y} + \tilde{w}\frac{\partial \tilde{u}}{\partial z} \end{aligned} \tag{2.4.8}$$

Einstein notation of the acceleration with i direction is

$$a_i = \frac{\partial \tilde{u}_i}{\partial t} + \tilde{u}_j \frac{\partial \tilde{u}_i}{\partial x_j} \left(= \frac{D\tilde{u}_i}{Dt}\right) \tag{2.4.9}$$

$\frac{D}{Dt}$ is called the material derivative. The second term in Eq. (2.4.9) is non-linear, which is called a convection term. Although the acceleration of the mass point corresponds with the first term in Eq. (2.4.9) called local acceleration, the convection term is added in the fluid acceleration.

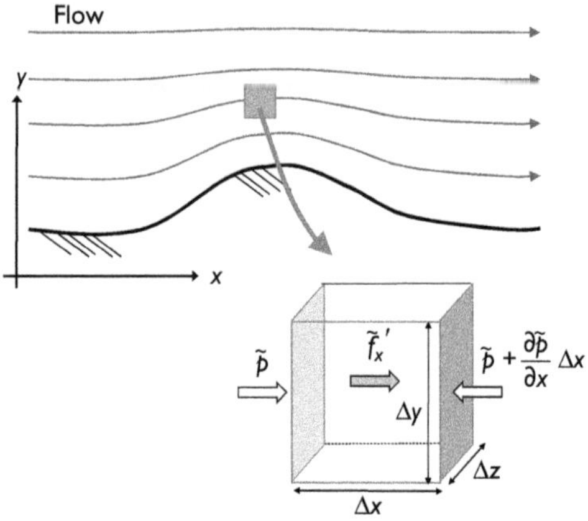

Figure 2.13 Forces acting on fluid parcel in *x* direction.

2.4.3.2 Derivation of Euler equations

FIG. 2.13 focuses on streamwise motion of the small fluid parcel in the three-dimensional current. What kind of forces act on the fluid parcel with x direction? Here, let's consider the perfect fluid for simpler discussion. Therefore, we need not consider a friction force related to the viscosity. The pressure and total external forces are $\tilde{p}$, $\tilde{f}'_x$, respectively. When the pressure on the upstream side is given by $\tilde{p}(x, y, z)$, that on the downstream side is expressed by $-\tilde{p}(x + dx, y, z) = -\tilde{p}(x, y, z) - \frac{\partial \tilde{p}}{\partial x}dx$. Hence, the total of them is $-\frac{\partial \tilde{p}}{\partial x}dx$. Since the area on which pressure acts is $dxdy$, the total pressure force acting on the fluid parcel is expressed by $-\frac{\partial \tilde{p}}{\partial x}dxdydz$. When there are some external forces, the total force that accelerates the fluid parcel with the x direction is $-\frac{\partial \tilde{p}}{\partial x}dxdydz + \tilde{f}'_x$.

The mass of the fluid parcel is $\rho dxdydz$. When the acceleration in x direction is given by (2.4.8), Newton's second law introduces the expression of $(\rho dxdydz) \times \left(\frac{\partial \tilde{u}}{\partial t} + \tilde{u}\frac{\partial \tilde{u}}{\partial x} + \tilde{v}\frac{\partial \tilde{u}}{\partial y} + \tilde{w}\frac{\partial \tilde{u}}{\partial z}\right) = -\frac{\partial \tilde{p}}{\partial x}dxdydz + \tilde{f}'_x$. Replacing the outer force by the volumetric force, $\tilde{f}_x = \tilde{f}'_x/(\rho dxdydz)$, the momentum equation of perfect fluid in the x direction is obtained. Those in y and z directions can be derived in the same way. They are as follows:

$$(x \text{ direction})\ \frac{\partial \tilde{u}}{\partial t} + \tilde{u}\frac{\partial \tilde{u}}{\partial x} + \tilde{v}\frac{\partial \tilde{u}}{\partial y} + \tilde{w}\frac{\partial \tilde{u}}{\partial z} = -\frac{1}{\rho}\frac{\partial \tilde{p}}{\partial x} + \tilde{f}_x \tag{2.4.10}$$

$$(y\ \text{direction})\ \frac{\partial \tilde{v}}{\partial t} + \tilde{u}\frac{\partial \tilde{v}}{\partial x} + \tilde{v}\frac{\partial \tilde{v}}{\partial y} + \tilde{w}\frac{\partial \tilde{v}}{\partial z} = -\frac{1}{\rho}\frac{\partial \tilde{p}}{\partial y} + \tilde{f}_y \tag{2.4.11}$$

$$(z\ \text{direction})\ \frac{\partial \tilde{w}}{\partial t} + \tilde{u}\frac{\partial \tilde{w}}{\partial x} + \tilde{v}\frac{\partial \tilde{w}}{\partial y} + \tilde{w}\frac{\partial \tilde{w}}{\partial z} = -\frac{1}{\rho}\frac{\partial \tilde{p}}{\partial z} + \tilde{f}_z \tag{2.4.12}$$

Their Einstein notation is written by

$$\frac{\partial \tilde{u}_i}{\partial t} + \tilde{u}_j\frac{\partial \tilde{u}_i}{\partial x_j} = -\frac{1}{\rho}\frac{\partial \tilde{p}}{\partial x_i} + \tilde{f}_i \quad (i, j = 1, 2, 3) \tag{2.4.13}$$

They are called Euler equations.

2.5 NAVIER STOKES EQUATIONS

2.5.1 Momentum Equation of Viscous Fluid

Adding the viscous force in Euler equations gives the momentum equations for viscous fluids. They are called Navier Stokes equations. When there is a local velocity variation around the small fluid parcel, it is distorted, contracted, expanded and rotated. The viscous stress arises with these fluid deformations. Here, two types of the viscous stress are considered, i.e., shear stress and normal stress. The following section explains them, acting on the two-dimensional fluid parcel.

2.5.2 Viscous Shear Stress

FIG. 2.14 shows a two-dimensional fluid parcel with side lengths Δx and Δy with unit width. The velocity on the lower side AB is given by $\tilde{u}$ and that on the higher side DC is given by $\tilde{u} + \Delta\tilde{u}$. Then the shear stress τ_{yx} arises with the x direction, and it distorts the fluid parcel. Here, assuming that the side DC moves the distance Δl in parallel with the x axis, the strain angle is defined as $\Delta\alpha_1$. Since the AD length is Δy, the strain angle is written by $\Delta\alpha_1 = \tan^{-1}\frac{\Delta l}{\Delta y}$. When $\Delta\alpha_1$ is very small, a following relation is formed.

$$\Delta\alpha_1 \simeq \frac{\Delta l}{\Delta y} = \frac{(\tilde{u} + \Delta\tilde{u})\Delta t - \tilde{u}\Delta t}{\Delta y} = \frac{\Delta\tilde{u}\Delta t}{\Delta y} \leftrightarrow \frac{\Delta\alpha_1}{\Delta t} = \frac{\Delta\tilde{u}}{\Delta y} \tag{2.5.1}$$

Hence, taking limitations, $\Delta t \to 0, \quad \Delta y \to 0$, (2.5.1) is rewritten by

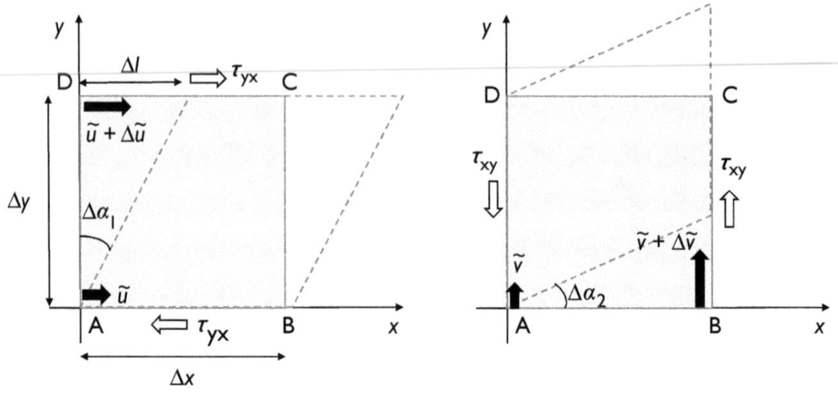

Figure 2.14 Distortion of two-dimensional fluid parcel.

$$\frac{\partial \alpha_1}{\partial t} = \frac{\partial \tilde{u}}{\partial y} \tag{2.5.2}$$

Newton's viscosity law defines the shear stress as

$$\tau = \mu \frac{\partial \tilde{u}}{\partial y} \tag{2.5.3}$$

Both Eqs. (2.5.2) and (2.5.3) introduce

$$\tau = \mu \frac{\partial \alpha_1}{\partial t} \tag{2.5.4}$$

This suggests that the shear stress is proportional to the time differential of the strain angle.

The velocity on the left-side AD is given by $\tilde{v}$ and that on the right-side BC is given by $\tilde{v} + \Delta\tilde{v}$. Then the shear stress τ_{xy} arises with the y direction. Then the strain angle is defined as $\Delta\alpha_2$. The similar relation to Eq. (2.5.4) is obtained as follows:

$$\frac{\partial \alpha_2}{\partial t} = \frac{\partial \tilde{v}}{\partial x} \tag{2.5.5}$$

When we assume that the fluid parcel is not rotated, the occurrence of τ_{yx} induces mechanically τ_{xy} to prevent a rotational moment production, as shown in FIG. 2.15(a). Since the moment about point A is zero, the shear stresses with horizontal and vertical directions become the same, i.e.,

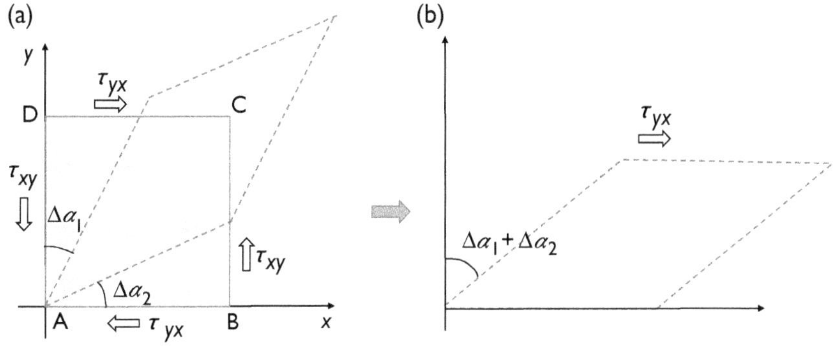

Figure 2.15 Composition of horizontal and vertical shear stresses.

$(\tau_{yx}\Delta x) \times \Delta y = (\tau_{xy}\Delta y) \times \Delta x \quad \rightarrow \quad \tau_{yx} = \tau_{xy}$. Total strain angle becomes $\Delta\alpha_1 + \Delta\alpha_2$, as shown in FIG. 2.15(b). When $\Delta\alpha_1$ and $\Delta\alpha_2$ are assumed very small, the strain angle corresponding to τ_{yx} can be thought of as $\Delta\alpha_1 + \Delta\alpha_2$. Hence, the shear stress is expressed by using the velocity gradient as follows:

$$\tau_{yx} = \tau_{xy} = \mu\frac{\partial(\alpha_1 + \alpha_2)}{\partial t} = \mu\left(\frac{\partial\tilde{u}}{\partial y} + \frac{\partial\tilde{v}}{\partial x}\right) \tag{2.5.6}$$

2.5.3 Viscous Normal Stress

As shown in FIG. 2.16, consider a small fluid element that is a square with a side of Δh. The depth is set to 1 for convenience. Suppose that this parcel ABCD is stretched in the x direction to become AB'C'D. In this deformation, a rhombus inside of the parcel is also stretched with the x direction. It results in the shear stress τ exerts on each side of the rhombus, as shown FIG. 2.16. Let us formulate a force balance about a triangle AEG using a strain angle γ. Here, the x and y components of viscous normal stresses acting on AEG are defined as τ_{xx} and τ_{yy}, respectively. Then a following relation holds.

x direction: x component of τ times contact area equals τ_{xx} times contact area.

y direction: y component of τ times contact area equals τ_{yy} times contact area.

In summary,

$$x \text{ direction: } \frac{\tau}{\sqrt{2}} \times \frac{\sqrt{2}}{2}\Delta h = \tau_{xx} \times \frac{\Delta h}{2} \quad \Leftrightarrow \quad \tau = \tau_{xx}\,(\text{tensile}) \tag{2.5.7}$$

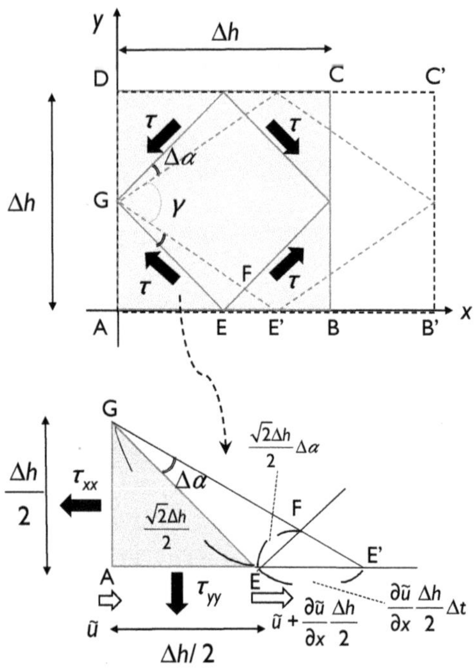

Figure 2.16 Viscous normal stress and internal strain.

$$y \text{ direction: } \frac{\tau}{\sqrt{2}} \times \frac{\sqrt{2}}{2}\Delta h = -\tau_{yy} \times \frac{\Delta h}{2} \quad \Leftrightarrow \quad -\tau = \tau_{yy}\,(\text{compression}) \tag{2.5.8}$$

are obtained. Note that Eq. (2.5.8) shows stretching in the x direction induces the compression force in the y direction with an equal strength.

Since the velocity difference between points A and E is expressed by $\frac{\partial \tilde{u}}{\partial x}\frac{\Delta h}{2}$ in FIG. 2.16, the travel distance from point E to point E' is given as $\frac{\partial \tilde{u}}{\partial x}\frac{\Delta h}{2}\Delta t$. Furthermore, the length of EF is $\frac{\sqrt{2}}{2}\Delta h \Delta\alpha$. When the strain angle is very small, the EE'F can be assumed to be an isosceles triangle with a small angle. That is to say, $\frac{\sqrt{2}}{2}\Delta h\Delta\alpha \simeq \frac{1}{\sqrt{2}}\frac{\partial \tilde{u}}{\partial x}\frac{\Delta h}{2}\Delta t$. Therefore, taking the zero limitation of Δt, i.e., $\Delta t \to 0$, $\frac{\partial \alpha}{\partial t} = \frac{1}{2}\frac{\partial \tilde{u}}{\partial x}$ is obtained. Due to $\Delta\gamma = -2\Delta\alpha$ (a minus sign is needed because the increase of α reduces γ), a following form is introduced.

$$\frac{\partial \gamma}{\partial t} = -\frac{\partial \tilde{u}}{\partial x} \tag{2.5.9}$$

The shear stress is the product of fluid viscosity and the time differential of strain angle as indicated in Eq. (2.5.4). Hence,

$$\tau = -\mu\frac{\partial \gamma}{\partial t} = \mu\frac{\partial \tilde{u}}{\partial x} \tag{2.5.10}$$

is obtained, in which the minus sign is needed because the strain angle is not α but γ. The normal stress $-\mu\frac{\partial \tilde{v}}{\partial y}$ exerts with the x direction due to spatial variation of vertical velocity $\tilde{v}$, as mentioned previously. The minus sign corresponds to Eq. (2.5.8). Since the sum of this stress and Eq. (2.5.10) is equal to τ_{xx},

$$\tau_{xx} = \mu\frac{\partial \tilde{u}}{\partial x} - \mu\frac{\partial \tilde{v}}{\partial y} = 2\mu\frac{\partial \tilde{u}}{\partial x} - \mu\left(\frac{\partial \tilde{u}}{\partial x} + \frac{\partial \tilde{v}}{\partial y}\right) = 2\mu\frac{\partial \tilde{u}}{\partial x} \tag{2.5.11}$$

is obtained using the continuity. It results in

$$\tau_{xx} = 2\mu\frac{\partial \tilde{u}}{\partial x}, \quad \tau_{yy} = 2\mu\frac{\partial \tilde{v}}{\partial y} \tag{2.5.12}$$

2.5.4 Total Viscous Stress

The x component of viscous stress exerting on the small fluid parcel can be expressed by using the above result. Here, the pulling direction is defined by positive, and the compressive direction is defined by negative. The normal stress exerts on sides AD and BC, as shown in FIG. 2.17. They are written by $-\tau_{xx}\Delta y$ for side AD and $\left(\tau_{xx} + \frac{\partial \tau_{xx}}{\partial x}\Delta x\right)\Delta y$ for side BC. In contrast, x components of the shear stress are written by $-\tau_{yx}\Delta x$ for side AB and $\left(\tau_{yx} + \frac{\partial \tau_{yx}}{\partial y}\Delta y\right)\Delta x$ for side DC, respectively. When these are added together,

$$\left(\frac{\partial \tau_{xx}}{\partial x} + \frac{\partial \tau_{yx}}{\partial y}\right)\Delta x\Delta y = \left(\frac{\partial}{\partial x}\left(2\mu\frac{\partial \tilde{u}}{\partial x}\right) + \frac{\partial}{\partial y}\mu\left(\frac{\partial \tilde{u}}{\partial y} + \frac{\partial \tilde{v}}{\partial x}\right)\right)\Delta x\Delta y$$
$$= \mu\left(\frac{\partial^2 \tilde{u}}{\partial x^2} + \frac{\partial^2 \tilde{u}}{\partial y^2} + \frac{\partial}{\partial x}\left(\frac{\partial \tilde{u}}{\partial x} + \frac{\partial \tilde{v}}{\partial y}\right)\right)\Delta x\Delta y = \mu\left(\frac{\partial^2 \tilde{u}}{\partial x^2} + \frac{\partial^2 \tilde{u}}{\partial y^2}\right)\Delta x\Delta y$$

is obtained. In the same way, the total of y components of the normal stress and the shear stress is expressed by $\left(\frac{\partial \tau_{yy}}{\partial y} + \frac{\partial \tau_{xx}}{\partial x}\right)\Delta x\Delta y = \cdots = \mu\left(\frac{\partial^2 \tilde{v}}{\partial x^2} + \frac{\partial^2 \tilde{v}}{\partial y^2}\right)\Delta x\Delta y$.

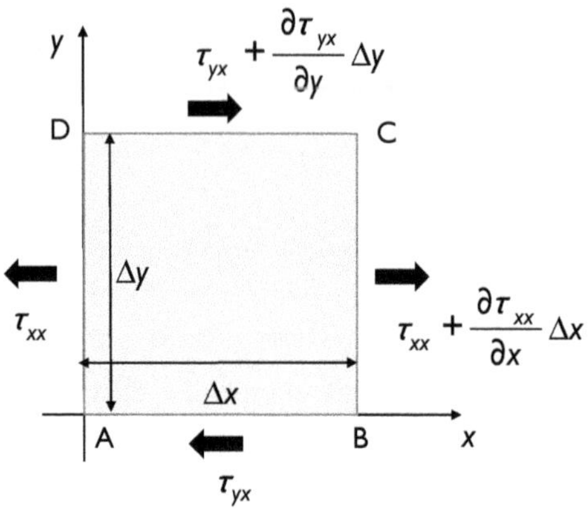

Figure 2.17 x components of normal stress and shear stress.

If we extend this result to three-dimensional fluid parcel with depth Δz, the total viscous stresses composed of the normal stress and the shear stress are expressed in the following forms:

$$(x \text{ component}) : \mu\left(\frac{\partial^2 \tilde{u}}{\partial x^2}+\frac{\partial^2 \tilde{u}}{\partial y^2}+\frac{\partial^2 \tilde{u}}{\partial z^2}\right)\Delta x \Delta y \Delta z \tag{2.5.13}$$

$$(y \text{ component}) : \mu\left(\frac{\partial^2 \tilde{v}}{\partial x^2}+\frac{\partial^2 \tilde{v}}{\partial y^2}+\frac{\partial^2 \tilde{v}}{\partial z^2}\right)\Delta x \Delta y \Delta z \tag{2.5.14}$$

$$(z \text{ component}) : \mu\left(\frac{\partial^2 \tilde{w}}{\partial x^2}+\frac{\partial^2 \tilde{w}}{\partial y^2}+\frac{\partial^2 \tilde{w}}{\partial z^2}\right)\Delta x \Delta y \Delta z \tag{2.5.15}$$

2.5.5 Navier Stokes Equations

Dividing viscous stresses in Eqs. (2.5.13) to (2.5.15) by $\rho \Delta x \Delta y \Delta z$,

$$(x \text{ component}) : \nu\left(\frac{\partial^2 \tilde{u}}{\partial x^2}+\frac{\partial^2 \tilde{u}}{\partial y^2}+\frac{\partial^2 \tilde{u}}{\partial z^2}\right) \tag{2.5.16}$$

$$(y \text{ component}) : \nu\left(\frac{\partial^2 \tilde{v}}{\partial x^2}+\frac{\partial^2 \tilde{v}}{\partial y^2}+\frac{\partial^2 \tilde{v}}{\partial z^2}\right) \tag{2.5.17}$$

$$(z \text{ component}): \nu\left(\frac{\partial^2 \tilde{w}}{\partial x^2} + \frac{\partial^2 \tilde{w}}{\partial y^2} + \frac{\partial^2 \tilde{w}}{\partial z^2}\right) \tag{2.5.18}$$

are obtained, in which the viscosity was replaced with the kinetic viscosity $\nu \equiv \mu/\rho$.

Hence, the Einstein notation of the viscous stress of i component is $\nu\left(\frac{\partial^2 \tilde{u}_i}{\partial x_j \partial x_j}\right)$. Adding this term to the right-hand side of Euler equation expressed by 2.4.13, the Navier Stokes (N-S) equation is derived as follows:

$$\frac{\partial \tilde{u}_i}{\partial t} + \tilde{u}_j \frac{\partial \tilde{u}_i}{\partial x_j} = -\frac{1}{\rho}\frac{\partial \tilde{p}}{\partial x_i} + \nu \frac{\partial^2 \tilde{u}_i}{\partial x_j \partial x_j} + \tilde{f}_i \quad (i, j = 1, 2, 3) \tag{2.5.19}$$

The second term on the left-hand side in 2.5.19 is nonlinear. This is called the convection term, which makes the fluid motion more complicated. When the external force is known, the number of unknown variables is four in $\tilde{u}$, $\tilde{v}$, $\tilde{w}$ and $\tilde{p}$ in the case of a three-dimensional field. The number of equations is also four, i.e., three components of the N-S equation and the continuity equation of Eq. (2.4.3). Therefore, the unknown variables are likely to be solved mathematically. However, we cannot solve them except in special cases, e.g., Couette flow, Hagen Poiseuille flow, etc.

The convection term is related to the cascade process, as explained in section 2.2.1. We assume that a spatial distribution of instantaneous velocity is expressed as $\tilde{u} = A\sin(kx)$, in which A is the amplitude and k is the wave number. Substituting this into the convection term in the one-dimensional N-S equation, i.e., $\tilde{u}(\partial\tilde{u}/\partial x)$, gives $\tilde{u}\frac{\partial \tilde{u}}{\partial x} = kA\sin(kx) \times A\cos(kx) = \frac{A^2 k}{2}\sin(2kx)$. Hence, the output waveform has twice the wave number of the input wave form. This suggests that the vortex scale becomes smaller due to the convection term.

2.5.6 Viscosity and Kinetic Viscosity

What is the physical difference between viscosity μ and kinetic viscosity ν? Here are two cups in which water and honey are filled, respectively. If you stir each of two beakers with water and honey with a spoon, the sticky honey requires more power. Then, the fluid velocity near the spoon becomes larger than the surrounding fluids. This velocity difference in space induces the shear stress, which is expressed as Eq. (2.5.3) of Newton's viscosity law. A proportional coefficient multiplier on the velocity gradient is a viscous coefficient, which is the physical characteristic value. The viscosity means the efficiency of how far the friction influences.

In contrast, a reduction of fluid velocity is considered, the fluid density is an important factor. The heavier fluid is harder to be decelerated compared to the lighter fluid. There is a related following example. When we brake a running small car and a running large truck using the same barking force, the truck requires a longer braking distance than the small car. Therefore, the fluid density must be considered to evaluate the influence of the friction force on the fluid velocity. In this situation, we do not use viscosity; we use kinetic viscosity ν.

It should be noted that the viscosity is larger in water than in air, whereas the kinetic viscosity is larger in air than in water. If you paddle the water and the air by your hand, you feel much larger resistance in water than in air. In other words, water with a large viscosity is easier to transmit external force than air. By contrast, when you paddle by the same force, you can make a faster stream in the air with a larger kinetic viscosity.

2.6 REYNOLDS-AVERAGED NAVIER STOKES EQUATIONS

2.6.1 Momentum Equation of Viscous Fluid

It is noted that the Navier Stokes equation expressed by 2.5.19 describes instantaneous fluid motion. Since the instantaneous velocity varies in time, as shown in FIG. 2.18, it is inconvenient for the quantitative evaluation of the fluid characteristics. To overcome this, the instantaneous value is decomposed into the time-mean value and the turbulence component as follows:

$$\tilde{u}_i(t) = U_i + u_i(t) \tag{2.6.1}$$

$$\tilde{p}(t) = P + p(t) \tag{2.6.2}$$

in which tilde means the instantaneous value and the capital letter and small letter mean time mean value and turbulence component, respectively. This procedure is called the Reynolds decomposition.

After substitution of Eqs. (2.6.1) and (2.6.2) into 2.5.19, the instantaneous components are removed. Then, the time mean operations in

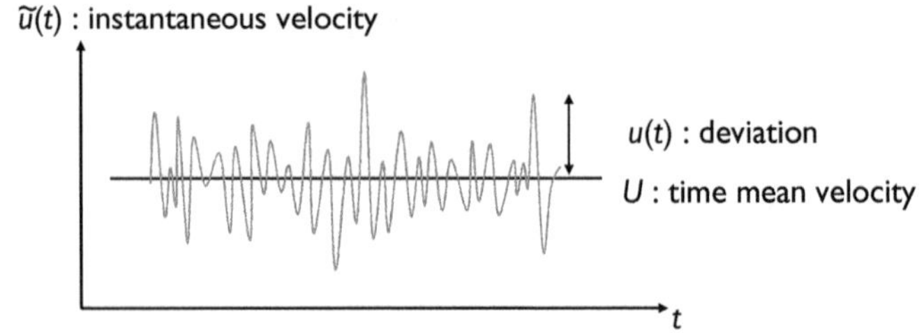

Figure 2.18 Time variation of instantaneous velocity and time mean velocity.

each term introduce a following form. Detailed mathematical expansion is descried in appendix A-1.

$$\frac{\partial U_i}{\partial t} + U_j \frac{\partial U_i}{\partial x_j} = -\frac{1}{\rho}\frac{\partial P}{\partial x_i} + \nu \frac{\partial^2 U_i}{\partial x_j \partial x_j} + \frac{\partial(-\overline{u_i u_j})}{\partial x_j} + F_i \quad (i, j = 1, 2, 3) \tag{2.6.3}$$

in which F_i is time mean external force, and over bar means the time mean operation. Eq. (2.6.3) is called the Reynolds averaged Navier Stokes (RANS) equation, which solves time mean values, i.e., U_i and P. However, the fourth term on the right-hand side is a product term of turbulence components and, therefore, Eq. (2.6.3) is not closed. This term is called the Reynolds stress, which is generally modeled for practical uses, as explained in section 3.5.

Chapter 3

2-D Open-Channel Turbulence

3.1 INTRODUCTORY REMARKS

This chapter specifically describes the turbulence structure in the straight 2-D open-channel flow with a smooth, flat bottom. It is well known that mean velocity and turbulence statistics varies significantly depending on the depth direction. The turbulence is generated near the bottom wall and it is conveyed toward the free-surface dissipating the turbulent kinetic energy. First, the concept of the boundary layer and related classical theories are explained, including the linear stability theories in turbulence generation. Detailed description of the governing equations and physical meaning of Reynolds stress are followed, and the vertical distributions of mean velocity and turbulence are explained based on previous studies. In the latter part, the turbulence properties in wave number space, including the Kolmogorov −5/3 law, are discussed. Finally, the interesting effects of the free surface on the turbulence structure are introduced.

3.2 BOUNDARY LAYER THEORY

3.2.1 What Is Boundary Layer?

When the viscous fluid flows around a solid body, the low-speed region is formed near the solid surface due to the viscous friction. This is called the boundary layer, which Prandtl first discussed. The mainstream over the boundary layer is not influenced by the solid surface. Prandtl simplified Navier Stokes equations to derive the governing equations about such a two-layer stream structure. It is important to consider how we define the boundary layer thickness and how this develops in the streamwise direction. The boundary layer thickness is closely related to the velocity profile, and Blasius gave a mathematical solution of the velocity profile. Karman proposed an integral formula of relation among streamwise variation of the mainstream velocity, wall shear stress and boundary layer thickness. The flow motion tends to be unstable in the boundary layer with the large

DOI: 10.1201/9781003112198-3

variation of the velocity profile. Hence, formation of the boundary layer contributes to turbulence generation or transition of laminar to turbulence.

3.2.2 Rayleigh Problem

The Rayleigh problem is one of the proper examples that help us to understand the physical meaning of the boundary layer formed near the solid wall. Here, think about two-dimensional flow fields, in which a flat plate with infinite length starts to move instantly in a constant speed U_0. The y axis is vertical to the plate, where the origin $y = 0$ is placed on the plate surface (FIG. 3.1).

The viscous friction acts on the plate surface, and the plate drags the fluid above it. This refers to $\tilde{u} = U_0 \quad (y = 0)$. In contrast, there is no influence apart from the plate, that is to say, $\tilde{u} = 0 \quad (y > 0)$. Then, how far does the viscous friction influence the plate surface? Let's consider this topic, analyzing a simplified N-S equation.

The unsteady term expressing the time-variation of local velocity cannot be deleted due to the unsteady phenomenon. The flow field is constant in horizontal direction x; we can think of $\partial/\partial x = 0$. Further, vertical fluid motion is assumed to be zero, $\tilde{v} = 0$. These conditions rewrite the N-S equation as follows:

$$\frac{\partial \tilde{u}}{\partial t} = \nu \frac{\partial^2 \tilde{u}}{\partial y^2} \tag{3.2.1}$$

in which the following boundary conditions are imposed.

$$\begin{aligned} t \le 0, &\quad \tilde{u} = 0\,(y \ge 0) \\ t > 0, &\quad u = U_0\,(y = 0) \\ t > 0, &\quad u = 0\,(y = \infty) \end{aligned}$$

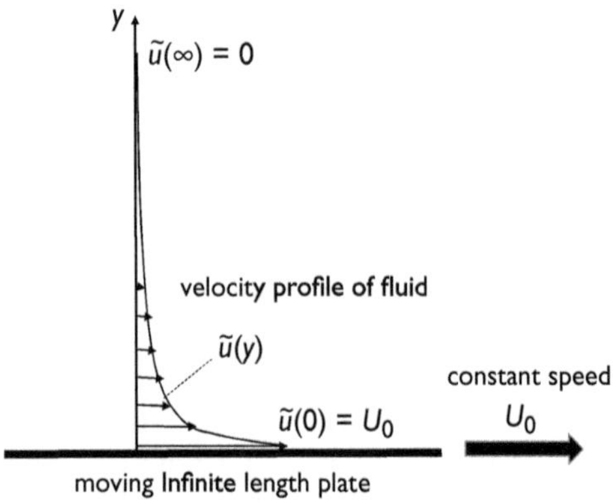

Figure 3.1 Velocity profile observed over the moving plate.

Note that the velocity is a function of two variables, y and t, i.e., $\tilde{u}(y, t)$. In order to connect these two variables, a following equation is introduced:

$$\eta = \frac{y}{2\sqrt{\nu t}} \tag{3.2.2}$$

Further, the velocity is normalized by using U_0.

$$f(\eta) = \frac{\tilde{u}}{U_0} \tag{3.2.3}$$

It implies that $\tilde{u}$ distributes in space and time coordinate $(y - t)$ is represented by a single function f and a single variable η.

The next step is to introduce the differential equation of f. Eq. (3.2.2) yields $\frac{\partial \eta}{\partial t} = -\frac{y}{4\sqrt{\nu}} t^{-3/2}$, $\frac{\partial \eta}{\partial y} = \frac{1}{2\sqrt{\nu t}}$.

By using them, the left-side hand term and the right-hand side term of (3.2.1) are rewritten in the following way, respectively:

$$\begin{aligned}
\frac{\partial \tilde{u}}{\partial t} &= U_0 \frac{\partial f}{\partial t} = U_0 \frac{df}{d\eta}\frac{\partial \eta}{\partial t} = -U_0 \frac{f' y}{4\sqrt{\nu}} t^{-3/2}, \\
\nu \frac{\partial^2 \tilde{u}}{\partial y^2} &= \nu U_0 \frac{\partial}{\partial y}\left(\frac{\partial f}{\partial y}\right) = \nu U_0 \frac{\partial}{\partial y}\left(\frac{df}{d\eta}\frac{\partial \eta}{\partial y}\right) = \nu U_0 \frac{1}{2\sqrt{\nu t}}\frac{\partial}{\partial y}\left(\frac{df}{d\eta}\right) \\
&= \nu U_0 \frac{1}{2\sqrt{\nu t}}\frac{\partial}{\partial \eta}\left(\frac{df}{d\eta}\right)\frac{\partial \eta}{\partial y} = \nu U_0 \left(\frac{1}{2\sqrt{\nu t}}\right)^2 f'' = \frac{U_0}{4t} f''
\end{aligned}$$

Here, the first-order differential form of f is $f' = \frac{\partial f}{\partial \eta} = \frac{df}{d\eta}$. Therefore, Eq. (3.2.1) is reformed to

$$-U_0 \frac{f' y}{4\sqrt{\nu}} t^{-3/2} = \frac{U_0}{4t} f'' \leftrightarrow \frac{y}{\sqrt{\nu t}} f' + f'' = 0 \leftrightarrow f'' + 2\eta f' = 0 \tag{3.2.4}$$

Eq. (3.2.4) is the second-order differential equation which can be solved under the boundary conditions, $f(0) = 1$, $f(\infty) = 0$. First, Eq. (3.2.4) is transformed like (3.2.4) $\leftrightarrow 2\eta = \frac{-f''}{f'} = -\frac{d(\ln f')}{d\eta}$. When this is integrated, it becomes $-\eta^2 = \ln f' + C_1 \leftrightarrow f' = e^{-\eta^2 - C_1} \leftrightarrow f' = C_2 e^{-\eta^2}$, in which C_1 and C_2 are integral constants. When integrated again in the range from 0 to η, we obtain

$$f(\eta) = C_2 \int_0^{\eta} e^{-\eta^2} d\eta + C_3 \tag{3.2.5}$$

in which C_3 is an integral constant. The B.C. $f(0) = 1$ gives $C_3 = 1$. Another B.C. $f(\infty) = 0$ leads $0 = C_2\left(\frac{\sqrt{\pi}}{2}\right) + 1$. Then, $C_2 = \frac{-2}{\sqrt{\pi}}$ is given by using the following Gauss's integral formula

$$\int_{-\infty}^{\infty} e^{-\alpha x^2} dx = \sqrt{\frac{\pi}{\alpha}}, \quad \int_{0}^{\infty} e^{-\alpha x^2} dx = \frac{1}{2}\sqrt{\frac{\pi}{\alpha}} \tag{3.2.6}$$

It results in $f(\eta) = -\frac{2}{\sqrt{\pi}} \int_0^{\eta} e^{-\eta^2} d\eta + 1$. From this, the space-time distribution of the velocity is as follows:

$$\tilde{u}(y, t) = U_0(1 - erf(\eta)) \tag{3.2.7}$$

where $erf(\eta) = \frac{2}{\sqrt{\pi}} \int_0^{\eta} e^{-\eta^2} d\eta$.

FIG. 3.2 shows the sketch of solution curves. They mean that the velocity reduction zone near the plate grows up over time.

They approach $\tilde{u} = 0$ with $y \to \infty$. Thus, we can define the height where the viscous friction influences, choosing conveniently $y = \delta$ corresponding to $\tilde{u}/U_0 = 0.5\%$. When we choose $\eta = 2$, $erf(2) = 0.9953$, it results in $\tilde{u}/U_0 = 0.5\%$. In this case, a following form is obtained:

$$\delta = 4\sqrt{\nu t} \tag{3.2.8}$$

This form suggests that the influence distance of the viscosity effect due to the plate depends on the kinematic viscosity and time scale. Viscous effect diffuses far away from the plate with the time. Eq. (3.2.1) is the same form

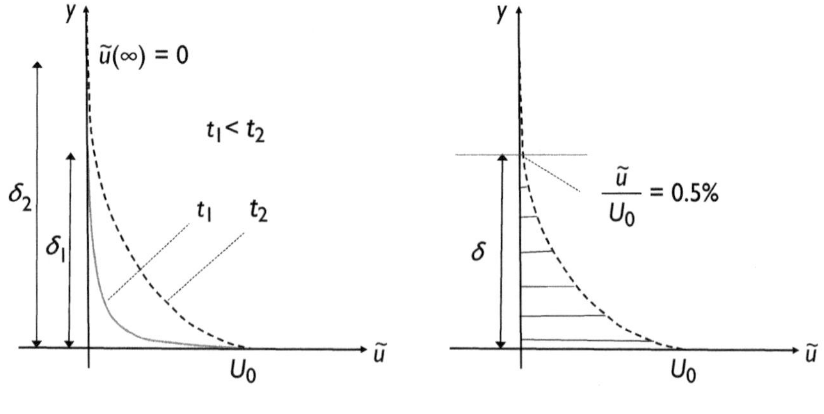

Figure 3.2 Time variation of solution curve and convenient definition of boundary layer thickness.

as the heat transfer equation, and thus the diffusion of the viscous influence region is expected to have the same mechanism observed in the heat and mass diffusions.

Let's move on to another situation, in which the long plate is placed in the uniform flow. The plate is parallel to the current direction. Although the flow velocity is uniform upstream of the plate, the velocity is reduced near the plate surface. The local velocity reduction is more remarkable due to the wall friction discussed previously, as the distance from the upstream-side edge of the plate becomes larger. The height δ where the velocity reduction appears is called the boundary layer.

A mild flow pattern is generally observed like laminar flow near the upstream side on the plate. Here we focus on the laminar boundary layer. Although this is a steady phenomenon, it varies in streamwise direction x. The above discussion gives $\delta \sim \sqrt{\nu t}$.

Considering that t is the time required for the fluid to travel downstream by the distance x from the board edge with a constant velocity U_∞, the following function of x is obtained.

$$\delta \sim \sqrt{\nu t} = \sqrt{\nu \frac{x}{U_\infty}} \tag{3.2.9}$$

3.2.3 Definitions of Boundary Layer

Blasius gave a following relation using the boundary layer equation assuming laminar condition.

$$\delta \cong 5.0\sqrt{\nu \frac{x}{U_\infty}} = 5.0x\,\mathrm{Re}_x^{-1/2} \tag{3.2.10}$$

The flow becomes more unstable even in the laminar boundary layer as it travels downstream. Then, it transitions to the turbulent boundary layer. A spatially increased rate in the turbulent boundary layer is quite different from that observed in the laminar boundary layer, as explained in section 3.3.

The boundary layer thickness δ is closely related to the skin friction and momentum transfer between the boundary layer and the free stream (FIG. 3.3). Thus, it is very important to predict it reasonably.

By the way, how do we evaluate δ from the existing mean velocity profile such as measured data or mathematically expressed one? Here, this section covers three kinds of definitions. Given the mean velocity profile near the bottom wall, we are able to evaluate quantitatively in one of the following ways, in which U_∞ is the oncoming or free stream velocity, and $U(y)$ is the vertical profile of the time-mean streamwise velocity without consideration of the turbulence disturbance.

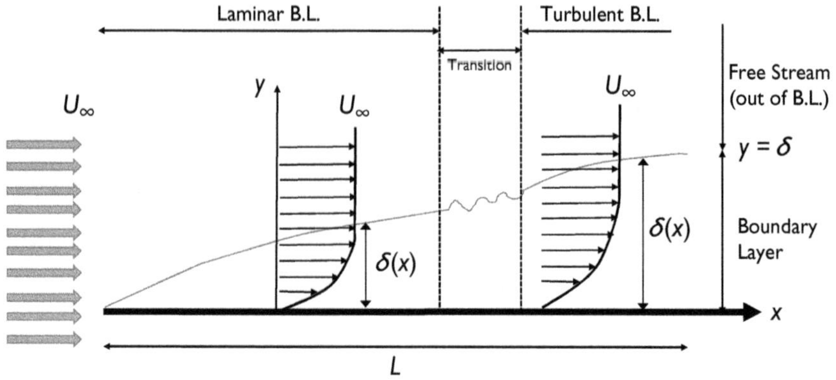

Figure 3.3 Development of boundary layer on the flat plate.

3.2.3.1 99% thickness

To use an elevation where the mean velocity is the 99% value of the free stream is the simplest method, although there is no physical meaning. The boundary layer thickness evaluated in this method is represented as δ_{99} in this book (FIG. 3.4).

3.2.3.2 Displacement thickness

In the perfect fluid without viscous stress, the fluid is not decelerated by the wall friction.

Consequently, $U(y)$ is a constant, i.e., U_∞ in the vertical direction including to wall region. However, in the viscous fluid, the non-slip condition is imposed, i.e., $U(0) = 0$, and the flow velocity is reduced only in the shaded area, as shown in FIG. 3.5. When the velocity distribution is considered to be constant up to the wall surface, in order to match the discharge flux of a certain cross section with that considering viscous reduction, only the diagonal rectangular part needs to be excluded from the

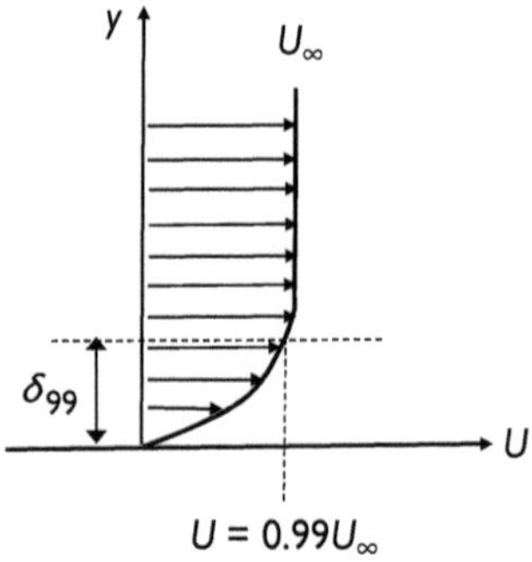

Figure 3.4 99% thickness.

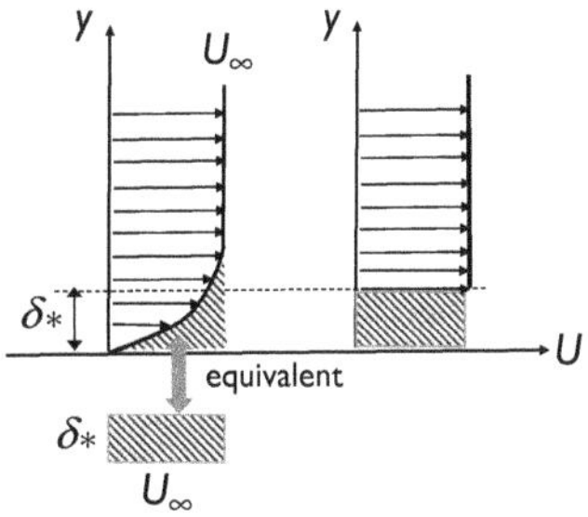

Figure 3.5 Displacement thickness.

flow velocity distribution. A height of the rectangle should be removed, called a displacement thickness. The boundary layer thickness in this method is given as δ^* here.

Since the area of two regions indicated by diagonal lines are just same.

$$U_\infty \delta^* = \int_0^\infty (U_\infty - U(y))\,dy \quad \leftrightarrow \delta^* = \int_0^\infty (U_\infty - U(y))\,dy / U_\infty \tag{3.2.11}$$

3.2.3.3 Momentum thickness

A momentum thickness θ is defined using the reduction of momentum flux per unit time in the same manner as the displacement thickness. The momentum is defined as a product of mass flux per unit time and velocity. Considering unit width in the spanwise direction, the mass flux passing a cross-section per unit time is given by $\rho U(y)dy$. Hence, the momentum flux is $\rho U(y)^2 dy$. We express the reduction of momentum thickness by the bottom friction at elevation y, as $\rho U(y)(U_\infty - U(y))$. The definition of θ assumes that the integral in the range from the bottom to infinity equals the area of the rectangle consisting of sides ρU_∞^2 and θ. Therefore, they are given by

$$\rho U_\infty^2 \theta = \int_0^\infty \rho U(y)(U_\infty - U(y))\,dy \quad \leftrightarrow \theta = \int_0^\infty U(y)(U_\infty - U(y))\,dy / U_\infty^2 \tag{3.2.12}$$

As the boundary layer thickness δ is unknown, integral range is mathematically chosen from $y = 0$ to ∞ in order to calculate the displacement thickness or the momentum thickness. When the discrete data of $U(y)$

measured in experiments is applied, the upper limit of the integral range needs to be cut off at a proper elevation y_p much larger than δ. A numerical integration is able to provide the values of δ^* and θ using Eqs. (3.2.11) or (3.2.12) with the range from $y = 0$ to y_p. If y_p is not chosen in the free-stream layer which has a constant velocity, the calculation result may be meaningless.

3.2.4 Boundary Layer Equations

The boundary layer thickness δ is much larger than the plate length L (or the traveling distance). Simplifying the N-S equations by using this scaling property, the boundary layer equations are introduced.

Two dimensional N-S equations and the continuity equation are as follows:

1. N-S equation for streamwise direction (x)

$$\frac{\partial \tilde{u}}{\partial t} + \tilde{u}\frac{\partial \tilde{u}}{\partial x} + \tilde{v}\frac{\partial \tilde{u}}{\partial y} = -\frac{1}{\rho}\frac{\partial \tilde{p}}{\partial x} + \nu\left(\frac{\partial^2 \tilde{u}}{\partial x^2} + \frac{\partial^2 \tilde{u}}{\partial y^2}\right) \tag{3.2.13}$$

2. N-S equation for vertical direction (y)

$$\frac{\partial \tilde{v}}{\partial t} + \tilde{u}\frac{\partial \tilde{v}}{\partial x} + \tilde{v}\frac{\partial \tilde{v}}{\partial y} = -\frac{1}{\rho}\frac{\partial \tilde{p}}{\partial y} + \nu\left(\frac{\partial^2 \tilde{v}}{\partial x^2} + \frac{\partial^2 \tilde{v}}{\partial y^2}\right) \tag{3.2.14}$$

3. Continuity equation

$$\frac{\partial \tilde{u}}{\partial x} + \frac{\partial \tilde{v}}{\partial y} = 0 \tag{3.2.15}$$

On the surface of the flat plate ($y = 0$), $\tilde{u} = \tilde{v} = 0$ (No-slip condition) is imposed as a boundary condition. In contrast, $\tilde{u} = U_\infty$, $\tilde{v} = 0$ is imposed in the infinite elevation, $y = \infty$. Several scaling factors are used to consider the developing boundary layer problem, as shown in FIG. 3.3. They are listed below.

U_∞: free stream velocity (typical velocity scale in x direction)
L: plate length or traveling distance from the leading edge (typical length scale in x)
δ: boundary layer thickness (typical length scale in y)

The result of the Rayleigh problem as explained in section 3.2.2 led to

$$\delta \sim \sqrt{\nu \frac{x}{U_\infty}} = x\,\mathrm{Re}_x^{-1/2} \tag{3.2.16}$$

Substituting $x = L$ into Eq. (3.2.16), $\frac{\delta}{L} \sim \mathrm{Re}_L^{-1/2}$ is obtained. Therefore, as the flow goes downstream, $\delta \ll L$ is obtained.

First, focus on the continuity equation, in which the boundary condition $\tilde{v} = 0$ at $y = 0$ is imposed. Integration from $y = 0$ to some elevation within the boundary layer expresses instantaneous vertical velocity.

$$\tilde{v} = -\int_0^y \frac{\partial \tilde{u}}{\partial x} dy \tag{3.2.17}$$

This result yields a following scaling relation:

$$\tilde{v} \sim U_\infty \delta / L \sim U_\infty \mathrm{Re}_L^{-1/2} \tag{3.2.18}$$

Thus, the vertical velocity is smaller, as the Reynolds number increases.

Therefore, scaling characteristics in the two-dimensional boundary layer are concluded in the following:

$$\begin{aligned} &\tilde{u} \sim U_\infty \\ &\tilde{v} \sim U_\infty \delta / L \\ &x \sim L \\ &y \sim \delta \end{aligned}$$

These allow to simplify N-S equations using smaller terms compared to others. Here, we consider within the boundary layer and the free-stream layer separately.

1. N-S equation within the boundary layer (x-direction)

$$\text{(convection terms)} \quad \tilde{u}\frac{\partial \tilde{u}}{\partial x} \sim \frac{U_\infty^2}{L}, \quad \tilde{v}\frac{\partial \tilde{u}}{\partial y} \sim \frac{U_\infty^2}{\delta} \tag{3.2.19}$$

$$\text{(viscous terms)} \quad \frac{\partial^2 \tilde{u}}{\partial x^2} \sim \frac{U_\infty}{L^2}, \quad \frac{\partial^2 \tilde{u}}{\partial y^2} \sim \frac{U_\infty}{\delta^2} \tag{3.2.20}$$

Focusing on Eq. (3.2.20), the relation of $\frac{\partial^2 \tilde{u}}{\partial x^2} \ll \frac{\partial^2 \tilde{u}}{\partial y^2}$ is able to delete $\frac{\partial^2 \tilde{u}}{\partial x^2}$ due to $L^2 \gg \delta^2$. Hence, the N-S equation is rewritten by

$$\frac{\partial \tilde{u}}{\partial t} + \tilde{u}\frac{\partial \tilde{u}}{\partial x} + \tilde{v}\frac{\partial \tilde{u}}{\partial y} = -\frac{1}{\rho}\frac{\partial \tilde{p}}{\partial x} + \nu\frac{\partial^2 \tilde{u}}{\partial y^2} \tag{3.2.21}$$

2. N-S equation within the boundary layer (y-direction)

The instantaneous vertical velocity is scaled by $\tilde{v} \sim U_\infty \delta / L$, which is very small. Thus, after terms other than the pressure gradient term are deleted, it is simplified to a following form:

$$0 = -\frac{1}{\rho}\frac{\partial \tilde{p}}{\partial y} \tag{3.2.22}$$

This implies that the pressure is constant in the vertical direction. In the free-stream layer over the boundary layer, there is no variation of the current structure with the vertical direction. Hence, the pressures in the two layers meet, and it is found that the pressure is propagated from the free stream to the boundary layer.

Momentum equations of the boundary are given by Eq. (3.2.21) and (3.2.22).

3. N-S equation in free-stream layer

We may consider only the streamwise direction in the free layer, where perfect fluid is assumed. In this layer, the current is considered potential flow without viscosity. Further, unidirectional flow is assumed, and it results in $\tilde{u} = U_\infty$, $\nu = 0$, $\partial/\partial y = 0$.

Hence, the N-S equation for the streamwise direction is replaced by

$$\frac{\partial U_\infty}{\partial t} + U_\infty \frac{\partial U_\infty}{\partial x} = -\frac{1}{\rho}\frac{\partial \tilde{p}}{\partial x} \tag{3.2.23}$$

Eqs. (3.2.21) to (3.2.23) are called boundary layer equations. The N-S equation in the x-direction, Eq. (3.2.13), is an elliptic differential equation including second-order differential of x and y. In contrast, the boundary layer equation, Eq. (3.2.21), is a parabolic partial differential equation. The latter has an advantage that a solution depends on only the upstream condition.

They are applied not only at the laminar boundary layer but also the turbulent boundary layer. However, instantaneous velocities must be considered within the boundary layer. When time-mean velocities are considered for practical purposes, it is likely to cause significant gaps between theory and reality. In the case of application to the turbulent boundary layer, rational measures including Reynolds decomposition, etc. should be taken into account.

3.2.5 Blasius's Solution

Blasius solved the velocity profile in the laminar boundary layer using the boundary layer equation, assuming conditions of steady, laminar and no pressure gradient. The coordinate system is shown in FIG. 3.6.

The boundary layer Eq. (3.2.21) is simplified to

$$U\frac{\partial U}{\partial x} + V\frac{\partial U}{\partial y} = \nu\frac{\partial^2 U}{\partial y^2} \tag{3.2.24}$$

The two coordinate variables (x, y) are transformed to (ξ, η) using Eq. (3.2.9).

$$\xi = x, \quad \eta = \frac{y}{\delta} = y\sqrt{\frac{U_\infty}{\nu x}} \tag{3.2.25}$$

In order to reduce the number of variables, we introduce the stream function $\Psi(\xi, \eta)$, defined by

$$\Psi(\xi, \eta) = \sqrt{\nu x U_\infty} f(\eta) \tag{3.2.26}$$

How is this definition introduced? At first, we assume that the velocity profile is expressed as a function $g(\eta)$, i.e., $\frac{U}{U_\infty} = g(\eta)$. Second, after $U = \frac{\partial \Psi}{\partial y}$ is integrated, $\Psi = \int U dy + c(x)$ is obtained. Here Eq. (3.2.25) gives $d\eta = dy\sqrt{\frac{U_\infty}{\nu x}}$; thus, the stream function is expanded in the following way.

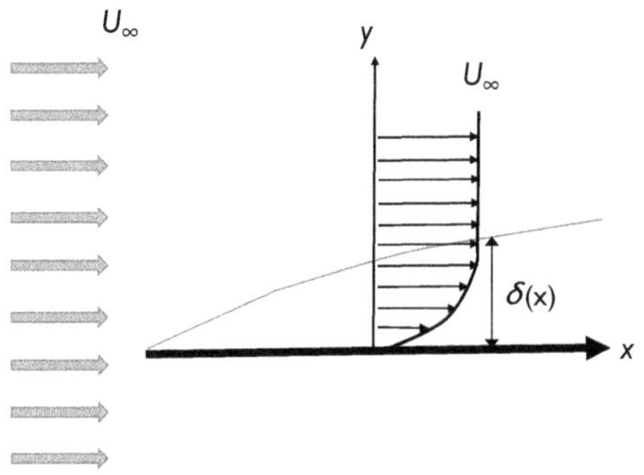

Figure 3.6 Coordinate of laminar boundary layer.

$$\Psi = \int U dy + c(x) = \int U_\infty g(\eta) dy + c(x) = U_\infty \sqrt{\frac{\nu x}{U_\infty}} \int g(\eta) d\eta + c(x)$$
$$= \sqrt{U_\infty \nu x} \int g(\eta) d\eta + c(x)$$

It results in $V = -\frac{\partial \Psi}{\partial x} = -\frac{\partial}{\partial x}(\sqrt{U_\infty \nu x} \int g(\eta) d\eta) - \frac{\partial}{\partial x} c(x)$.

The no-slip condition gives $U = V = 0$ at $y = 0 (\eta = 0)$, that is, $g(0) = 0$ and $V(\eta = 0) = -\frac{\partial}{\partial x}(\sqrt{U_\infty \nu x} \int g(0) d\eta) - \frac{\partial}{\partial x} c(x) = -\frac{\partial}{\partial x} c(x)$. Hence, $c(x) = 0$. Finally, $\Psi = \sqrt{U_\infty \nu x} \int g(\eta) d\eta$ is obtained. Here, defining $f(\eta) = \int g(\eta) d\eta$, Eq.(3.2.26) is introduced.

The velocity profile in the laminar boundary layer is theoretically introduced by solving the boundary layer equation expressed with $f(\eta)$. See appendix 2, which describes the detailed calculation process.

3.2.6 Karman's Integral Equation of Momentum

Karman's integral equation of momentum is a significant formula expressing the relation between the velocity profile and the bottom friction. The boundary layer equation in a streamwise component can be rewritten through Reynolds decomposition as follows:

$$\frac{\partial U}{\partial t} + U\frac{\partial U}{\partial x} + V\frac{\partial U}{\partial y} = -\frac{1}{\rho}\frac{\partial P}{\partial x} + \nu\frac{\partial^2 U}{\partial y^2} - \frac{\partial \overline{uv}}{\partial y},$$

in which the normal Reynolds stress term is neglected due to $\frac{\partial - \overline{uu}}{\partial x} \ll \frac{\partial - \overline{uv}}{\partial y}$.

Here is a height γ, which is above the boundary layer. After integral from $y = 0$ to γ,

$$\int_0^\gamma \frac{\partial U}{\partial t} dy + \int_0^\gamma U\frac{\partial U}{\partial x} dy + \int_0^\gamma V\frac{\partial U}{\partial y} dy$$
$$= -\int_0^\gamma \frac{1}{\rho}\frac{\partial P}{\partial x} dy + \nu\int_0^\gamma \frac{\partial^2 U}{\partial y^2} dy + \int_0^\gamma \frac{\partial - \overline{uv}}{\partial y} dy \tag{3.2.27}$$

Summation of the second and third terms on the right-hand side is expressed as

$$\nu\int_0^\gamma \frac{\partial^2 U}{\partial y^2} dy + \int_0^\gamma \frac{\partial - \overline{uv}}{\partial y} dy = \nu\left[\frac{\partial U}{\partial y}\right]_0^\gamma + [-\overline{uv}]_0^\gamma$$
$$= 0 - \nu\left[\frac{\partial U}{\partial y}\right]_0 + 0 - 0 = -\tau_w/\rho.$$

It should be noted that Reynolds stress is zero on the plate surface due to the no-slip condition. However, the velocity gradient $\partial U/\partial y$ is not zero, and it contributes to the wall friction. Newton's law gives $\tau_w = \mu \partial U/\partial y$.

Above the boundary layer, ($y > \delta$), $\partial U/\partial y = 0$ is used due to uniform flow condition. Further, there is no turbulence generation because of $\partial U/\partial y = 0$, and thus Reynolds stress is also not produced. Substituting the continuity equation, $V = -\int_0^y \frac{\partial U}{\partial x} dy$, the third term on the left-hand side is transformed in the following way:

$$\int_0^\gamma V\frac{\partial U}{\partial y}dy = -\int_0^\gamma \left(\frac{\partial U}{\partial y}\int_0^y \frac{\partial U}{\partial x}dy\right)dy = -\left[U\int_0^y \frac{\partial U}{\partial x}dy\right]_0^\gamma + \int_0^\gamma U\frac{\partial U}{\partial x}dy$$
$$= -U_\infty \int_0^\gamma \frac{\partial U}{\partial x}dy + \int_0^\gamma U\frac{\partial U}{\partial x}dy$$

Therefore, Eq. (3.2.27) is rewritten as

$$\int_0^\gamma \frac{\partial U}{\partial t}dy + \int_0^\gamma 2U\frac{\partial U}{\partial x}dy - U_\infty \int_0^\gamma \frac{\partial U}{\partial x}dy = -\int_0^\gamma \frac{1}{\rho}\frac{\partial P}{\partial x}dy - \frac{\tau_w}{\rho}$$

Removing the first term on right-hand side of the boundary layer equation in the free layer, Eq. (3.2.23), with $\tilde{p} = P$

$$\begin{aligned} &\int_0^\gamma \frac{\partial (U_\infty - U)}{\partial t}dy - \int_0^\gamma 2U\frac{\partial U}{\partial x}dy + U_\infty \int_0^\gamma \frac{\partial U}{\partial x}dy + U_\infty \int_0^\gamma \frac{\partial U_\infty}{\partial x}dy = \frac{\tau_w}{\rho} \\ &\leftrightarrow \int_0^\gamma \frac{\partial (U_\infty - U)}{\partial t}dy - \int_0^\gamma \left(2U\frac{\partial U}{\partial x} - U_\infty\frac{\partial U}{\partial x} - U_\infty\frac{\partial U_\infty}{\partial x}\right)dy = \frac{\tau_w}{\rho} \end{aligned} \tag{3.2.28}$$

is obtained. Further, the integrand in the second term of the left side is transformed in the following form:

$$\begin{aligned} 2U\frac{\partial U}{\partial x} - U_\infty\frac{\partial U}{\partial x} - U_\infty\frac{\partial U_\infty}{\partial x} &= \frac{\partial U^2}{\partial x} - \frac{\partial U_\infty U}{\partial x} + U\frac{\partial U_\infty}{\partial x} - U_\infty\frac{\partial U_\infty}{\partial x} \\ &= \frac{\partial \{U(U - U_\infty)\}}{\partial x} - (U_\infty - U)\frac{\partial U_\infty}{\partial x} \end{aligned}$$

Hence, Eq. (3.2.28) is rewritten as follows:

$$\int_0^{\gamma} \frac{\partial (U_\infty - U)}{\partial t} dy - \int_0^{\gamma} \frac{\partial}{\partial x}\{U(U - U_\infty)\} dy + \int_0^{\gamma} (U_\infty - U)\frac{\partial U_\infty}{\partial x} dy = \frac{\tau_w}{\rho}$$

Here, we can change the upper limit of the integral range from γ to ∞, since the integrant becomes zero for $y \geq \gamma$. We can change the operation order of the differential and the integral, because x, y, t are independent. Thus,

$$\frac{\partial}{\partial t}\int_0^{\infty} (U_\infty - U) dy + \frac{\partial}{\partial x}\int_0^{\infty} \{U(U_\infty - U)\} dy + \frac{\partial U_\infty}{\partial x}\int_0^{\infty} (U_\infty - U) dy = \frac{\tau_w}{\rho}.$$

This form can be expressed as

$$\frac{\partial U_\infty \delta^*}{\partial t} + \frac{\partial U_\infty{}^2 \theta}{\partial x} + U_\infty \delta^* \frac{\partial U_\infty}{\partial x} = \frac{\tau_w}{\rho} \tag{3.2.29}$$

This is called Karman's integral equation of momentum, which connects the streamwise variation of the uniform velocity and the wall shear stress. Of particular significance is that this equation can be applied to both the laminar and turbulent flows.

3.3 TURBULENCE GENERATION

3.3.1 Transition of Laminar Boundary Layer to Turbulent Boundary Layer

There are many theoretical, experimental, and numerical works on turbulent generation. A disturbance in the laminar boundary layer appears and after then it grows into a two-dimensional fluctuation called Tollmien–Schlichting waves. Further, turbulent spots like spike and spanwise vortices are produced and they grow into a three-dimensional turbulence structure (see e.g., Robinson 1991). Although a mainstream in the river is fully developed turbulent flow, turbulence and related vortices are newly produced in the zones where the velocity varies locally, e.g., the confluence region, drop zone and mainstream/embayment boundary. This section focuses on the transition of a laminar boundary layer to a turbulent

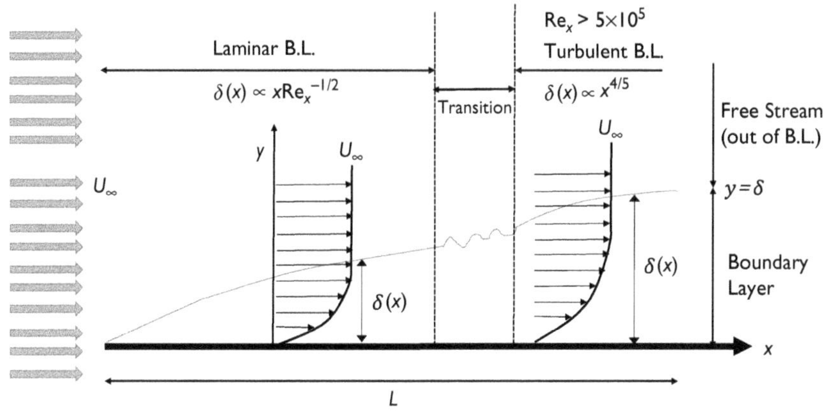

Figure 3.7 Laminar and turbulent boundary layers.

one and the shear instability that is often observed in the compound open-channel flows.

The laminar boundary layer (LBL) is formed near the upstream-side edge of the plate as previously described. Small oscillations and fluctuations appear a little downstream due to an external small disturbance. A part of them grows up and become turbulence further downstream with the condition of $Re_x (= U_\infty x/\nu) > 5 \times 10^5$, as shown in FIG. 3.7.

What are characteristics of the turbulence in the boundary layer? Other than visual impressions such as many vortices, shapeless violent motions and random fluctuation, the physical properties are listed as follows:

1. The boundary thickness, δ, in the TBL is developing proportional to $x^{4/5}$. In contrast, the development of δ in the LBL is proportional to $x\,Re_x^{-1/2}$. This fact means that δ grows downstream more rapidly in the TBL than in the LBL.
2. The velocity profile in the TBL is a logarithmic type and the velocity gradient near the bottom is much larger than in the LBL introduced by Blasius, as shown in FIG. 3.8.

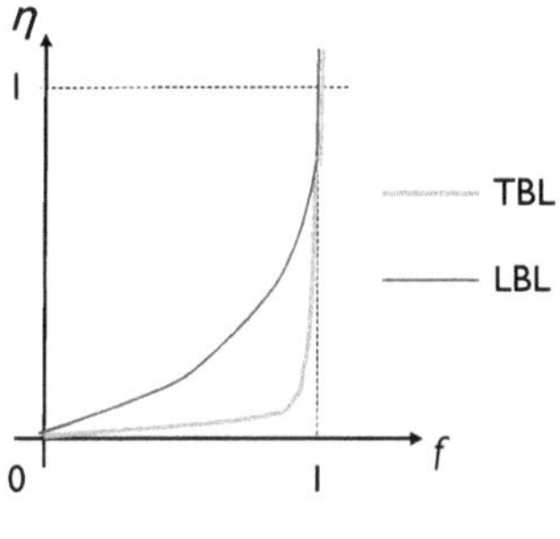

Figure 3.8 Typical velocity profiles in laminar and turbulent boundary layers.

3. It is known that the relationship of friction coefficient and Reynolds number is also different for the LBL and the TBL.

3.3.2 Development of Turbulent Boundary Layer

This section introduces the developing property of the boundary layer thickness in the TBL, i.e., the relation of $\delta \propto x^{4/5}$. First, a power law is applied to the velocity profile as follows:

$$\frac{U(y)}{U_\infty} = \left(\frac{y}{\delta}\right)^{1/n} \tag{3.3.1}$$

in which $n = 7$ is often used in the boundary layer flow on the flat plate and the open-channel flow (1/7 power law). In order to apply Eq. (3.3.1) to Karman's integral formula, we express the displacement thickness and the momentum thickness by the scaling function $\phi(\eta) = U/U_\infty = \eta^{1/n}$, in which $\eta = y/\delta$.

The displacement thickness can be explored as follows:

$$\delta^* = a_2\delta = \int_0^1 (1 - \phi)\,d\eta\cdot\delta = \int_0^1 (1 - \eta^{1/n})\,d\eta\cdot\delta$$

$$= \left[\eta - \frac{n}{n+1}\eta^{\frac{n+1}{n}}\right]_0^1 \cdot \delta = \frac{\delta}{n+1}$$

in the case of $n = 7$,

$$\delta^* = \frac{\delta}{8} \tag{3.3.2}$$

The momentum thickness can be explored as follows:

$$\theta = a_1\delta = \int_0^1 \phi(1 - \phi)\,d\eta\cdot\delta = \int_0^1 (\eta^{1/n} - \eta^{2/n})\,d\eta\cdot\delta$$

$$= \left[\frac{n}{n+1}\eta^{\frac{n+1}{n}} - \frac{n}{n+2}\eta^{\frac{n+2}{n}}\right]_0^1 \cdot\delta = \frac{n\delta}{(n+1)(n+2)}$$

in the case of $n = 7$,

$$\theta = \frac{7}{72}\delta \tag{3.3.3}$$

Second, we consider the steady current over the flat plate with zero attacking angle, in which background uniform velocity is U_∞, Karman's integral equation is rewritten as

$$U_\infty{}^2 \frac{d\theta(x)}{dx} = \frac{\tau_0(x)}{\rho} \tag{3.3.4}$$

Substituting Eq. (3.3.3) into (3.3.4) gives

$$\frac{\tau_0(x)}{\rho U_\infty{}^2} = \frac{7}{72}\frac{d\delta}{dx} \tag{3.3.5}$$

Here, a following equation is derived from the Blasius's experimental formula for the pipe flow. This is applied to the boundary layer on the flat plate.

$$\frac{\tau_0(x)}{\rho U_\infty{}^2} = 0.0225\left(\frac{U_\infty \delta}{\nu}\right)^{-1/4} \tag{3.3.6}$$

Eliminating $\tau_0(x)/\rho U_\infty{}^2$ from Eqs. (3.3.5) and (3.3.6) gives

$$0.0225\left(\frac{U_\infty \delta}{\nu}\right)^{-1/4} = \frac{7}{72}\frac{d\delta}{dx} \quad \leftrightarrow \int_0^x \left(\frac{0.0225 \times 72}{7}\right)\left(\frac{U_\infty}{\nu}\right)^{-1/4} dx = \int_0^\delta \delta^{1/4} d\delta$$

$$\leftrightarrow \delta = 0.387\left(\frac{U_\infty}{\nu}\right)^{-1/5} x^{4/5} + C.$$

When $\delta = 0$ is given at $x = x_t$

$$\delta = 0.387\left(\frac{U_\infty}{\nu}\right)^{-1/5} (x - x_t)^{4/5} \tag{3.3.7}$$

is introduced.

Note that we assume the turbulent boundary layer thickness is zero at $x = x_t$, which appears in the transition from LBL to TBL. However, the boundary layer thickness is not really zero. Hence, δ at $x = x_t$ is evaluated with a reference value calculated by a following equation for the LBL

$$\delta \cong 5.0\sqrt{\nu \frac{x}{U_\infty}} = 5.0x\,\mathrm{Re}_x{}^{-1/2} \tag{3.3.8}$$

Although it is difficult to define strictly x_t, the combination of Eqs. (3.3.7) and (3.3.8) reveals the development process of the boundary layer.

3.3.3 Turbulence Generation

a. How to evaluate growth of disturbance

Why does laminar flow become unstable in the boundary layer? Why is the state of the laminar flow not maintained while traveling downstream? The Orr-Sommerfeld equation contributed significantly to the solution of this problem. Now, consider how to evaluate the growth of disturbance in the boundary layer under the following assumptions:

1. Spatial variation of streamwise velocity is very small, i.e., $\partial U/\partial x \simeq 0$,
2. Mean vertical velocity V is negligible compared to U, i.e., $V \simeq 0$
3. Spatial variation of pressure is also small, i.e., $\partial P/\partial x \simeq 0,\ \ \partial P/\partial y \simeq 0$.

Then, the instantaneous velocity is decomposed into the mean and turbulence components in the following:

$$\begin{aligned} \tilde{u}(x, y, t) &= U(y) + u(x, y, t) \\ \tilde{v}(x, y, t) &= v(x, y, t) \\ \tilde{p}(x, y, t) &= P(y) + p(x, y, t) \end{aligned} \tag{3.3.9}$$

They are substituted into the N-S equations and then second-order terms are ignored. For example, the convection term in the x component of the N-S equation can be expressed as follows:

$$\tilde{u}\frac{\partial \tilde{u}}{\partial x} = (U + u)\frac{\partial (U + u)}{\partial x} \simeq (U + u)\frac{\partial u}{\partial x} \simeq U\frac{\partial u}{\partial x}$$

Exploring also other terms in the same way, the transport equations of the turbulence components are obtained from N-S equations and continuity equation.

$$\begin{aligned} &\frac{\partial u}{\partial t} + U\frac{\partial u}{\partial x} + v\frac{\partial U}{\partial y} + \frac{1}{\rho}\frac{\partial p}{\partial x} = \nu\nabla^2 u \\ &\frac{\partial v}{\partial t} + U\frac{\partial v}{\partial x} + \frac{1}{\rho}\frac{\partial p}{\partial y} = \nu\nabla^2 v \\ &\frac{\partial u}{\partial x} + \frac{\partial v}{\partial y} = 0 \end{aligned} \tag{3.3.10}$$

It should be noted that they are linear equations with an advantage for theoretical exploration.

When small disturbance occurs in the stream, how they grow or decay? We assume that the disturbance is a group of sine waves with various wave lengths and periods. Hence, it can be expressed by a following form:

$$a_1 \sin(\alpha_1 x - \beta_1 t) + a_2 \sin(\alpha_2 x - \beta_2 t) + a_3 \sin(\alpha_3 x - \beta_3 t) + \cdots + a_n \sin(\alpha_n x - \beta_n t) \tag{3.3.11}$$

in which a_n is amplitude, $a_n = 2\pi/L_n$ (L_n is wavelength) and $\beta_n = 2\pi/T_n$ (T_n: period).

When superimposed disturbance is a solution of Eq. (3.3.10), an individual component of Eq. (3.3.11) also becomes a solution since Eq. (3.3.10) is linear. We focus on a representative component of Eq. (3.3.11), $a \sin(\alpha x - \beta t)$. Introducing the complex number $z = e^{i\theta} = \cos\theta + i\sin\theta$, $\sin\theta$ refers the variation in the imaginary axis in FIG. 3.9. Thus, the wave component is rewritten as $ae^{i(\alpha x - \beta t)} = ae^{i\alpha(x-ct)}$. While $z = e^{i\theta}$ orbits a circle with the radius of one, $\sin\theta$ reciprocates along the imaginary axis.

Assuming that turbulence components u, v, p fluctuate with same wave number, the period due to the external disturbance, they are expressed as follows:

$$\begin{aligned} u(x, y, t) &= \tilde{u}(y)e^{i\alpha(x-ct)} \\ v(x, y, t) &= \tilde{v}(y)e^{i\alpha(x-ct)} \\ p(x, y, t) &= \tilde{p}(y)e^{i\alpha(x-ct)} \end{aligned} \tag{3.3.12}$$

in which $\tilde{u}$, $\tilde{v}$, $\tilde{p}$ refers the amplification.

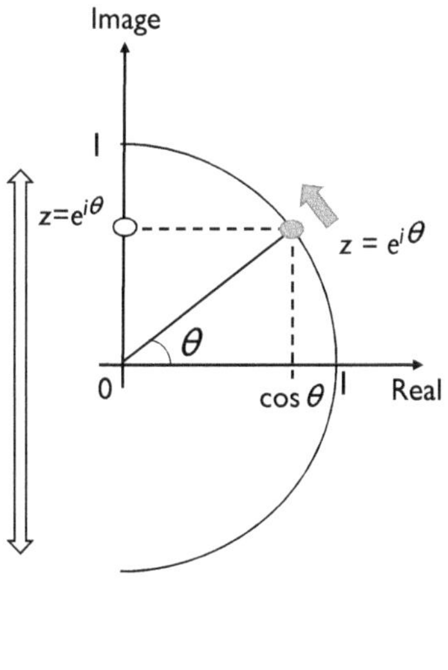

Figure 3.9 Complex plane for expression of a wave component.

Let's focus on the wave component $e^{i\alpha(x-ct)}$ again. If α is the real number and $c = c_r + ic_i$ is the complex number, then $e^{i\alpha(x-ct)} = e^{\alpha c_i t} e^{i\alpha(x-c_r t)}$. $e^{i\alpha(x-c_r t)}$ means a periodical variation and $e^{\alpha c_i t}$ means the developing rate of the amplitude. $\alpha > 0$ is the wave number. Hence, the time growth of disturbance is summarized by the sign of c_i as follows:

$c_i > 0$: disturbance develops and it results in turbulence generation
$c_i < 0$: disturbance decreases without turbulence generation
$c_i = 0$: neutral, where the amplitude of disturbance is constant

b. Orr-Sommerfeld equation

In order to normalize Eq. (3.3.10) using the free stream U_∞ and the displacement thickness δ^*, the following normalized variables are considered.

$$\hat{x} = x/\delta^*, \quad \hat{y} = y/\delta^*, \quad \hat{U} = u/U_\infty, \quad \hat{u} = u/U_\infty, \quad \hat{v} = v/U_\infty,$$
$$\hat{p} = p/(\rho U_\infty{}^2), \quad \hat{t} = tU_\infty/\delta^*$$

Substituting them to the top equation in Eq. (3.3.10) gives a following normalized expression:

$$\frac{\partial(U_\infty \hat{u})}{\partial(\delta^* \hat{t}/U_\infty)} + U_\infty \hat{U}\frac{\partial(U_\infty \hat{u})}{\partial(\delta^* \hat{x})} + U_\infty \hat{v}\frac{\partial(U_\infty \hat{u})}{\partial(\delta^* \hat{y})} = -\frac{1}{\rho}\frac{\partial(\rho U_\infty{}^2 p')}{\partial(\delta^* \hat{x})} + \nu\left(\frac{\partial^2(U_\infty \hat{u})}{\partial(\delta^* \hat{x})^2} + \frac{\partial^2(U_\infty \hat{u})}{\partial(\delta^* \hat{y})^2}\right)$$

$$\leftrightarrow \frac{\partial \hat{u}}{\partial \hat{t}} + \hat{U}\frac{\partial \hat{u}}{\partial \hat{x}} + \hat{v}\frac{\partial \hat{U}}{\partial \hat{y}} = -\frac{\partial \hat{p}}{\partial \hat{x}} + \frac{1}{\text{Re}}\nabla^2 \hat{u} \tag{3.3.13}$$

The middle and bottom in Eq. (3.3.10) are also normalized in the same way as follows:

$$\frac{\partial \hat{v}}{\partial \hat{t}} + \hat{U}\frac{\partial \hat{v}}{\partial \hat{x}} = -\frac{\partial \hat{p}}{\partial \hat{y}} + \frac{1}{\text{Re}}\nabla^2 \hat{v} \tag{3.3.14}$$

$$\frac{\partial \hat{u}}{\partial \hat{x}} + \frac{\partial \hat{v}}{\partial \hat{y}} = 0 \tag{3.3.15}$$

in which $\text{Re} = U_\infty \delta^*/\nu$.

Turbulence components obey the continuity equation of Eq. (3.3.15), and they can be expressed by the stream function.

$$\hat{u} = \frac{\partial \psi}{\partial y'}, \quad \hat{v} = -\frac{\partial \psi}{\partial x'} \tag{3.3.16}$$

Substituting Eq. (3.3.16) into Eqs. (3.3.13) and (3.3.14), we can differentiate them by y, x, respectively. Then, eliminating the pressure gradient term, a following equation is introduced:

$$\frac{\partial}{\partial t}\frac{\partial^2 \psi}{\partial x^2} + \frac{\partial}{\partial t}\frac{\partial^2 \psi}{\partial y^2} + U\frac{\partial^3 \psi}{\partial x \partial y^2} + U\frac{\partial^3 \psi}{\partial x^3} = \frac{\partial^2 U}{\partial y^2}\frac{\partial \psi}{\partial x} + \frac{1}{\mathrm{Re}}\Delta^2 \psi \tag{3.3.17}$$

We omit the symbol hat after Eq. (3.3.17), in order to avoid complications. Assuming that the stream function also has the same fluctuation as u, v, it is given by

$$\psi(x, y, t) = \phi(y) e^{i\alpha(x-ct)} \tag{3.3.18}$$

Substitution of Eq. (3.3.18) to (3.3.17) results in

$$\frac{1}{\mathrm{Re}}(D^2 - \alpha^2)^2 \phi(y) = i\alpha\left[(U - c)(D^2 - \alpha^2)\phi(y) - \frac{d^2U}{dy^2}\phi(y)\right] \tag{3.3.19}$$

where $D = d/dy$. Eq. (3.3.19) is called the Orr-Sommerfeld equation, which is able to evaluate the development of the fluctuations.

c. Rayleigh's instability theory

A following simple form called the Rayleigh equation is introduced, taking the limit of Re ($\mathrm{Re} \to \infty$) in the Orr-Sommerfeld equation.

$$\phi'' - \left(\alpha^2 + \frac{U''}{U - c}\right)\phi = 0 \tag{3.3.20}$$

in which the prime refers the derivative with respect to y.

When Eq. (3.3.20) is multiplied with ϕ^*, which is a conjugate function of ϕ, and integrated for the interval y_1 to y_2, a following equation is obtained using the partial integration, $\int_{y1}^{y2} \phi''\phi^* dy = [\phi''\phi^*]_{y1}^{y2} - \int_{y1}^{y2} \phi'\phi'^* dy$.

$$\int_{y1}^{y2}\left\{|\phi'|^2+\left(\alpha^2+\frac{U''(U-c)^*}{|U-c|^2}\right)\right\}|\phi|^2dy=0 \tag{3.3.21}$$

Introducing $c = c_r + ic_i$, the imaginary part of (Eq. 3.3.21) is expressed as

$$c_i\int_{y1}^{y2}\frac{U''}{|U-c|^2}|\phi|^2dy=0 \tag{3.3.22}$$

For Eq. (3.3.22) to hold in the case of $U'' \neq 0$, $c_i = 0$ is required, that means the disturbance stability is neutral. In contrast, while $c_i \neq 0$, the integral is zero, which means that the sign of U'' varies in the integral interval. That is to say, $U(y)$ must have an inflectional point. It leads to the following conclusion.

An inflection point must exist for the disturbance to grow. Note that this is not a necessary and sufficient condition. Although the existence of the inflection point doesn't mean the development of the disturbance, it does not occur without the inflection point. This is a claim of Rayleigh's instability theory.

Let us imagine the confluence of two rivers with different speeds or the compound channel, where the surface velocity profile varies in a lateral direction, like a tangent-hyperbolic curve. The inflectional point appears and streamwise current becomes unstable, in which horizontal vortices generated periodically are often observed.

3.4 GOVERNING EQUATION

3.4.1 Open-Channel and Boundary Layer

The free-stream layer is situated over the boundary layer. Since the height of the free-stream layer is much larger than that of the boundary layer, it is not interested. In contrast, the open channel should consider the restriction of the water depth. Assuming that lateral variation of the stream can be ignored, let us focus on the two-dimensional open-channel flow that we observe from the side of the flume. A standard open-channel consists of the smooth, flat bottom and side walls. Since it is tilted a small angle with respect to the horizontal plain, the streamwise current is driven by the gravity force. When water flows from the upstream side of the channel, the bottom friction reduces the velocity near the bottom. It results in formation of the boundary layer, as shown in FIG. 3.10. The boundary layer thickness increases in the streamwise direction, similar to what was as observed on the flat plate in section 3.2. When the boundary layer height reaches the free surface, it is expected for the flow to be fully developed. Hence, the open-channel flow is thought to be a kind of boundary layer

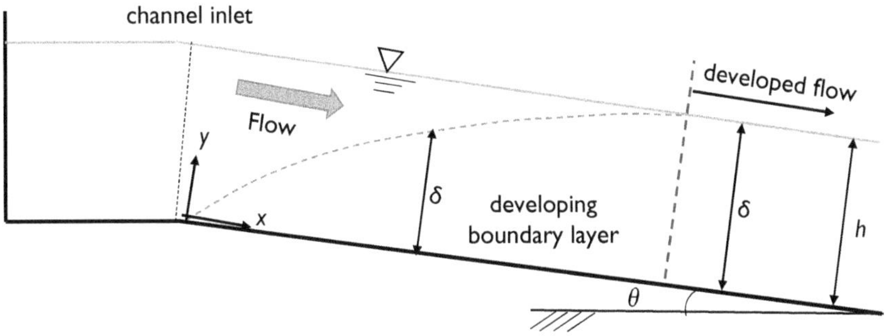

Figure 3.10 The boundary layer formed in the 2-D open-channel flow.

that occupies the whole depth region. The distance from the channel entrance where the flow is fully developed reported by previous studies is about 50 to 100 times as large as the water depth.

3.4.2 2-D Governing Equation in Open-Channel Flow

The external force is gravity, which must be considered in the open channel. Steady RANS equation in the streamwise direction is as follows:

$$U\frac{\partial U}{\partial x} + V\frac{\partial U}{\partial y} = g\sin\theta - \frac{\partial}{\partial x}\left(\frac{P}{\rho}\right) + \frac{\partial}{\partial x}(-\overline{u^2}) + \frac{\partial}{\partial y}(-\overline{uv}) + \nu\nabla^2 U \quad (3.4.1)$$

Assuming the uniform condition in the streamwise direction and the vertical velocity is zero, Eq. (3.4.1) is simplified:

$$0 = g\sin\theta + \frac{\partial}{\partial y}\left(-\overline{uv} + \nu\frac{\partial U}{\partial y}\right) \quad (3.4.2)$$

in which the second term refers the variation in the depth direction of the shear stress $\frac{\tau(y)}{\rho} = \left(-\overline{uv} + \nu\frac{\partial U}{\partial y}\right)$ consists of the Reynolds stress and the viscous shear stress. Hence, this equation means that the gravity force driving the streamwise current matches the shear stress acting as the friction. After the integral of Eq. (3.4.2) with the interval from an arbitrary elevation to the free surface $y = h$, a following form is obtained:

$$0 = \int_y^h g\sin\theta + \frac{\partial}{\partial y}\left(-\overline{uv} + \nu\frac{\partial U}{\partial y}\right)dy \leftrightarrow 0 = (h - y)gI_e + \frac{\tau(h) - \tau(y)}{\rho} \quad (3.4.3)$$

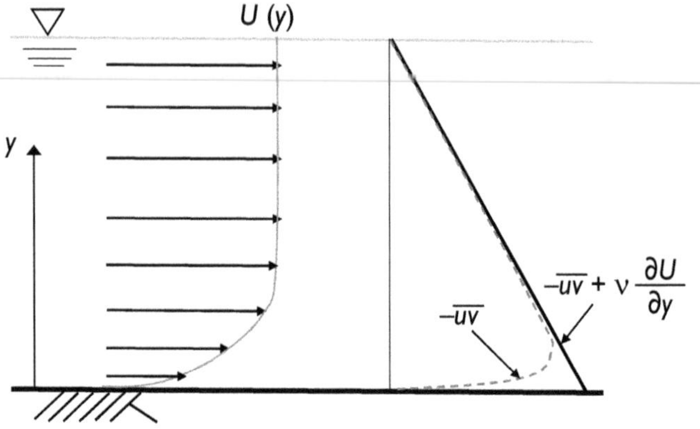

Figure 3.11 Profile of shear stress in the open-channel flow.

in which the energy slope $I_e = \sin\theta$. Since the shear stress in the free-surface $\tau(h)$ is almost, zero, Eq. (3.4.3) can be simplified:

$$\frac{\tau(y)}{\rho} = (h - y)gI_e = (1 - y/h)ghI_e = U_*^2(1 - y/h) \tag{3.4.4}$$

in which $U_* \equiv \sqrt{ghI_e}$.

This suggests that the shear stress has the following linear profile, as shown in FIG. 3.11.

$$\frac{\tau}{\rho} = -\overline{uv} + \nu\frac{\partial U}{\partial y} = U_*^2(1 - y/h) \tag{3.4.5}$$

In contrast, Reynolds stress decays significantly near the bottom. It is comparable with the linear total shear stress $\tau(y)$, except the bottom region, where viscous force is reduced due to small velocity gradient $\frac{\partial U}{\partial y} \simeq 0$.

3.5 MODELING OF SHEAR STRESS

3.5.1 Reynolds Stress

Reynolds stress is a kind of force related to the time variation of momentum, i.e, force product. FIG. 3.12(a) shows the vertical mixing of the momentum in the two-dimensional turbulent boundary layer. Here, the coordinate system is expressed by using streamwise direction x and vertical direction y. Reynolds stress could be expressed by $-\overline{uv}$. The velocity gradient, $\partial U/\partial y$, is large near the bottom and it induces the flow instability accompanied by small fluctuations and spanwise vortices.

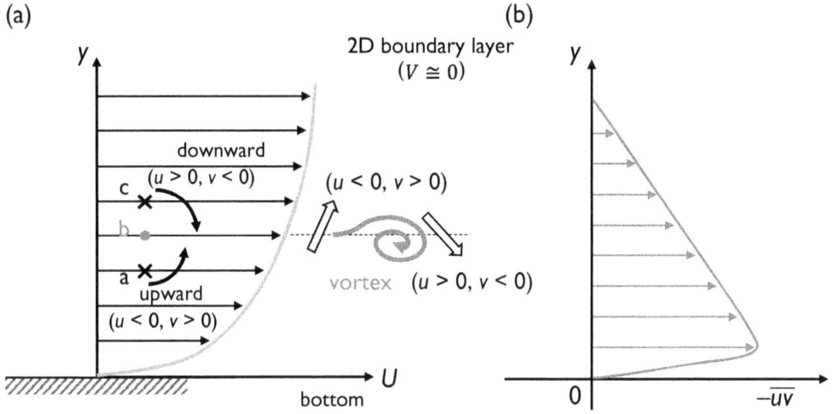

Figure 3.12 Mean velocity and Reynolds stress in 2-D open-channel flow: (a) momentum exchange by the vortex and (b) profile of Reynolds stress.

Then we focus on point "b" situated in the boundary layer. When upward currents happen suddenly in point "a" closer the bottom than point "b", this fluid parcel reaches the elevation of "b", which has lower speed than the mean velocity observed in point "b". This is because the mean velocity is lower in point "a" than in point "b". Therefore, it is considered that deviations of velocity components ($u < 0$, $v > 0$) are observed while the low-speed upward current passes at point "b", in which the vertical time-averaged velocity is $V \cong 0$. In contrast, when the downward currents happen suddenly at point "c", which is farther from the bottom compared to "b", deviations of velocity components ($u > 0$, $v < 0$) are detected because the high-speed downward currents passes at point "b". Making a connection with a spanwise vortex, these downward and upward currents appear in the front and rear ends of the vortex. The sign of the product of the velocity deviations is negative; it results in $-\overline{uv} > 0$. Actually, the Reynolds stress profile has a positive linear curve except for the near bottom region. This property was recognized by the many experiments and numerical simulations (FIG. 3.12(b)).

Let's consider the vertical mixing of the momentum. Before the mathematical procedure, we assume the vertical mean velocity is much smaller than the sreamwise one; that is to say, $V = 0$. Furthermore, time averaged of velocity deviations (u, v) is zero.

The mass flux passing an arbitrary horizontal plane par unit time is given by $\rho\tilde{v} = \rho(V + v) = \rho v$. Since instantaneous velocity is decomposed by $\tilde{u} = U + u$, the momentum passing vertically through the horizontal plane is $\rho\tilde{v}\,\tilde{u} = \rho v(U + u)$. Both positive and negative momentum flux are possible. We can calculate the fluctuation of the vertical momentum flux after the time-averaging integral of the momentum flux with the long time period gives

$$\frac{1}{T}\int_0^T \rho v(U+u)vdt = \frac{1}{T}\int_0^T \rho vUdt + \frac{1}{T}\int_0^T \rho uvdt$$
$$= \frac{\rho U}{T}\int_0^T vdt + \frac{\rho}{T}\int_0^T uvdt = 0 + \rho\overline{uv} = \rho\overline{uv} \tag{3.5.1}$$

This fact means that the Reynolds stress certainly corresponds to the variation of the momentum flux.

3.5.2 Mixing Length Model

Reynolds stress is a second-order term, and thus it is comparatively hard to measure or calculate. Then it is modeled by using the mean velocity component, which is easily obtained.

Let's consider two points, A and B, with spacing l near the bottom, as shown in FIG. 3.13. Point C is situated in the middle of them. The instantaneous high-speed downward current ($u > 0, \quad v < 0$) originated from point B is observed at point C and, conversely, the low-speed upward current ($u < 0, \quad v > 0$) from point A is detected at point C. These coherent events yield Reynolds stress. When the distances of C-B and C-A increase, the instantaneous upward and downward flows does not arrive at point C. Assuming the spatial range influenced by these coherent events, such a characteristic length is called the mixing length l. This also means the influence range of the vortex is like a vortex diameter.

Plandtl assumed the following three relations:

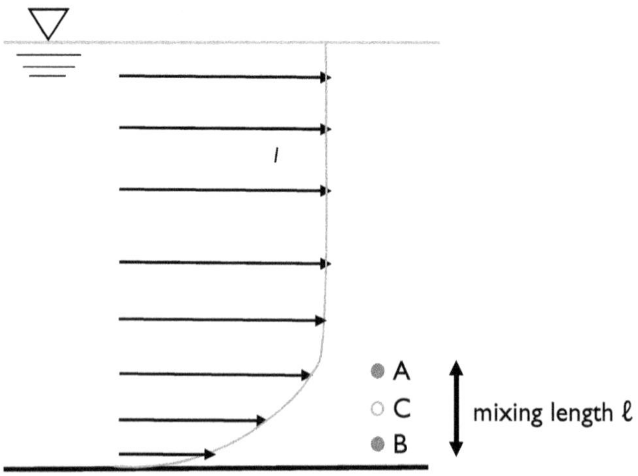

Figure 3.13 Mixing length near the bottom.

$$|u| \propto |v|$$
$$|u| \propto |\Delta U|$$
$$\Delta U \propto \frac{\partial U}{\partial y} l$$

ΔU is a velocity gap between the two points of which the height is different. Hence, a following scaling is introduced.

$$uv \propto uu \propto (\Delta U^2) \propto \left(l \frac{\partial U}{\partial y} \right)^2 \tag{3.5.2}$$

It results in the expression of the Reynolds stress using the gradient of mean streamwise velocity.

$$-\overline{uv} = l^2 \left(\frac{\partial U}{\partial y} \right)^2 \tag{3.5.3}$$

The square of mixing length implies proportional constant. Although l is still a vague physical value, the form of Eq. (3.5.3) has a good visibility for further theoretical expansion and is advantageous for the practical applications.

3.5.3 Eddy Viscosity Model

Shear friction stress could be expressed by Newton's viscosity law in the following form. The detailed derivation is explained in section 2.5.2.

$$\frac{\tau_{ij}}{\rho} = \nu \left(\frac{\partial \tilde{u}_i}{\partial x_j} + \frac{\partial \tilde{u}_j}{\partial x_i} \right) \tag{3.5.4}$$

It should be noted that this form is applied even to the instantaneous velocity fields. As mentioned previously, Reynolds stress contributes significantly to the momentum exchange in the shear flows with a normal gradient of streamwise velocity. Hence, Reynolds stress can be considered as a kind of shear stress, and it is expressed as follows in the same way as the above form.

$$-\overline{u_i u_j} = \nu_t \left(\frac{\partial U_i}{\partial x_j} + \frac{\partial U_j}{\partial x_i} \right) \tag{3.5.5}$$

in which ν_t is called eddy viscosity. Eq. (2.6.3) is rewritten by this assumption as Eq. (3.5.6).

$$\frac{\partial U_i}{\partial t} + U_j \frac{\partial U_i}{\partial x_j} = F_i - \frac{1}{\rho}\frac{\partial P}{\partial x_i} + (\nu + \nu_t)\frac{\partial^2 U_i}{\partial x_j \partial x_j} \quad (i, j = 1,\ 2,\ 3) \tag{3.5.6}$$

This expression implies that the momentum diffusion due to turbulence is concentrated on the eddy viscosity. It seems to make it possible to calculate the turbulence flow using only mean velocity components. However, how do we evaluate the eddy viscosity, ν_t? Unfortunately, Eq. (3.5.6) is not still closed, and further modeling of the eddy viscosity is required.

Many researchers suggested several turbulence modeling of the eddy viscosity. The most popular one is the $k - \varepsilon$ model proposed by Jones & Launder (1972), which expressed the eddy viscosity by using the turbulent kinetic energy, k, and the turbulence energy dissipation rate, ε, in the following form:

$$\nu_t = C_\mu \frac{k^2}{\varepsilon} \tag{3.5.7}$$

C_μ is a parameter, and $C_\mu = 0.09$ evaluated based on the benchmark tests is often used.

The behaviors of the bottom turbulence statistics are significantly influenced by the viscosity, and then a damping function f_μ is used as follows:

$$\nu_t = C_\mu f_\mu \frac{k^2}{\varepsilon} \tag{3.5.8}$$

When f_μ is equal to 1.0, the viscous effect by the wall disappears, and this is equal to the standard $k - \varepsilon$ model. Although a damping function, f_μ, has been investigated by many researchers, e.g., Jones & Launder (1972), Patel et al. (1985), and Yang & Shih (1993) for boundary layers and pipe flows, the behavior of f_μ near the bottom with the bottom normal coordinate y should be considered. Taylor expansions of turbulence characteristics near the wall, i.e., $y \cong 0$, are summarized in the following proportional relations, e.g., see Patel et al. (1985):

$$-\overline{uv} \propto y^3, \quad k \propto y^2, \quad \varepsilon \propto y^0, \quad U \propto y^1 \quad \text{at} \quad y \cong 0 \tag{3.5.9}$$

In the uniform flow, Eq. (3.5.5) is simplified to

$$-\overline{uv} = \nu_t \left(\frac{\partial U}{\partial y} \right) \tag{3.5.10}$$

Eqs. (3.5.9) and (3.5.10) give $\nu_t \propto y^3$. As a result, it is found from Eq. (3.5.8) that $f_\mu \propto y^{-1}$ is required near the bottom wall. On the basis of previous findings, f_μ is expressed reasonably by the van Driest–type function, as suggested by Abe et al. (1992) and Park & Sung (1997). Nezu & Sanjou (2006) proposed the following damping function, f_μ, from the wall to the outer layer in the open-channel turbulence.

$$f_\mu = 2\xi^{1/2}\{1 - \exp(-0.4y^+)\}^{1/2}(1 + R_t^{-1/2}) \propto y^{-1} \quad (0 \le \xi \equiv y/h \le 1/4) \tag{3.5.11}$$

where h is the water depth, $y^+ = yU_*/\nu = R_*\xi$ is the vertical coordinate (wall unit) normalized by the inner variables of the friction velocity U_* and the kinematic viscosity ν and $R_* \equiv hU_*/\nu$ and $R_t \equiv (k^2/\varepsilon)\nu$ is Reynolds numbers based on the friction velocity and the turbulence, respectively.

3.6 TIME MEAN VELOCITY PROFILE

3.6.1 Time Mean Velocity Profile

Substitution of Eq. (3.5.3) into Eq. (3.4.4) gives a following profile of the time mean velocity:

$$\frac{dU^+}{dy^+} = \frac{2(1-\xi)}{1 + \sqrt{1 + 4l^{+2}(1-\xi)}} \tag{3.6.1}$$

in which $\xi = y/h$ ($\xi = 0$ in the bottom, $\xi = 1$ in the free-surface), $U^+ = U/U_*$, $y^+ = yU_*/\nu$ and $l^+ = lU_*/\nu$.

3.6.2 Law of the Wall

The regions in $\xi = y/h < 0.2$ and $\xi = y/h > 0.2$ are called the inner layer and the outer layer, respectively, as shown in FIG. 3.14. Particularly, the velocity varies much larger in the inner layer near the bottom, and the velocity profile formula there is called the law of the wall.

Due to $\xi \ll 1$ near the bottom, Eq. (3.6.1) is reduced to

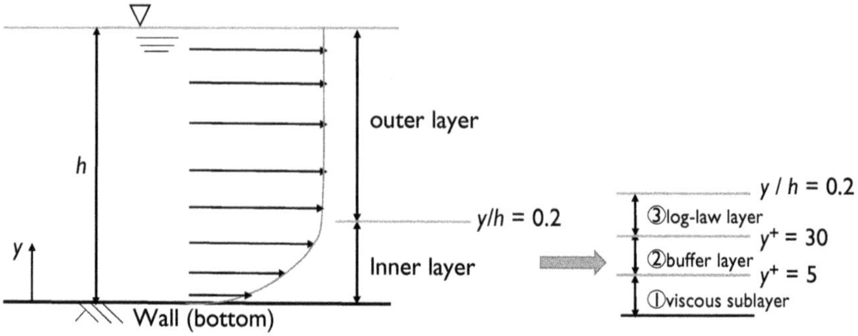

Figure 3.14 Outline of mean velocity profile.

$$\frac{dU^+}{dy^+} = \frac{2}{1 + \sqrt{1 + 4l^{+2}}} \tag{3.6.2}$$

Assuming that the normalized mixing length l^+ increases farther from the bottom because of no generation of large-scale vortex, l^+ is often expressed by $l^+ = \kappa y^+$. κ is the von Karman constant and $\kappa = 0.412$ is proposed by Nezu & Rodi (1986). When a sudden reduction of the mixing length near the bottom is considered carefully, l^+ is expressed by $l^+ = \kappa y^+\{1 - \exp(-y^+/26)\}$ with a damping function.

3.6.3 Inner Layer

The inner layer is divided into three sublayers, as follows:

1. Viscous sublayer ($y^+ \le 5$)

This is the region nearest the bottom wall, and it can be influenced by the viscous stress. When we consider $l^+ \ll 1$ in the viscous sublayer, Eq. (3.6.2) is reduced to

$$\frac{dU^+}{dy^+} = 1 \quad \leftrightarrow \quad U^+ = y^+ \tag{3.6.3}$$

Therefore, the linear profile is obtained in the viscous sublayer without any model parameters. An accurate non-intrusive velocimetry such as LDA is required for the measurements in the viscous sublayer, because the thickness of the viscous sublayer is very thin.

2. Buffer layer

This is the transition between the viscous sublayer and the log law layer. Generally, the velocity profile in this layer is calculated by the numerical integral of Eq. (3.6.1).

3. Log law layer ($30 \le y^+, \quad y < 0.2h$)

This is situated in the upper zone in the inner layer. When we consider $l^+ \gg 1$, Eq. (3.6.1) is rewritten by

$$\frac{dU^+}{dy^+} = \frac{1}{l^{+2}} = \frac{1}{\kappa y^+} \quad \rightarrow \quad U^+ = \frac{1}{\kappa} \ln y^+ + A \tag{3.6.4}$$

A is an integral constant, and $A = 5.29$ is a world standard value proposed by Nezu & Rodi (1986).

3.6.4 Outer Layer

The velocity profile is found to be deviated from the log low in the outer layer. The deviation $w(\xi)$ depends on Reynolds number and, thus, the profile is corrected by the wake function in the following form:

$$U^+ = \frac{1}{\kappa} \ln y^+ + A + w(\xi), \quad w(\xi) = \frac{2\Pi}{\kappa} \sin^2\left(\frac{\pi}{2}\xi\right) \tag{3.6.5}$$

The relation of Π and Re is obtained by the Nezu & Rodi (1986) curve, which is shown in FIG. 3.15. Π is almost zero with a low Reynolds number flow, and increases as Re becomes larger. It reaches a constant $\Pi = 0.2$ with $Re^* \equiv U_* h/\nu \simeq 10000$.

FIG. 3.16 shows the zone classification in the normalized velocity profile. The low Re number flows have the log-law expressed by Eq. (3.6.3) in the

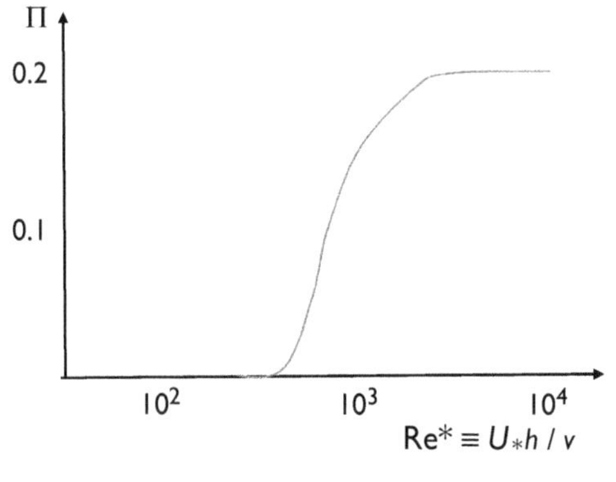

Figure 3.15 Outline of Nezu–Rodi curve.

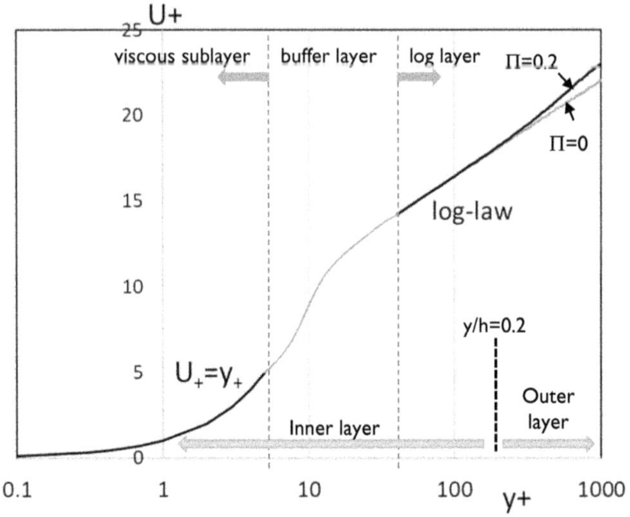

Figure 3.16 Zone classification in velocity profile normalized by inner variables.

whole outer layer, including near the free surface. In contrast, in the larger Re number flows, the velocity profile is shifted upward in the outer layer. This property is expected in uniform open-channel flows with different scales, e.g., the laboratory flume, brook and large river.

3.7 TURBULENCE STRUCTURE IN OPEN-CHANNEL

3.7.1 Profiles of Turbulence Statistics

Eq. (3.4.5) leads to the following profile of normalized Reynolds stress:

$$-\overline{uv}/U_*^2 = (1 - y/h) - \nu\frac{\partial U}{\partial y}/U_*^2 \tag{3.7.1}$$

The second term on the right-hand side refers the viscous effect. Actually, Reynolds stress approaches zero near the wall. In contrast, the viscous effect becomes weaker as the distance from the wall increases due to the very small velocity gradient, as shown in FIG. 3.17. Turbulence intensities $u'_i = \sqrt{\overline{u_i u_i}}$ are defined by normal components of the Reynolds stress tensors, $\overline{uu}$, $\overline{vv}$, $\overline{ww}$, which are related to the turbulent kinetic energy. The normal components can be measured by one velocity component velocimetry unlike the shear component, such as $\overline{uv}$, which requires two-component-type velocimetry.

Following semi-theoretical formulae, the turbulence intensities are introduced assuming the energy equilibrium of turbulence generation and

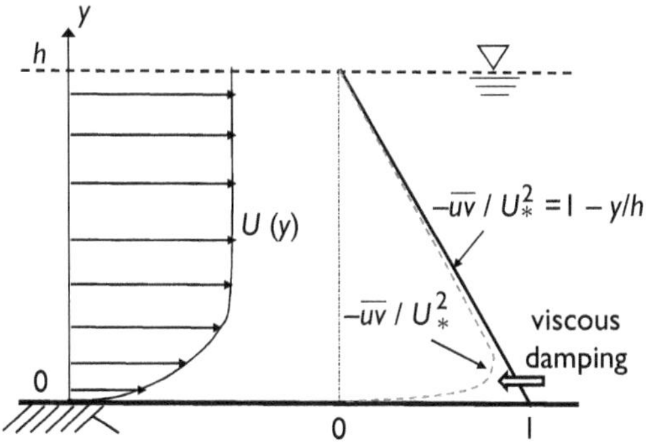

Figure 3.17 Reynolds stress profile in two-dimensional open-channel flow.

dissipation (Nezu & Nakagawa 1993). They fit well with measured data except for near the bottom and near the free surface.

$$\frac{u'}{U_*^2} = 2.30\exp\left(-\frac{y}{h}\right) \quad (0.1 < y/h < 1.0) \tag{3.7.2}$$

$$\frac{v'}{U_*^2} = 1.27\exp\left(-\frac{y}{h}\right) \quad (0.1 < y/h < 0.9) \tag{3.7.3}$$

$$\frac{w'}{U_*^2} = 1.63\exp\left(-\frac{y}{h}\right) \quad (0.1 < y/h < 1.0) \tag{3.7.4}$$

They are non-linear functions that decrease monotonically from the bottom to the free surface. The anisotropic property appears significantly in the bottom and free-surface boundaries, in which the turbulence in vertical component reduction is more remarkable than those in streamwise and spanwise components. It should be noted that the measured velocity data includes influence of the free-surface disturbance in the super critical flows. That is to say, the wave signal induced by the free surface is superimposed on the turbulence intensity. Therefore, they must be separated properly to analyze the turbulence structure correctly.

3.7.2 Transport Equations of Mean Flow and Turbulent Kinetic Energy

Transport equations of mean flow and turbulent kinetic energy have a common term with different signs. This implies that a part of kinetic energy

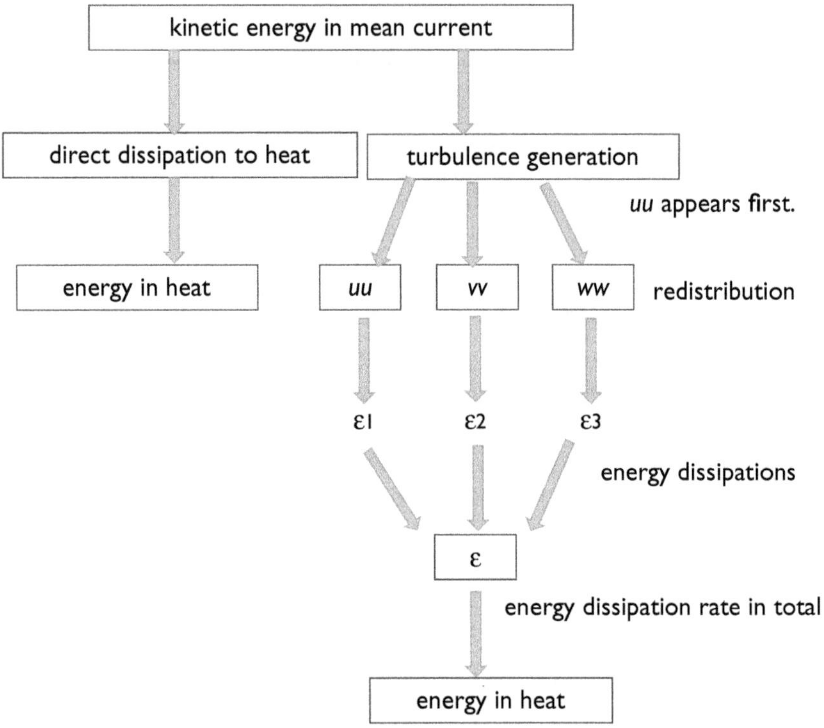

Figure 3.18 Generation and dissipation of turbulence energy.

in the mean current is used for turbulence generation. In the laminar flow, the mean current energy is directly dissipated to heat (FIG. 3.18). The detailed mechanism is explained by transport equations.

A following transport equation of Reynolds stress $R_{ij} = \overline{u_i u_j}$ is derived from the N-S equations:

$$\frac{\partial}{\partial t} R_{ij} + U_k \frac{\partial R_{ij}}{\partial x_k} = G_{ij} + \Pi_{ij} + \frac{\partial}{\partial x_k} (J_{T\ ijk} + J_{P\ ijk} + J_{V\ ijk}) - \varepsilon_{ij} \tag{3.7.5}$$

in which and $i, j = 1, 2$ for a two-dimensional field and $i,\ \ j,\ \ k = 1,\ \ 2,\ \ 3$ for a three-dimensional field. Each term is as follows:

Turbulence generation term: $G_{ij} = -R_{ik} \frac{\partial U_j}{\partial x_k} - R_{jk} \frac{\partial U_i}{\partial x_k}$

Pressure strain term: $\Pi_{ij} = \overline{\frac{p}{\rho} \left(\frac{\partial u_i}{\partial x_j} + \frac{\partial u_j}{\partial x_i} \right)}$

Diffusion terms:

Velocity diffusion: $J_{T\ ijk} = -\overline{u_i u_j u_k}$

Pressure diffusion: $J_{P\ ijk} = -(\overline{pu_i}\delta_{jk} + \overline{pu_j}\delta_{ik})$

Viscous diffusion: $J_{V\ ijk} = \nu \dfrac{\partial R_{ij}}{\partial x_k}$

Turbulent energy dissipation: $\varepsilon_{ij} = 2\nu \overline{\dfrac{\partial u_i}{\partial x_k}\dfrac{\partial u_j}{\partial x_k}}$

We consider a diagonal component of the Reynolds stress tensor in the two-dimensional field ($i, j, k = 1, 2$) for easy understanding. The subscripts 1 and 2 refer respectively to the streamwise direction x and vertical y. The transport equations of R_{11} and R_{22} are as follows:

1. Streamwise normal stress $R_{11} = \overline{u_1 u_1} (= \overline{uu})$

$$\frac{\partial}{\partial t}R_{11} + U_1\frac{\partial R_{11}}{\partial x_1} + U_2\frac{\partial R_{11}}{\partial x_2} = -R_{11}\frac{\partial U_1}{\partial x_1} - R_{12}\frac{\partial U_1}{\partial x_2} - R_{11}\frac{\partial U_1}{\partial x_1} - R_{12}\frac{\partial U_1}{\partial x_2} + 2\overline{p\frac{\partial u_1}{\partial x_1}} + diffs. - 2\nu\left(\overline{\frac{\partial u_1}{\partial x_1}\frac{\partial u_1}{\partial x_1}} + \overline{\frac{\partial u_1}{\partial x_2}\frac{\partial u_1}{\partial x_2}}\right) \tag{3.7.6}$$

in which *diffs.* means the diffusion term. The expression using the $x - y$ coordinate is

$$\frac{\partial}{\partial t}\overline{uu} + U\frac{\partial \overline{uu}}{\partial x} + V\frac{\partial \overline{uu}}{\partial y} = \underbrace{-2\overline{uu}\frac{\partial U}{\partial x} - 2\overline{uv}\frac{\partial U}{\partial y}}_{\text{Generation term}} + \underbrace{2\overline{p\frac{\partial u}{\partial x}}}_{\text{Pressure strain term}} + diffs. - 2\nu\left(\overline{\frac{\partial u}{\partial x}\frac{\partial u}{\partial x}} + \overline{\frac{\partial u}{\partial y}\frac{\partial u}{\partial y}}\right) \tag{3.7.7}$$

2. Vertical normal stress $R_{22} = \overline{u_2 u_2} (= \overline{vv})$

$$\frac{\partial}{\partial t}\overline{vv} + U\frac{\partial \overline{vv}}{\partial x} + V\frac{\partial \overline{vv}}{\partial y} = \underbrace{-2\overline{uv}\frac{\partial V}{\partial x} - 2\overline{vv}\frac{\partial V}{\partial y}}_{\text{Generation term}} + \underbrace{2\overline{p\frac{\partial v}{\partial y}}}_{\text{Pressure strain term}} + diffs. - 2\nu\left(\overline{\frac{\partial v}{\partial x}\frac{\partial v}{\partial x}} + \overline{\frac{\partial v}{\partial y}\frac{\partial v}{\partial y}}\right) \tag{3.7.8}$$

In the uniform flow with no variation in the streamwise component, i.e., $\partial/\partial x = 0$, the generation term in $\overline{uu}$ can be approximated to be $-\ 2\overline{uv}\frac{\partial U}{\partial y}$, and the vertical one can be approximated to be 0 because of $V \ll U$. The generation term in turbulence means the energy supply from the mean current, as explained later. That is, only the $\overline{uu}$ component is given directory the energy by the mean current. Note that the energy is also redistributed to the $\overline{vv}$ component through the pressure strain term. When the streamwise turbulence component receives the energy from the mean current, the streamwise turbulence is promoted or maintained. Then, a strain rate $\partial u/\partial x$ appears locally, and a vertical strain rate $\partial v/\partial y$ is also produced through the continuity equation in turbulence components, i.e., $\partial v/\partial y = -\partial u/\partial x$. This fact suggests that when a positive streamwise strain rate $\partial u/\partial x < 0$ occurs, the negative vertical strain rate $\partial v/\partial y > 0$ occurs. If the correlation with the pressure fluctuation is high, it means that the energy of $\overline{uu}$ is reduced in Eq. (3.7.7), and the energy of $\overline{vv}$ is increased in Eq. (3.7.8). This is the mechanism of energy redistribution, which can be extended to the three-dimensional field. The turbulence is finally converted to heat.

Defining the energy in the mean current as $K = \frac{1}{2}U_i U_i$, the corresponding transport equation is given by

$$\frac{\partial}{\partial t}K + U_j\frac{\partial K}{\partial x_j} = -G - \nu\frac{\partial U_i}{\partial x_j}\frac{\partial U_i}{\partial x_j} + \frac{\partial}{\partial x_j}\left(-U_i R_{ij} - \frac{P}{\rho}U_i\delta_{ij} + \nu\frac{\partial K}{\partial x_j}\right) \quad (3.7.9)$$

We can see the generation term with a minus sign on the right-hand side. This implies that a part of the kinetic energy in the mean current is transferred to the turbulence component. A flowchart about energy transport in the open-channel turbulent flow is summarized in FIG. 3.18.

3.7.3 Energy Equilibrium in the Depth Direction

The turbulent kinetic energy $k = \frac{1}{2}\overline{u_i u_i} = \frac{1}{2}(\overline{u_1 u_1} + \overline{u_2 u_2} + \overline{u_3 u_3})$ (for a three-dimensional field) is the sum of diagonal components in the Reynolds stress tensor. Therefore, the total sum of the diagonal components in Eq. (3.7.5) leads to the following form:

$$\frac{\partial}{\partial t}k + U_j\frac{\partial k}{\partial x_j} = G + \frac{\partial}{\partial x_j}(J_{T\,j} + J_{P\,j} + J_{V\,j}) - \varepsilon \quad (3.7.10)$$

in which each term is specifically expressed as follows:

Turbulence generation term: $G = -R_{ij}\frac{\partial U_i}{\partial x_j}$

Diffusion terms:

Velocity diffusion: $J_{T\,j} = -\frac{1}{2}\overline{u_i u_i u_j}$
Pressure diffusion: $J_{P\,j} = -\overline{\frac{p}{\rho} u_j}$
Viscous diffusion: $J_{V\,j} = \nu \frac{\partial k}{\partial x_j}$
Energy dissipation term: $\varepsilon = \nu \overline{\frac{\partial u_i}{\partial x_j} \frac{\partial u_i}{\partial x_j}}$

The pressure strain term (pressure diffusion term) in Eq. (3.7.5) disappears in Eq. (3.7.10), and it is found to contribute to only the energy transfers among the components of normal Reynolds stress.

G is the turbulence generation rate due to the energy supply from the mean current and, in contrast, ε means the turbulent energy dissipation rate including energy supply to surrounding smaller vortices and conversion to heat. When the turbulence generation and dissipation are balanced, i.e., $G \simeq \varepsilon$, this state is called energy equilibrium, in which the turbulence does not increase or decrease. (3.7.2) to (3.7.4) are semi-theoretical formulae assuming the energy equilibrium except for near the bottom and the free-surface. FIG. 3.19 shows outlines of depth profiles of turbulence generation and energy dissipation terms reported by experiments and numerical simulations. The proper balance is lost between G and ε in the bottom and the free-surface boundaries. The energy equilibrium is observed except for these boundaries. The energy budget has the following properties at the bottom and the free surface, respectively.

(bottom)
The turbulence generation exceeds the energy dissipation in this layer ($G > \varepsilon$). An existence of turbulence source like the velocity shear $\frac{\partial U}{\partial y}$ causes a local non-equilibrium. A part of the turbulent energy that is not dissipated is transported toward the free surface.

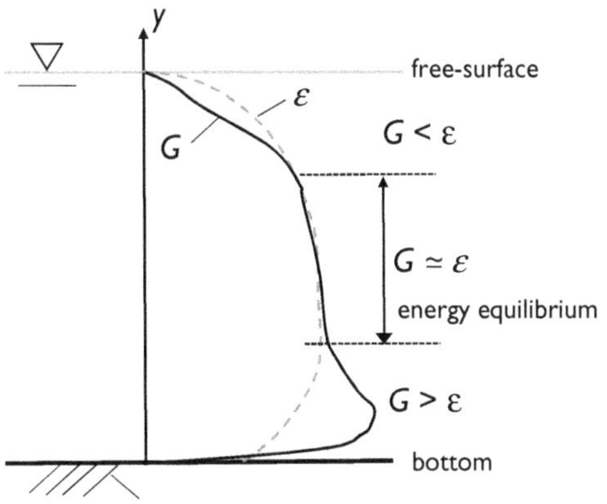

Figure 3.19 Outline of turbulence generation and energy dissipation terms in the depth direction.

(free surface)
The velocity shear becomes small and, as a result, the production of turbulence is insufficient ($G < \varepsilon$). The turbulent kinetic energy is diffused from the lower layer to make up for the shortage.

3.8 FREQUENCY ANALYSIS AND CASCADE PROCESS

3.8.1 Frequency and Wave Number Fields

Turbulence flows include various scale vortices as described in previous sections with the statistical analysis. However, it is unclear how much kinetic energy a vortex with a certain length scale possesses and how multiple vortices of different sizes affect each other.

We look down a free-surface flow in a straight river, as shown in FIG. 3.20. We can see it in practice that transverse disturbance occurs locally in the mainstream. This is a kind of turbulence; those maximum scales are expected to reach a river width in lateral and a water depth in vertical.

The direct measurement at a single point by a velocimetry provides time-series of instantaneous velocity. In contrast, simultaneous multi-point measurements using particle image velocimetry (PIV) are able to obtain spatial distribution of instantaneous velocity components, as shown in FIG. 3.21(a). They are useful information in physical space that are easy to understand. However, the distribution of turbulence fluctuation is unknown in the wave space of FIG. 3.21(b).

3.8.2 Correlation Coefficient and Spectrum

3.8.2.1 Correlation coefficient

It is expected that there is some relationship between the velocity fluctuations at two points included in the same vortex. A following cross-correlation coefficient is often used to quantitatively evaluate how well two velocity signals match in the $x - y$ plane.

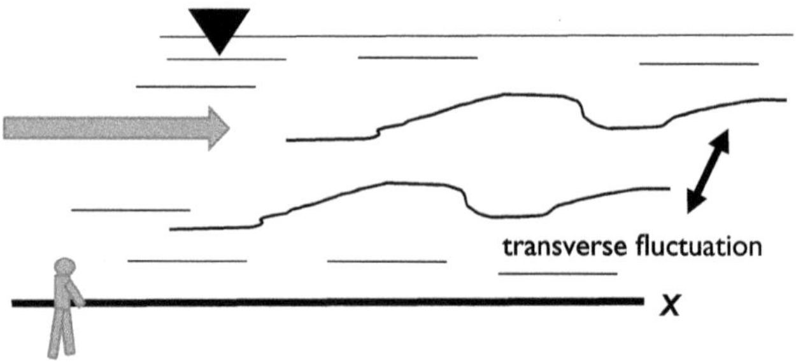

Figure 3.20 Free-surface flow.

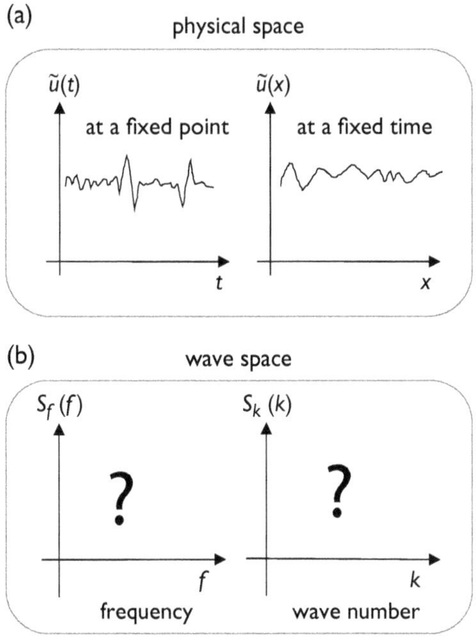

Figure 3.21 Comparison of physical and frequency spaces: (a) variation of instantaneous velocity in space and time; (b) energy distribution of turbulence against frequency and wave number.

$$R_{uu}(x, y, \Delta x, \Delta y) = \frac{\overline{u(x, y)u(x + \Delta x, y + \Delta y)}}{u'(y, z)u'(x + \Delta x, y + \Delta y)} \tag{3.8.1}$$

This means the two-dimensional cross correlation of turbulence fluctuations in the streamwise direction is between the points (x, y) and $(x + \Delta x, y + \Delta y)$. The numerator means the correlation value and the denominator means the product of turbulence intensities at two points for normalization. When the same point is chosen as the two reference points, R_{uu} becomes one.

FIG. 3.22 shows an experimental setup in the open-channel flume submerged vegetation models, in which a high-speed camera and laser-light sheet are used for the measurement of the velocity correlation in the $x - y$ plane. Time series of multi-velocity signals are obtained simultaneously. The reference point "x" is fixed and another point is moved. FIG. 3.23 shows a distribution of R_{uu} evaluated by the measured data, where streamwise and vertical positions are normalized by the vegetation height h and the vegetation width B_v. It results in that a peak appears at the reference point and the cross-correlation coefficient decreases moving farther away from the reference.

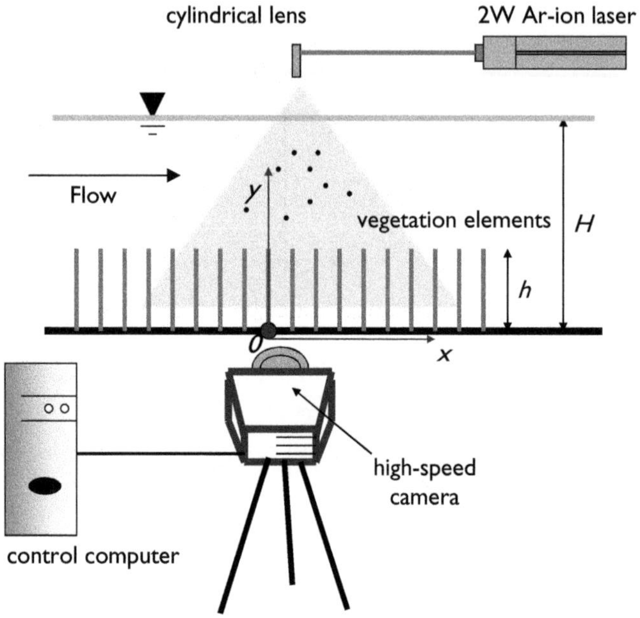

Figure 3.22 Experimental setup for simultaneous velocity measurement system.

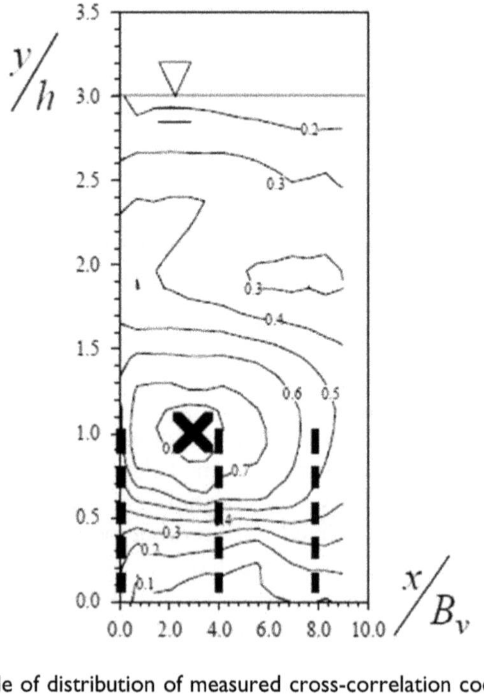

Figure 3.23 Example of distribution of measured cross-correlation coefficient.

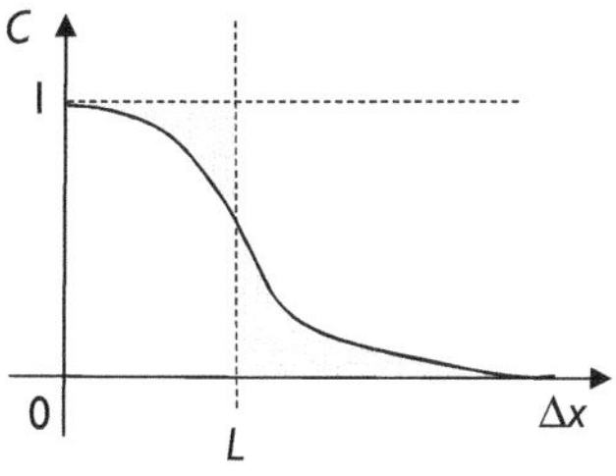

Figure 3.24 Autocorrelation function profile and integral scale.

The numerator in Eq. (3.8.1) is defined as a cross-correlation function.

$$C(x, y, \Delta x, \Delta y) = \overline{u(x, y)u(x + \Delta x, y + \Delta y)} \quad (3.8.2)$$

Focusing on only the streamwise direction, $C(x, \Delta x) = \overline{u(x, y)u(x + \Delta x, y)}$ is expressed.

When a time-series of velocity signal in a single point is considered, the correlation between the velocity fluctuations at two different times is also calculated. This is called the autocorrelation function, given as

$$C(x, \Delta x) = \overline{u(x, t)u(x, t + \Delta t)} \quad (3.8.3)$$

in which Δt is the variable time delay. Further, the space-time correlation function, which varies both of the time delay and spatial gap, is also analyzed for the evaluation of the size and period of the coherent vortices in the open-channel turbulence. The fluid is expected to move in a unified fashion within a region with $C(x, \Delta x) = 1$. We consider a rectangular of height $C = 1$ with the same area as the autocorrelation function profile, as shown in FIG. 3.24, in which the area of two dotted zones are equal. A width of the rectangular is defined as L. This is called the integral scale, which means the length of mean size vortex. L is expressed mathematically by

$$L = \int_0^\infty C(x, \Delta x)d(\Delta x) \quad (3.8.4)$$

3.8.2.2 Spectrum

How much kinetic energy does an individual vortex with a certain size in frequency and wave number possesses? Let us consider a time-series of fluctuation in instantaneous velocity $u(t)$. When we assume that $u(t)$ is a periodic function having a period $2L$, it is expressed by the following Fourier series:

$$u(t) = \frac{a_0}{2} + \sum_{n=1}^{\infty} \{a_n \cos(\pi nt/L) + b_n \sin(\pi nt/L)\} \tag{3.8.5}$$

This is rewritten by

$$\begin{aligned} u(t) &= \tfrac{a_0}{2} \\ &+ a_1 \cos(\pi t/L) + b_1 \sin(\pi t/L) \\ &+ a_2 \cos(2\pi t/L) + b_2 \sin(2\pi t/L) \\ &+ a_3 \cos(3\pi t/L) + b_3 \sin(3\pi t/L) \\ &+ \cdots \end{aligned} \tag{3.8.6}$$

Second, third and fourth lines in Eq. (3.8.6) have periods $2L$, L, $\frac{2}{3}L$, respectively.

In contrast, for $u(t)$ with a period $2mL$,

$$\begin{aligned} u(t) &= \tfrac{a_0}{2} \\ &+ a_1 \cos(\pi t/mL) + b_1 \sin(\pi t/mL) \\ &+ a_2 \cos(2\pi t/mL) + b_2 \sin(2\pi t/mL) \\ &+ a_3 \cos(3\pi t/mL) + b_3 \sin(3\pi t/mL) \\ &+ \cdots \end{aligned} \tag{3.8.7}$$

is expressed. Eq. 3.8.7 implies that the fundamental frequency becomes smaller as m increases. Hence, when the period of $u(t)$ has infinity, i.e., $L \to \infty$, discrete components becomes continuous and mathematical sign Σ changes to $\int$ in Eq. (3.8.5).

A function with a period $2L$, u_L, is expressed as follows:

$$u_L(t) = \frac{a_0}{2} + \sum_{n=1}^{\infty} \{a_n \cos(\pi nt/L) + b_n \sin(\pi nt/L)\} \tag{3.8.8}$$

in which $a_n = \frac{1}{L}\int_{-L}^{L} u_L(r)\cos\frac{n\pi r}{L}dr$ and $b_n = \frac{1}{L}\int_{-L}^{L} u_L(r)\sin\frac{n\pi r}{L}dr$.

It is rewritten by

$$
\begin{aligned}
u_L(t) &= \frac{1}{2L}\int_{-L}^{L} u_L(r)\,dr + \frac{1}{L}\sum_{n=1}^{\infty}\left(\cos\frac{n\pi t}{L}\int_{-L}^{L} u_L(t)\cos\frac{n\pi r}{L}dr \right. \\
&\qquad \left. + \sin\frac{n\pi t}{L}\int_{-L}^{L} u_L(t)\sin\frac{n\pi r}{L}dr\right) \\
&= \frac{1}{2L}\int_{-L}^{L} u_L(r)\,dr + \frac{1}{\pi}\sum_{n=1}^{\infty}\Delta\omega\left(\cos\omega_n t\int_{-L}^{L} u_L(t)\cos\omega_n r dr \right. \\
&\qquad \left. + \sin\omega_n t\int_{-L}^{L} u_L(t)\sin\omega_n r dr\right)
\end{aligned}
\tag{3.8.9}
$$

in which $\omega_n = \frac{n\pi}{L}$, $\Delta\omega = \omega_{n+1} - \omega_n = \frac{\pi}{L}$. Defining $u_\infty(t) = \lim_{L\to\infty}\frac{n\pi}{L}u_L(t)$, $u_\infty(t)$ is no longer periodic. Hence, $u_\infty(t)$ is replaced by a time series of velocity fluctuation $u(t)$.

Taking the limit of L in Eq. (3.8.9),

$$
u(t) = \frac{1}{\pi}\sum_{n=1}^{\infty}\Delta\omega\left(\cos\omega_n t\int_{-\infty}^{\infty} u(t)\cos\omega_n r dr + \sin\omega_n t\int_{-\infty}^{\infty} u(t)\sin\omega_n r dr\right)
\tag{3.8.10}
$$

is obtained. ω_n is rewritten by ω because of $\Delta\omega \to 0$, i.e., ω_n is no longer discrete, when $L \to \infty$. Therefore, a following form is introduced:

$$
\lim_{\Delta\omega\to 0}\sum_{n=1}^{\infty}\Delta\omega F(\omega_n) \to \int_0^{\infty} F(\omega)\,d\omega
\tag{3.8.11}
$$

Eq. (3.8.10) is rewritten using Eq. (3.8.11) as follows:

$$
\begin{aligned}
u(t) &= \frac{1}{\pi}\left(\int_0^{\infty}\cos\omega t d\omega\int_{-\infty}^{\infty} u(t)\cos\omega r dr + \int_0^{\infty}\sin\omega t d\omega\int_{-\infty}^{\infty} u(t)\sin\omega r dr\right) \\
&= \frac{1}{\pi}\left(\int_0^{\infty} dr\int_{-\infty}^{\infty} d\omega u(r)(\cos\omega t\cos\omega r + \sin\omega t\sin\omega r\right) \\
&= \frac{1}{\pi}\int_0^{\infty} d\omega\int_{-\infty}^{\infty} dr u(r)\cos\omega(t - r)
\end{aligned}
\tag{3.8.12}
$$

This is a Fourier formula. Eq. (3.8.12) is further expanded.

$$u(t) = \frac{1}{\pi}\int_0^\infty d\omega \int_{-\infty}^{\infty} dr u(r)\cos\omega(t-r) = \frac{1}{\pi}\int_0^\infty d\omega \int_{-\infty}^{\infty} dr u(r)\frac{1}{2}\left(e^{i\omega(t-r)} + e^{-i\omega(t-r)}\right)$$

$$= \frac{1}{2\pi}\int_{-\infty}^{\infty} d\omega \int_{-\infty}^{\infty} dr u(r) e^{i\omega(t-r)} = \frac{1}{2\pi}\int_{-\infty}^{\infty} d\omega \int_{-\infty}^{\infty} dr u(r) e^{-i\omega r} e^{i\omega t}$$

$$= \frac{1}{2\pi}\int_{-\infty}^{\infty}\left\{\int_{-\infty}^{\infty} u(r) e^{-i\omega r} dr\right\} e^{i\omega t} d\omega$$

Introducing $F(\omega) = \int_{-\infty}^{\infty} u(r)e^{-i\omega r}dr$, Eq. (3.8.12) is rewritten as

$$u(t) = \frac{1}{2\pi}\int_{-\infty}^{\infty} F(\omega) e^{i\omega t} d\omega \tag{3.8.13}$$

in which

$$F(\omega) = \int_{-\infty}^{\infty} u(r) e^{-i\omega r} dr = \int_{-\infty}^{\infty} u(t) e^{-i\omega t} dt \tag{3.8.14}$$

Eqs. (3.8.13) and (3.8.14) are inverse Fourier transform and Fourier transform, respectively.

Further, $F(\omega)$ is expressed as follows:

$$F(\omega) = |F(\omega)| e^{i\theta_\omega} \tag{3.8.15}$$

Eqs. (3.8.13) and (3.8.15) introduce

$$u(t) = \frac{1}{2\pi}\int_{-\infty}^{\infty} |F(\omega)| e^{i(\omega t + \theta_\omega)} d\omega \tag{3.8.16}$$

in which $|F(\omega)|$, $|F(\omega)|^2$, θ_ω means amplification, energy and phase of individual wave having a frequency ω. Here, a following function $S_f(\omega)$ is defined:

$$S_f(\omega) = \lim_{T\to\infty} \frac{|F(\omega)|^2}{T} \tag{3.8.17}$$

T is a sampling duration of velocity measurement. S_f means a time averaged energy. A following relation also holds:

$$\int_{-\infty}^{\infty} S_f(\omega)d\omega = \overline{u(t)^2} = k \tag{3.8.18}$$

This means that total area of energy distribution is equal to the turbulence kinetic energy k. Due to the relation of $\omega = 2\pi f$, Eq. (3.8.13) is rewritten by

$$u(t) = \frac{1}{2\pi}\int_{-\infty}^{\infty} F(f)e^{i2\pi ft}df \tag{3.8.19}$$

in which $F_f(f) = 2\pi \int_{-\infty}^{\infty} u(r)e^{-2i\pi fr}dr = 2\pi \int_{-\infty}^{\infty} u(t)e^{-2i\pi ft}dt$. Here, the frequency spectrum can be defined by

$$S_f(f) = \lim_{T\to\infty} \frac{|F_f(f)|^2}{T} \tag{3.8.20}$$

For one-dimensional spatial distribution of instantaneous velocity at a certain moment, the wave number spectrum is also given by

$$S_k(k) = \lim_{X\to\infty} \frac{|F_k(k)|^2}{X} \tag{3.8.21}$$

in which $F_k(k) = \int_{-\infty}^{\infty} u(x)e^{-ikx}dx$.

3.8.2.3 Wiener Khinchin theorem

The spectrum is associated with the autocorrelation function by following the Wiener Khinchin theorem.

$$C(\Delta t) = \int_{-\infty}^{\infty} S_f(\omega)e^{i\omega\Delta t}d\omega \tag{3.8.22}$$

This means that the autocorrelation function is obtained after the inverse Fourier transform of the frequency spectrum. By contrast, the frequency spectrum is calculated by the Fourier transform of the autocorrelation function.

3.8.2.4 Conversion between frequency spectrum and wave number spectrum

Taylor's frozen turbulence hypothesis permits a conversion between frequency spectrum and wave number spectrum. It is generally hard to evaluate the wave number spectrum $S_k(k)$, which requires simultaneous multi-points measurements. In contrast, the frequency spectrum $S_f(f)$ can be evaluated by time-series of an instantaneous velocity obtained in a single-point measurement, which is popular in experimental studies. Hence, $S_k(k)$ is converted from $S_f(f)$ in a following process.

Taylor's frozen turbulence hypothesis involves a vortex structure that is not changed in a turbulent flow while traveling downstream. Here, the vortex can be considered as a wave without any dispersion. The traveling distance of the vortex dx is expressed by the convection velocity U_c and the traveling time dt.

$$dx = U_c \times dt \tag{3.8.23}$$

A dispersion relation is given by using U_c in the following form:

$$U_c = \frac{\omega}{k} = \frac{2\pi f}{k} \tag{3.8.24}$$

Replacing k and x by ω and t in Eq. (3.8.14),

$$\begin{aligned} F_k(k) &= \int_{-\infty}^{\infty} u(x)e^{-ikx}dx = U_c \int_{-\infty}^{\infty} u(t)e^{-i\frac{\omega}{U_c}U_c t}dt = U_c \int_{-\infty}^{\infty} u(t)e^{-2i\pi f t}dt \\ &= \frac{U_c}{2\pi}F_f(f) \end{aligned} \tag{3.8.25}$$

is introduced. The wave number spectrum can be converted from the frequency spectrum by Eq. (3.8.25).

3.8.3 Kolmogorov's −5/3 Power Law

Kolmogorov (1941) proposed the –5/3 power law, applicable to various kinds of flow fields under the following assumptions.

1. Kinetic energy in mean flow is conveyed to the largest vortex at the initial stage whose wave number is k_0. The large vortex becomes smaller in size and smaller vortices appear as shown in FIG. 3.25. This is called the cascade process, which is promoted by the convection

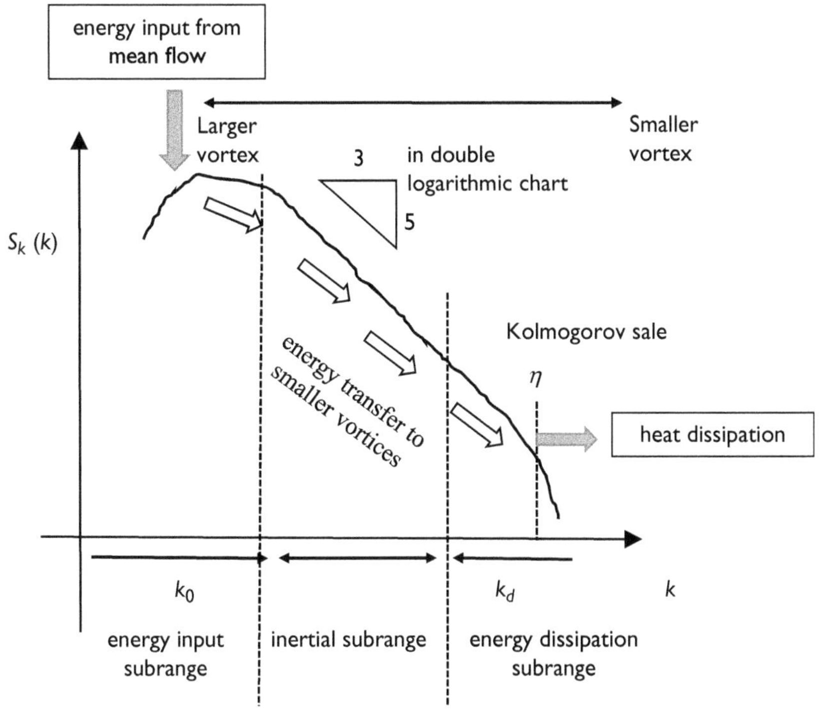

Figure 3.25 Distribution of wave number spectrum.

term in N-S equations. The smallest vortex with a wave number k_d is dissipated and converted to heat energy.

2. The cascade process, in which energy transfer process from largest vortex with length scale $1/k_0$ to smaller vortex with $1/k_d$, is controlled by the energy dissipation rate ε and kinetic viscosity ν.
3. Length and velocity scales of the largest vortex are given by $1/k_0$ and $(\varepsilon/k0)^{1/3}$, respectively. The Reynolds number is defined by $\mathrm{Re} = \varepsilon^{1/3}k_0{}^{-4/3}/\nu$. By contrast, the wave number of the smallest vortex is given by $k_d = (\varepsilon/\nu^3)^{4/3}$ and, thus, $\mathrm{Re} = (kd/k0)^{4/3}$ is introduced.
4. The third assumption suggests that k_d is much larger than k_0 ($k_0 \ll k_d$) in a large Reynolds number condition. It results in wider inertial subrange situated between the largest and the smallest vortices' scales, in which viscous effect is comparatively weak.
5. Direct energy transfer between different wave numbers with a large gap can be ignored.
6. The fifth assumption suggests how the largest vortex with $1/k_0$ receives energy from the mean flow does not influence the cascade process continued to the smallest vortex with $1/k_d$. Even if the largest vortex is anisotropic, the pressure strain term make vortices isotropic in the cascade process. This property is called local isotropy.

Kolmogorov's length and velocity scales are expressed by $\eta \equiv (\nu^3/\varepsilon)^{1/4}$ and $u \equiv (\nu\varepsilon)^{1/4}$, respectively. they mean scales of the minimum vortex influenced significantly by the viscosity. In the wave number region comparable to the Kolmogorov scales minimum vortex, the local isotropy holds. After dimension analysis, the wave number spectrum in such a region is expressed by

$$S_k(k) \sim u^2\eta \times \Upsilon(\eta k) \tag{3.8.26}$$

in which $\Upsilon(\eta k)$ is a non-dimensional function.

A wave number in the inertial subrange k exists between k_0 and k_d, i.e., $k_0 \ll k \ll k_d$. The viscous effect is not significant here, as mentioned above. When $\Upsilon(\eta k) = (\eta k)^{-5/3}$ is chosen, the viscous effect does not appear in Eq. (3.8.26). It results in

$$S_k(k) = \alpha\varepsilon^{2/3}k^{-5/3} \tag{3.8.27}$$

This is Kolmogorov's –5/3 power law, in which α is a dimensionless constant that is found experimentally to be between 0.5 and 1.5 (e.g., Qian (1996), Tsuji (2009)).

Eq. (3.8.27) is applicable to the measurement of the turbulent energy dissipation ε. First, the frequency spectrum $S_f(f)$ is calculated with the time-series of instantaneous velocity at a single point by the Fourier transform. Second, the wave number spectrum $S_k(k)$ is transformed by Eq. 3.8.25, in which the convection velocity U_c is evaluated by the time-space correlation analysis if there is a time-series data of simultaneous two-points velocity measurement, e.g., Nezu & Sanjou (2008). If it's not, the local time-mean velocity is substituted. Finally, the energy dissipation rate is obtained, fitting the best to the theoretical curve of Eq. (3.8.27) within the inertial subrange.

3.8.4 Inverse Cascade in Two-Dimensional Turbulence

A three-dimensional turbulence field includes the cascade process with the energy transfer from larger to smaller vortices, as mentioned above. However, the inverse cascade process which is the energy transfer from smaller to larger vortices is reported in two-dimensional turbulence flows. The positive dissipation rate $\varepsilon > 0$ means the normal cascade process, and in contrast the negative one $\varepsilon < 0$ means the inverse cascade process highlighted here.

A following structure function Dk_p considering the relation between two velocities at different points is introduced for further discussion:

$$Dk_p(l) = \langle \Delta\tilde{u}^p \rangle, \quad \Delta\tilde{u} = \tilde{u}(x + l) - \tilde{u}(x) \tag{3.8.28}$$

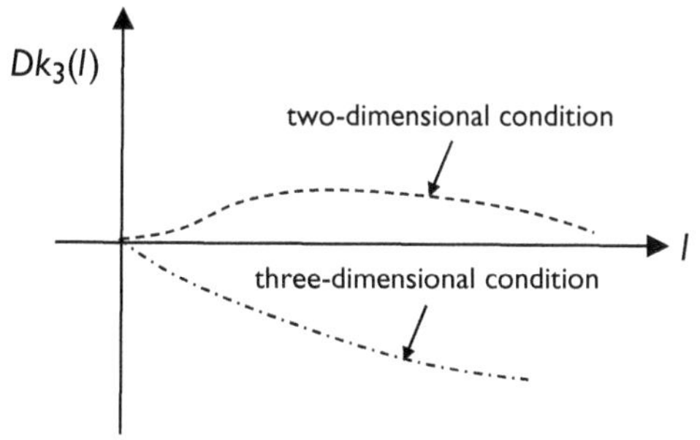

Figure 3.26 Distributions of structural function reported by Nikora et al. (2007).

in which angle brackets mean the spatial average and l is a distance between two reference points. One of the popular Kolmogorov's theories is a following 4/5 law on the structure function with $p = 3$ (see e.g., Rasmussen 1999).

$$Dk_3(l) = -\frac{4}{5}\varepsilon l \tag{3.8.29}$$

Nikora et al. (2007) showed the distribution of Dk_3 like FIG. 3.26. They measured velocities in horizontal 2-D and in 3-D flows generated in a wide, straight flume by PTV technique. In the case of deep condition, i.e., 3-D turbulence, Dk_3 is negative, irrespective of l and it results in that ε is always positive corresponding to the cascade process, whereas in the case of shallow conditions, i.e., 2-D turbulence, Dk_3 is positive, irrespective of l and it results in that ε is always negative, corresponding to the inverse cascade process. They reported that the inverse cascade process is observed in the wide width conditions when the aspect ratio B/H is larger than about 20.

3.9 COHERENT STRUCTURES

3.9.1 Coherent Turbulence

Theodorsen (1952) first anticipated that coherent structure may exist in turbulent boundary layer, and proposed a horseshoe vortex model to explain turbulence characteristics. The hairpin vortex is significantly related to low-speed streaks that are lifted up from the wall. Smith (1984) observed the coherent vortices with a hydrogen bubbles and explained them reasonably by the hairpin vortex model. Moin and Kim (1985) also reproduced numerically the hairpin vortices formation and their evolutional process. Their numerical results were in fairly good agreement with those obtained by dye visualization experiments. Robinson (1991) reviewed carefully these hairpin vortex properties and the relation with bursting phenomena.

Most earlier studies have targeted the inner layer near the wall at low Reynolds numbers. Because of the difficulties in visualization of high-speed flows, many uncertainties remain concerning hairpin vortex structures in the outer layer farther from the wall. Recent innovative developments in PIV techniques allow us to evaluate coherent vortices in the inner and outer layers more quantitatively. Nino & Garcia (1996) suggested that the coherent structures in the near-wall region play significant roles on entrainment of suspended sediments. Nagaosa & Handler (2003) conducted DNS to investigate dynamics of coherent vortices and their origin and development. Further, they educed hairpin-like coherent structure in detail by applying a second invariant of the velocity gradient tensor. Recently, Hurther et al. (2007) found that coherent structures appear even in the outer layer and influence the transport of turbulent energy. The intensive investigations on coherent structures in open-channel flows is necessary to consider not only the wall-turbulence effects but also the air/water interface effects as pointed out by Nezu (2005).

3.9.2 Bursting Phenomenon

Although turbulence seems like a random event, some characteristics called coherent structures are found. Particularly, the hairpin vortex and related bursting phenomena are generally observed near the flume bottom in the open-channel turbulence flows. Their three-dimensional properties could also be investigated in the both experiments and numerical simulations. Ejections and sweeps are observed near the bottom wall in FIG. 3.27. The ejection event means instantaneous low-speed upward current from the near the bottom. By contrast, the sweep event means instantaneous high-speed downward current toward the flume bottom. They play significant roles for the sedimentation, scouring, nutrient adsorption in bottom materials and momentum exchange between the lower and upper layers. In particular, the ejection accompanied by the spanwise vortex is called bursting.

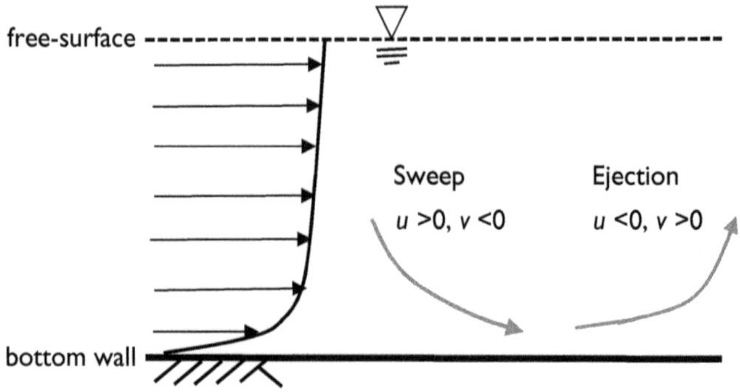

Figure 3.27 Ejection and sweep in open-channel turbulence.

The quadrant analysis is one of the popular methods to detect the coherent structures. This method requires time-series of two components' velocity data obtained simultaneously. By plotting the combination of two components' velocity fluctuations in the two-dimensional coordinate, contributions of the ejections and the sweeps are evaluated quantitatively. Random phenomenon without coherent motions makes a circle like the distribution shown in FIG. 3.28(a). When the velocity fluctuations include the bursting events, there is an inclined ellipse-like distribution, as shown in FIG. 3.28(b). The ejection and the sweep events occur in the second and fourth quadrants, respectively.

The following index is used to evaluate occurrence frequency and magnitude of organized motions with velocity fluctuations of (u, v).

$$RS_i = (\overline{uv})^{-1} \cdot \lim_{T\to\infty} \frac{1}{T} \int_0^T uvI_i dt \quad (i = 1 \sim 4) \tag{3.9.1}$$

When (u, v) appears in the ith quadrant with the magnitude condition, $|uv| \geq H|\overline{uv}|$, $I_i = 1$ is given. H is a threshold parameter called hole size. Removing random small fluctuations, we can detect characterized organized motions using the hole size; otherwise, $I_i = 0$ is given. The contribution rate in each quadrant is expressed as follows:

RS_1: outward interaction ($i = 1$)
RS_2: ejection ($i = 2$)
RS_3: inward interaction ($i = 3$)
RS_4: sweep ($i = 4$)

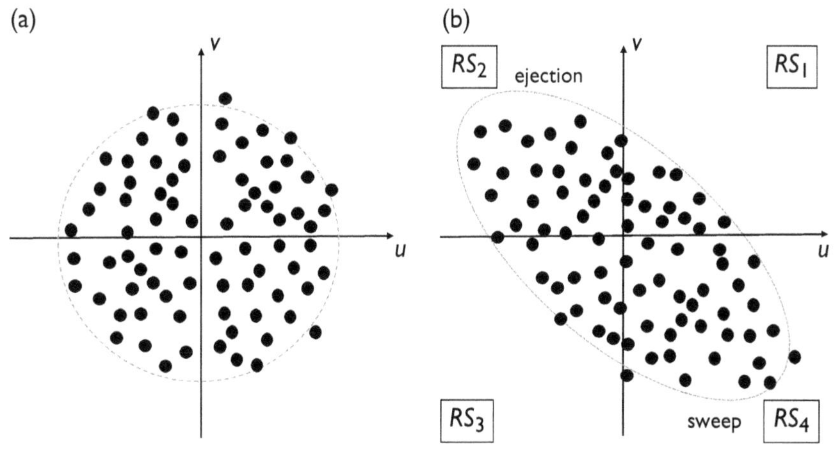

Figure 3.28 Examples of quadrant map for two components of instantaneous velocity fluctuations: (a) no coherent structure and (b) including coherent structure.

It is generally known that the ejections are more significant than the sweeps, i.e., $RS_4 < RS_2$ in the smooth bottom open-channel flows, whereas the sweep events are more significant than the ejections, i.e., $RS_2 < RS_4$ in the open channel with submerged vegetation. There still remain uncertainties about the relation between the bottom roughness and the coherent motions, and thus this topic is expected to be studied further.

Nakagawa & Nezu (1977) proposed the evaluation method of the mean period of the bursting, $\overline{T}_B$. First, they calculated the variation of contribution rate against the hole value. Then, the hole value corresponding to the half contribution rate calculated for $H = 0$ is defined as the threshold value (see section 8.4.1). The period of the bursting is calculated, counting the velocity fluctuation pair (u, v) over the threshold value.

They revealed the contribution of ejections and sweeps to Reynolds stress, and the mean period of the bursting event in open-channel turbulence is expressed by the outer variables of the water depth and maximum velocity, irrespective of the Froude number and Reynolds number as follows:

$$\overline{T}_B U_{\max}/h = 1.5 \sim 3 \qquad (3.9.2)$$

3.9.3 Streaks and Hairpin Vortex

Various physical and conceptual models of the coherent structures in the developing or developed turbulent boundary layer have been proposed by many researchers (e.g., Robinson 1991). Although there still remain uncertainties, an outline of them is briefly explained as shown in FIG. 3.29.

A two-dimensional T-S wave with a wave number with the streamwise direction is unstable and develops linearly in the transition from the laminar boundary layer to the turbulent boundary layer. The growth of disturbance in the T-S wave produces oblique waves and magnitude of velocity fluctuation varies with the spanwise direction, which is called a peak-valley structure (Asai & Nishioka 1989). Then, velocity disturbance becomes three dimensional. We consider that a spanwise vortex tube appears near the bottom wall in this situation. If the middle part of the tube is distorted downstream and stretched, it results in the axis of both sides in the tube as no longer along the spanwise direction. They are called legs, which compose longitudinal circulation. Hence, upward currents are generated between the legs. The middle of the vortex tube, called the head, is lifted up by the upward currents, and a Λ-vortex is formed. The head part has a local spanwise vorticity that promotes further turbulence motion, and thus, the Λ-vortex develops into the hairpin vortex tilted by a certain angle with respect to the wall surface. The sweeps occurs just downstream of the head. The ejections occur between the legs where the low-speed streak is formed. The coherent structure near the bottom can be reproduced by

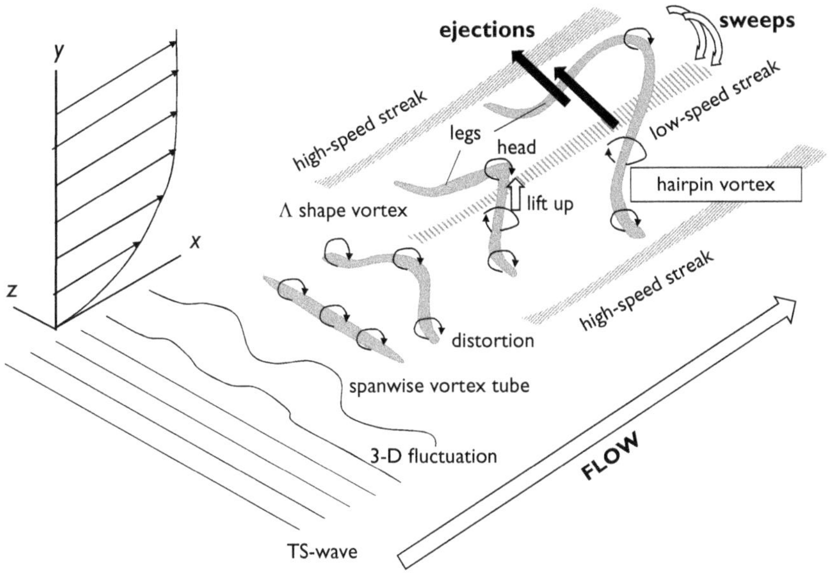

Figure 3.29 Generation of hairpin vortex near the bottom wall.

instability of the low-speed streak. This implies that the bottom turbulence has a mechanism of self-sustaining process, proposed by Waleffe (1997).

Kline et al. (1967) first observed the streak structure in the buffer layer of the developed turbulent boundary layer. The normalized spanwise spacing of streaks, $\Delta z^{+} \equiv \Delta z U_{*}/\nu$ is about 100, which is well accepted as a universal value, e.g., Brereton & Hwang (1994). Adrian et al. (2000) conducted PIV measurements of turbulent boundary layers, and suggested that the coherent structure is likely to be composed of packets, in which several hairpin vortices are lined up in the streamwise direction. Furthermore, they divided the boundary layer into three subzones with approximately uniform momentum, and revealed that the normalized thickness of three subzones depended on the Reynolds number, but the mean velocity in each subzone was almost independent of the Reynolds number. Tomkins & Adrian (2003) have conducted PIV measurements of horizontal plane at several heights throughout the buffer and logarithmic layers in order to reveal a 3-D structure. Their study strongly supported the vortex packet model proposed by Adrian et al. (2000), as shown in FIG. 3.30. Zhou et al. (1999) have predicted such packet structure by a direct numerical simulation (DNS). Handler et al. (1999) conducted direct numerical simulations (DNS) in fully developed turbulent open-channel flows, in which there exists a vertical profile of temperature, and they first proposed that hairpin vortices play a significant role on heat transfer. Sanjou & Nezu (2011a) has measured the generation of hairpin vortex and their development in the open-channel flows, as shown in FIG. 3.31. Jiang et al. (2020) investigated the three-dimensional structures of streaks and hairpin vortices by using a Tomo-PIV.

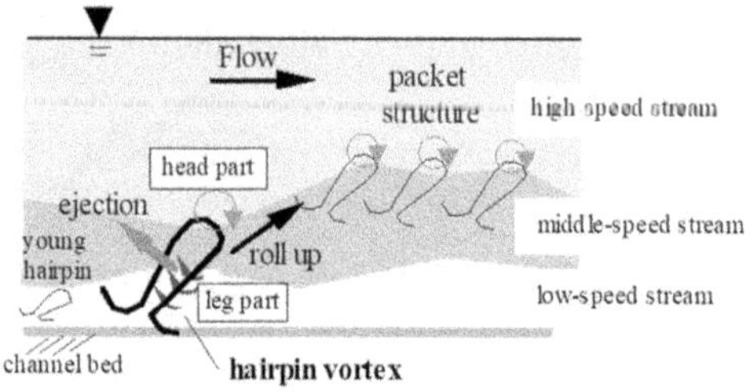

Figure 3.30 Packet structure in outer layer in the open channel (after Sanjou & Nezu 2011a).

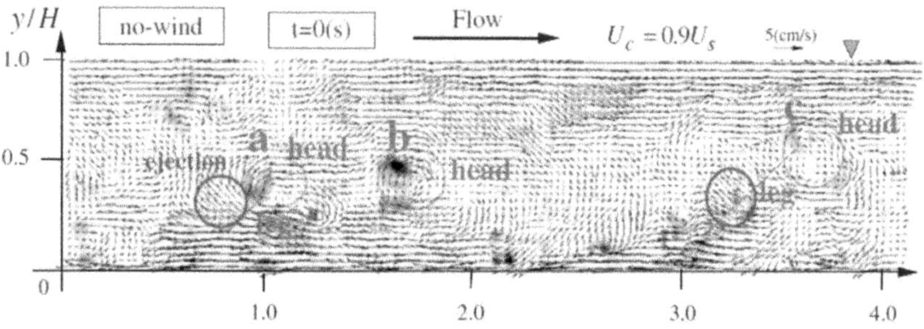

Figure 3.31 Head part of hairpin vortex accompanied with a bursting measured in the open-channel flow (after Sanjou & Nezu 2011a).

3.9.4 Boil

It is generally known that large-scale divergence of the free-surface velocity occurs periodically in natural rivers. When an upward fluid parcel originated in the bottom reaches the free surface, the surface rises locally like ebullition and it diffuses around. Therefore, this phenomenon is called a boil. Although it is difficult to measure in the laboratory flume, previous observation in rivers tell the following properties. First is that the larger bottom roughness supports a boil production more significantly. Second is that the diameter of the horizontal boil is about 30 to 60% of the water depth. Third is that periodicity is expressed by the characteristic velocity U and the water depth h, i.e., $7.6h/U$ (see Jackson 1976, Nezu & Nakagawa 1993). For example, the boil occurs every 7.6 seconds in a mild river with 1 m/s velocity and 1 m depth.

The boil is generated by the following three mechanisms:

1. Generation by sand dune

The uniform flow induces the moving bottom boundary in the gravel bed rivers. Particularly, in the case of dune formation, the shear layer appears, accompanied by the flow separation behind the crest (see section 7.5.2). The periodically separated vortices travel downstream. After attaching the bottom, they ascend to the free surface together with the highly concentrated sediments. But, the rising mechanism of the vortices is still unknown.

2. Generation by sand ribbon

Formation of sand ribbon along the stream promotes spiral-like secondary cells, as shown in FIG. 3.32(a). We can know the existence of the secondary cells observing the bubbles' motions in the free surface and the streamwise streaks in the bottom. The upward and downward currents appear between the neighboring secondary cells. They play important roles of vertical mass and momentum exchanges. Especially, it is expected a fluid parcel conveyed by the upward current develops into the boil in the free-surface, as shown in FIG. 3.32(b).

3. Generation by bursting

The periodicity of the boil is known to correspond with that of the bursting in a low Reynolds number condition. This means the free surface is locally varied in the timing of the ejections. However, this mechanism is limited to a low Reynolds number condition, and thus most boils are expected to be promoted by the dune and the sand ribbon in the natural rivers having a large Reynolds number.

3.10 FREE-SURFACE EFFECTS

3.10.1 Turbulence Damping in the Free Surface

Although the free surface is one of the boundaries as well as the walls, it is not rigid like the bottom surface. The velocity shear $\partial U/\partial y$ can be ignored and thus the turbulence generation is not significant there. The turbulence structure near the free surface is different from that near the bottom wall.

The turbulence intensities decrease monotonically toward the free surface except for near the boundaries. They are expressed by (3.7.2) to (3.7.4). Previous experimental results by Komori et al. (1982) suggest that the vertical component of turbulence intensity v' is dampened near the free surface due to the gravity and the surface tension force, as shown in FIG. 3.33. In contrast, streamwise and spanwise components u' and w' increase slightly. This property may be explained by the pressure strain term in Eq. (3.7.5). Because summation of the pressure strain terms in transport equations of u', v' and w' is zero, the reduction of the presser strain term in v' increases them in u' and w'. That is to say, there is a redistribution of the turbulent kinetic energy near the free surface where a part of the energy in v' is transferred to the energies in u' and w'.

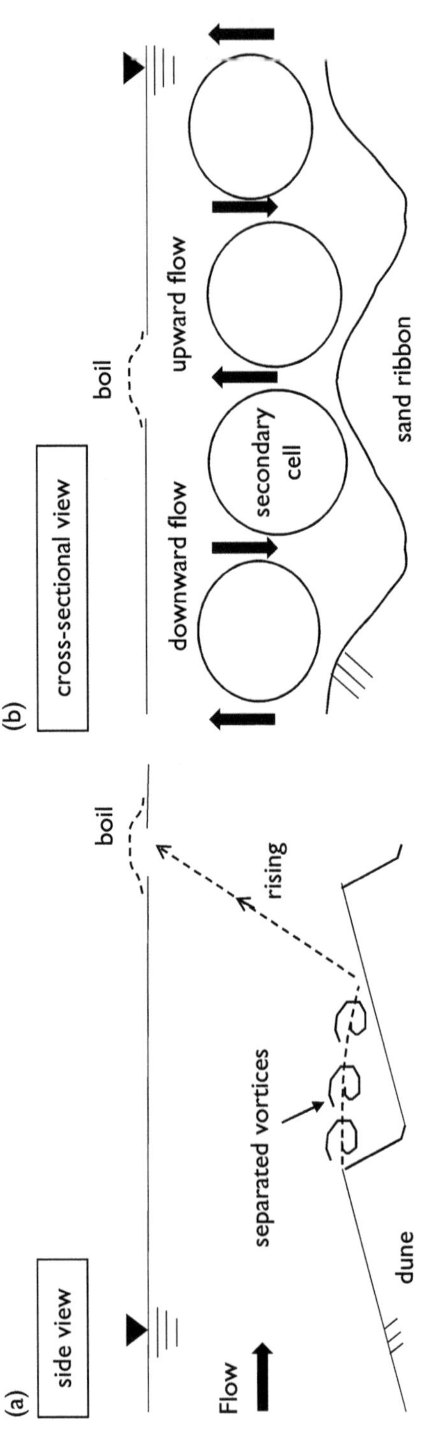

Figure 3.32 Examples of boil observed in rivers: (a) boil formed behind dune (side view) and (b) boil generated by secondary cells (cross-sectional view).

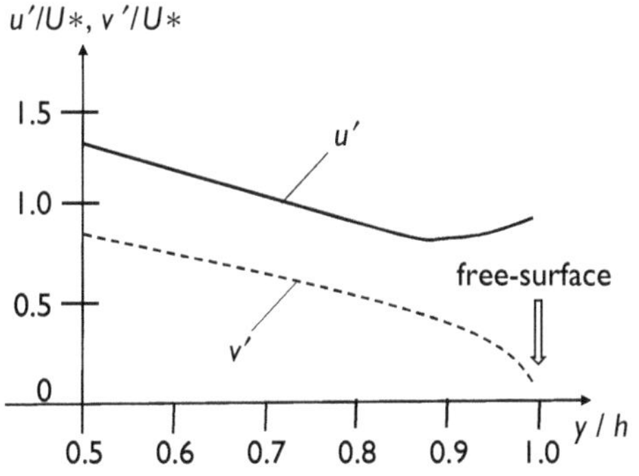

Figure 3.33 Distribution of turbulence intensities in the free-surface region.

Further, the generation term, the turbulence dissipation term and diffusion term are distributed near the free surface, as reported by Komori et al. (1982) (FIG. 3.34). The velocity shear becomes small approaching the free surface and the turbulence generation is not significant. The dissipation term has a negative value in the free-surface region and the turbulence energy is dissipated to heat. The diffusion term, which is positive, in the mid-depth region changes to negative in the free surface. Therefore, a positive diffusion and negative dissipation are balanced there.

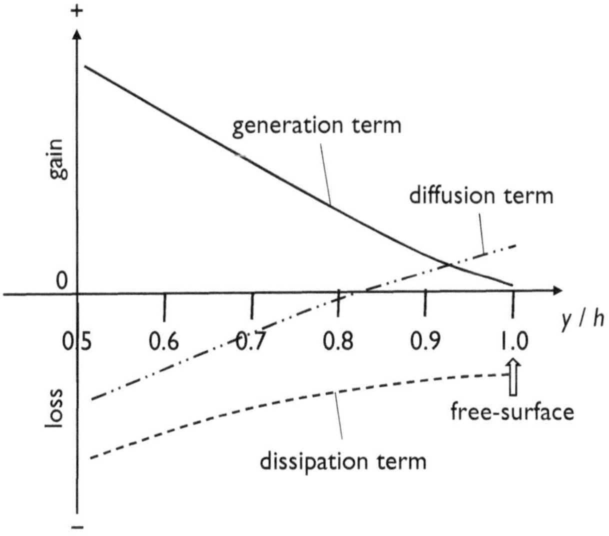

Figure 3.34 Distribution of turbulence generation, dissipation and diffusion terms.

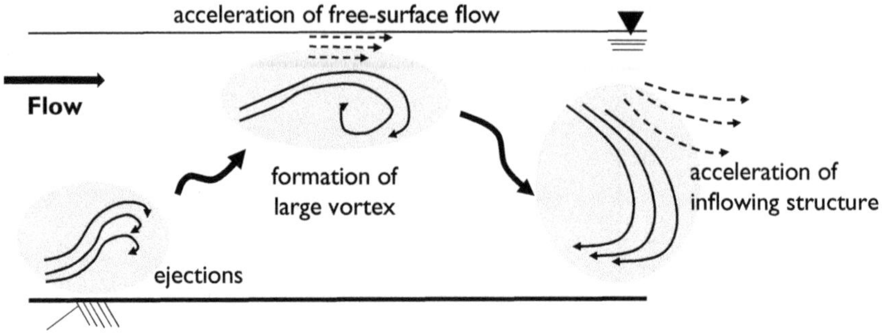

Figure 3.35 Phenomenological model for interaction of bursting and free surface suggested by Rashidi and Banerjee (1988).

3.10.2 Interaction of Bursting and Free Surface

Rashidi and Banerjee (1988) conducted flow visualization in open-channel flow using the oxygen bubble technique to investigate upward transport mechanism of the bottom-oriented bursting motion. FIG. 3.35 shows their conceptual model. As described in section 3.9, the bursting occurs periodically in the bottom wall, and the ejections having an upwelling instantaneous low momentum are observed. When it reaches near the free surface, the current over the bursting parcel is accelerated with locally enhanced velocity shear. After arrival at the free surface, the bursting parcel starts to descend with chaotic motion. Such an organized parcel moves to the bottom, accelerating around the fluids. This sequential process is called surface renewal, as explained in the section 4.3, which plays an important role in the gas and heat exchanges in the free surface.

3.10.3 Generation of Water Waves in Open-Channel Flow

There are small disturbances in the free surface of the high-speed open-channel flow even if under a no wind condition. What induces such a surface deformation accompanied with wave motion? Waves of the supercritical flow are induced by the internal flow instabilities, as theoretically explained by Jeffreys (1925), Vadernikov (1945), Iwasa (1955), etc.

Researches on the turbulence and the free-surface fluctuation are also in progress, regardless of the distinction between supercritical flow and subcritical flow. They pointed out that the bursting reaching the free surface yields the disturbances and then the gravity and surface tension act as restoring forces. It results in the gravity waves and capillary waves propagating in the free surface. However, it remains unknown how the free-surface deformation occurs, and a common explanation of this mechanism has not been obtained yet.

Savelsberg and de Water (2009) focused on the relation between the free-surface disturbance and shedding vortices behind a vertical cylinder. Their

PIV measurements concluded that the effect of turbulence on the free surface is overwhelmed by the capillary waves propagating away from the turbulence disturbance in the uniform isotropic field. In contrast, Guo & Shen (2010) reported that only 2.2% to 12.1% of the potential energy in the free surface contributes to the wave propagation. They concluded that the gravity response to the turbulence disturbance is important in surface deformation.

Although further studies including accurate numerical simulations are required to investigate detailed mechanisms, Nichols et al. (2016) suggested an interesting physical model of the free-surface deformation. They focused a free-surface swelling induced by the boil, and modeled by a convex cone, as shown in FIG. 3.36. The cone height reduces with time by the gravity and the surface tension, and it changes to the concave cone. Nichols et al. expressed the restoring force acting on the cone F_s in the following:

$$F_s = \frac{1}{12}\pi L_s^2 \rho g x_s + \pi L_s \gamma_s \sin\theta \simeq \frac{1}{12}\pi L_s^2 \rho g x_s + 2\pi\gamma_s x_s \tag{3.10.1}$$

in which L_s is deformation length, x_s is surface deformation height and γ_s is surface tension. When downward direction is positive, x_s is positive for the concave and negative for the convex. They considered the analogy of standard spring theory and, therefore, a spring stiffness is given by $K = \frac{1}{12}\pi L_s^2 \rho g + 2\pi\gamma_s$. The frequency is expressed by

$$f = \frac{1}{2\pi}\sqrt{\frac{K}{M_s}} \tag{3.10.2}$$

in which M_s is a mass. Note that M_s is not only the cone mass but includes the additional mass influenced by the free-surface deformation under the water. Neglecting the surface tension effect compared to the gravity, they obtained a following final form:

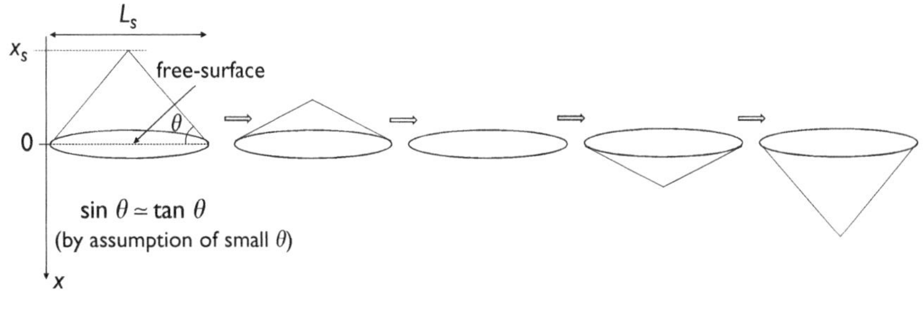

Figure 3.36 Surface deformation model using identified cone suggested by Nichols et al. (2016).

$$f = \frac{1}{2\pi}\sqrt{\frac{g}{(1.5N+1)\sigma}} \tag{3.10.3}$$

in which N is depth of influence factor and σ is root mean square of wave height. This model corresponded well with experimental data. In the near future, the generation mechanism and physical modeling will be discussed in more detail by comparison with the frequency around the turbulence frequency and consideration of the surface roughness effects.

3.11 BOTTOM ROUGHNESS EFFECTS

3.11.1 Equivalent Sand Roughness

The previous sections cover the flow over the flat and smooth bottom. However, the river bed is generally a rough surface that consists of sand grains as shown in Fig. 3.37. The time-mean velocity profile on the rough bottom changes from the smooth bottom due to an increase of the bottom friction. Nikuradse (1933) measured the velocity carefully and evaluated the friction factor in the unidirectional flow on the dense sand roughness. There are various types of elements, such as artificially created regular ones, random ones, and transparent types like submerged vegetation. The parameter that treats them with a unified standard is an equivalent sand roughness. The uniform grain size of sand used in Nikuradse's experiment is defined as the roughness, k_s. In any of roughness flows, try changing k_s, and when it matches the velocity distribution with the dense uniform sand roughness, k_s is the equivalent sand roughness peculiar of the target flow.

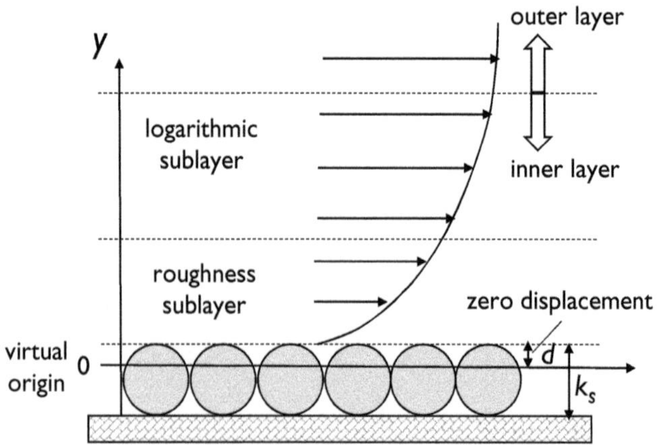

Figure 3.37 The regime of time-averaged velocity profile in the roughness flow.

3.11.2 Virtual Zero Level

For a flat and smooth bed flow, the origin in the depth direction should be on the bottom surface. On the other hand, it is difficult to set the origin in the roughness flow whose bottom surface height is uneven. Nikora et al. (2002) stated that the height at which a large-scale vortex felt as the bed surface is the virtual origin.

Furthermore, assuming that the virtual origin is located d below the top surface of the roughness, a following expression is introduced:

$$\left(\frac{\partial U}{\partial y'}\right)^{-1} = \kappa(y' + d)/U_* \tag{3.11.1}$$

in which y' is a vertical coordinate whose origin is the top surface of the roughness element. Generally, the virtual origin is located at the elevation shifted downward by d from the reference level like the top of the roughness element. d is determined so that the mean flow velocity profile fits the log law best. Although it varies depending on the researcher, it is said to be $d/k_s = 0.15$ to 0.3 (Nezu & Nakagawa 1993).

3.11.3 Mean Velocity Profile

FIG. 3.38 shows the outline of the time-mean velocity profile over the rough bottom, normalized by inner variables so that it is first the log law. We can see a downward drift ΔU^+ of the mean velocity profile below the smooth-bed case. From this finding, it is difficult to handle the rough surface and smooth beds in a unified manner with only the kinetic viscosity and the friction velocity.

Clauser (1954) proposed the following log law for the rough surface flows, considering downward displacement:

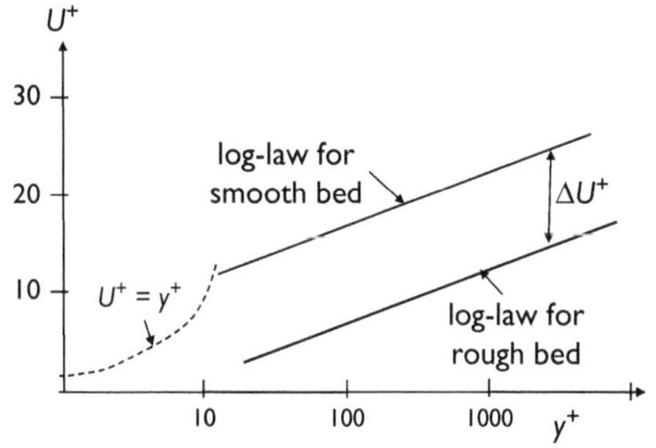

Figure 3.38 Comparison of normalized time-mean velocity profiles in smooth and rough beds.

$$U^+ = \frac{1}{\kappa}\ln y^+ + A - \Delta U^+ \tag{3.11.2}$$

When $\Delta U^+ = 0$, this matches the log law for the smooth bed. Hama (1954) experimentally showed that ΔU^+ could be represented by the function of k_s^+:

$$\Delta U^+ = \frac{1}{\kappa}\ln k_s^+ + A - B \tag{3.11.3}$$

in which B is a constant. Schlichting (1979) proved that $B = 8.5$ in the completely rough bed condition described later. Considering that this value can be applied to the open channel, the time-mean velocity profile for the completely rough bed flow can be expressed as follows by combining Eqs. (3.11.2) and (3.11.3):

$$U^+ = \frac{1}{\kappa}\ln \frac{y}{k_s} + 8.5 \tag{3.11.4}$$

The influence of the roughness is manly the inner layer, and the impact on the outer layer is limited (Raupach et al. 1991, Jimenez 2004). By adding the wake function of Eq. (3.6.4), we can see the following equation that can be applied to the outer layer:

$$U^+ = \frac{1}{\kappa}\ln \frac{y}{k_s} + 8.5 + w(\xi) \tag{3.11.5}$$

3.11.4 Rough Flow Regime

Hydraulic bed roughness is classified into the following three types according to the relationship between the thickness of the viscous sublayer and the roughness height (Nikuradse).

1. Hydraulically smooth bed

When the normalized roughness height, $k_s^+ = k_s U_*/\nu$, is larger than a lower threshold, k_{Smooth}^+, i.e., $k_s^+ < k_{Smooth}^+$, the roughness is buried in the viscous sublayer. Since it is strongly influenced by the viscosity, the roughness influence does not actually appear and it can be regarded as the same as the smooth surface.

2. Incompletely rough bed

When $k^+_{Smooth} < k_s^+ < k^+_{Rough}$, the influence of viscosity coexists with the roughness. ΔU^+ changes as a function of k_s^+, which is called the Nikuradse-type roughness function (e.g., Andersson et al. 2020).

3. Completely rough bed

When $k_s^+ > k^+_{Rough}$, the effect of viscosity disappears and only the roughness effect appears. The roughness function is logarithmic like Eq. (3.11.3). The larger the roughness k_s^+, the closer ΔU^+ is to a certain value. B also converges to 8.5.

Nikuradse set the thresholds for the uniform sand-grain roughness to $k^+_{Smooth} = 5$ and $k^+_{Rough} = 70$, respectively. These depend on the roughness type, and various values have been proposed by subsequent researchers (summarized in Table 2 of Kadivar et al. 2021). Many researchers are continuing to study not only the roughness functions that fit this regime, but also the functions called Colebrook-type roughness functions that do not depend on the regime division (e.g., White 1991, Flack & Schultz 2014).

3.11.5 Turbulence Structure

The effect of the roughness is small on the outer layer, and the profiles of turbulence statistics normalized by the friction velocity are not much different from the result of the smooth bed (e.g., Raupach et al. 1991, Di Cicca et al. 2018). On the other hand, it has been found that the turbulence structure near the roughness surface has different characteristics from the smooth surface due to the vortex generated by the individual roughness element.

Di Cicca et al. compared the probability density function (PDF) of streamwise velocity fluctuation at a height of about $y^+ = 30$ on the smooth and rough beds. The outline is shown in FIG. 3.39(a). While it becomes a weak negative skew in the smooth bed, it becomes a positive skew in the rough bed. As a result, as shown in FIG. 3.39(b), the skewness of the streamwise velocity component Sk_u varies signs with a dependency on the bottom roughness. These features are similar to the CFD result by Lee & Sung (2011). The CFD for the smooth bed shows that the skewness is positive in the viscous sublayer, but it is negative in most zones from the buffer layer to the outer layer. In contrast, it becomes positive in the roughness sublayer. This means that the sweep event, which is the downward high-speed current, is promoted more significantly on the rough surface bed than on the smooth bed except for near the bottom.

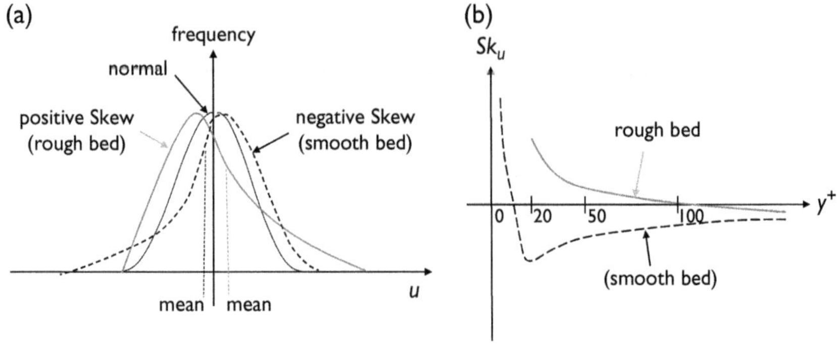

Figure 3.39 Distortion of probability density function (a) PDF of streamwise velocity fluctuation and (b) profile of the skewness of the streamwise velocity in the depth direction.

Raupachi (1981) concludes that sweeps dominate in the roughness sublayer and the thickness is several times the height of the roughness element. Nezu and Nakagawa (1993) show that the ratio of sweep intensity to ejection intensity increases with the increase of k_s^+. The effect of the bottom wall becomes smaller as the distance from the bottom increases, and the intensities of ejection and sweep are getting closer.

Also, there is a slight tendency for isotropy near the rough bed in the turbulent flow, as stated by Nezu and Nakagawa. The maximum of streamwise turbulence intensity u'/U_* is about 2.8 near the smooth bed, but it drops to about 2.0 on the rough surface. The turbulence is generated by the flow separation behind the roughness element accompanied by the sweep, unlike the smooth bed. The integral scale in the streamwise direction decreases as the roughness scale increases, which also contributes to the decrease of u'. On the other hand, v'/U_* increases a little near the rough bed. This is because the larger the roughness size, the weaker the vertical confinement due to the bottom wall. The tendency of turbulence isotropy in the roughness boundary layer is also reported by Smalley et al. (2002).

Further, such a different behavior in the turbulent motion between the rough and smooth beds may be induced by an amplitude modulation effect, in which the large-scale motion originated at the roughness element influences the small-scale motion (e.g., Squire et al. 2016, Di Cicca et al. 2018). Further development of related research is expected.

Chapter 4

Horizontal Open-Channel Turbulence

4.1 HORIZONTAL SHEAR INSTABILITY

4.1.1 Horizontal Unstable Phenomena in Open-Channel

Horizontal fluid dynamics is much more important than vertical one in large scale river and inundation flows. Horizontal velocity fluctuations or vertical vortices govern turbulence phenomena there. Periodical shedding vortices are generally observed behind bridge peer, groins and spur dikes and in the junction between the main channel and the floodplain. This kind of flows are often studied as horizontal two-dimensional current and a depth averaged governing equations are introduced. They have some advantages in theoretical expansion or in reduction of calculation cost compared to the three-dimensional analysis. Kimura et al. (1995) conducted stability analysis in the two-dimensional flow with lateral velocity shear using the depth-averaged equations. This section introduces briefly that linearization of the two-dimensional depth-averaged equation proposed by Kimura et al. and one example of results obtained by their stability analysis. Mathematical derivation of the depth-averaged equations is explained in detail in section 4.4.

4.1.2 Theoretical Background

Kimura et al. (1995) carried out the linear stability analysis using the depth-averaged horizontally two-dimensional Euler equation and the continuity equation as follows:

continuity equation

$$\frac{\partial \tilde{h}}{\partial t} + \frac{\partial \tilde{h}\tilde{u}}{\partial x} + \frac{\partial \tilde{h}\tilde{v}}{\partial y} = 0 \tag{4.1.1}$$

x component of the Euler equation

DOI: 10.1201/9781003112198-4

$$\frac{\partial \tilde{h}\tilde{u}}{\partial t} + \frac{\partial \tilde{h}\tilde{u}^2}{\partial x} + \frac{\partial \tilde{h}\tilde{u}\tilde{v}}{\partial y} + gh\frac{\partial \tilde{h}}{\partial x} = 0 \tag{4.1.2}$$

y component of the Euler equation

$$\frac{\partial \tilde{h}\tilde{v}}{\partial t} + \frac{\partial \tilde{h}\tilde{u}\tilde{v}}{\partial x} + \frac{\partial \tilde{h}\tilde{v}^2}{\partial y} + gh\frac{\partial \tilde{h}}{\partial y} = 0 \tag{4.1.3}$$

in which $\tilde{u}$, $\tilde{v}$ are depth-averaged instantaneous velocity components for x, y directions, respectively. $\tilde{h}$ is an instantaneous water depth. This analysis supposes that the mean streamwise velocity in the x direction has a transverse profile $U_0(y)$ expressed by a tangent hyperbolic function, as shown in FIG. 4.1.

The instantaneous values are decomposed into the mean value using subscript 0 and the deviation using a small letter. The mean of the lateral velocity is assumed to be zero.

$$\tilde{h} = h_0 + h, \quad \tilde{u} = U_0 + u, \quad \tilde{v} = v \tag{4.1.4}$$

Substituting Eq. (4.4.4) into Eqs. (4.4.1) to (4.4.3) with removal of more than second-order terms, Eqs. (4.4.1) to (4.4.3) are linearized as follows:

$$\frac{\partial h}{\partial t} + h_0\frac{\partial u}{\partial x} + U_0(y)\frac{\partial h}{\partial x} + h_0\frac{\partial v}{\partial y} = 0 \tag{4.1.5}$$

$$h_0\frac{\partial u}{\partial t} + h_0 U_0(y)\frac{\partial u}{\partial x} + h_0\frac{dU_0}{dy}v + gh_0\frac{\partial h}{\partial x} = 0 \tag{4.1.6}$$

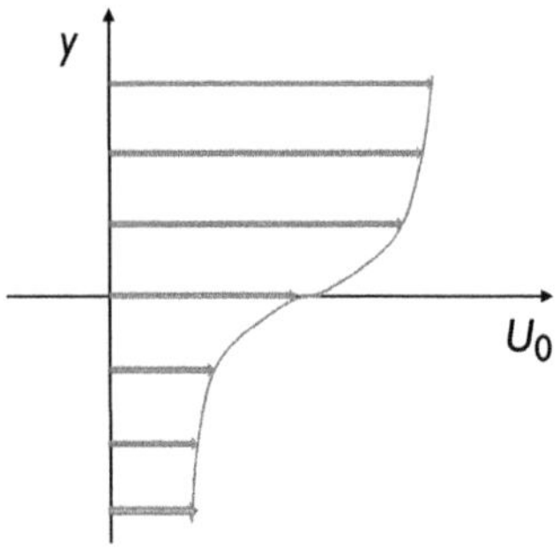

Figure 4.1 Mean streamwise velocity profile used for the stability analysis.

$$h_0 \frac{\partial v}{\partial t} + h_0 U_0(y) \frac{\partial v}{\partial x} + g h_0 \frac{\partial h}{\partial y} = 0 \tag{4.1.7}$$

Eqs. (4.1.5) to (4.1.7) obtain the linear equation for h:

$$\frac{\partial^3 h}{\partial t^3} + 3U_0 \frac{\partial^3 h}{\partial t^2 \partial x} + (3U_0^2 - g h_0) \frac{\partial^3 h}{\partial t \partial x^2} - g h_0 \frac{\partial^3 h}{\partial t \partial y^2}$$
$$+ U_0 (U_0^2 - g h_0) \frac{\partial^3 h}{\partial x^3} - g h_0 U_0 \frac{\partial^3 h}{\partial x \partial y^2} + 2 g h_0 \frac{dU_0}{dy} \frac{\partial^2 h}{\partial x \partial y} = 0 \tag{4.1.8}$$

Some normalized values are defined as follows.

$$\breve{x} = \frac{x}{h_0}, \quad \breve{y} = \frac{y}{l}, \quad \breve{t} = \frac{tU_\infty}{h_0}, \quad \breve{h} = \frac{h}{h_0}, \quad \breve{U}_0 = \frac{U_0}{U_\infty}, \quad F_{r0} = \frac{U_\infty}{\sqrt{g h_0}}$$

U_∞, l are characteristic velocity and length scales, respectively. The fluctuation of water depth is approximately expressed by

$$\breve{h} = \phi(\breve{y}) \exp[i(\breve{k}\breve{x} - \breve{\beta}\breve{t})] \tag{4.1.9}$$

in which $\phi(\breve{y})$ is a small amplification and $\breve{k} = k h_0$ and $\breve{\beta} = \frac{\beta h_0}{U_\infty}$. In the stability analysis for the spatial growth of disturbance, the wave number is decomposed into the real and imaginary parts, i.e., $\breve{k} = \breve{k}_r + i\breve{k}_i$. Hence, Eq. (4.1.9) is rewritten by

$$\breve{h} = \exp(-\breve{k}_i \breve{x}) \phi(\breve{y}) \exp[i(\breve{k}_r \breve{x} - \breve{\beta}\breve{t})] \tag{4.1.10}$$

This expression implies that the disturbance grows in space when $-\breve{k}_i \breve{x}$ is positive.

Substituting of Eq. (4.1.10) into the normalized form Eq. (4.1.8) introduces Eq. (4.1.11).

$$(\breve{k}\breve{U}_0 - \breve{\beta}) \frac{d^2\phi}{dy^2} - 2\breve{k} \frac{d\breve{U}_0}{dy} \frac{d\phi}{dy} + \left(\frac{l}{h_0}\right)^2 (\breve{k}\breve{U}_0 - \breve{\beta})$$
$$\times \left\{ F_{r0}^2 (\breve{k}\breve{U}_0 - \breve{\beta})^2 - \breve{k}^2 \right\} \phi = 0 \tag{4.1.11}$$

Solving Eq. (4.1.11) numerically, we can discuss some crucial stability properties. For example, Kimura et al. (1995) obtained a relationship

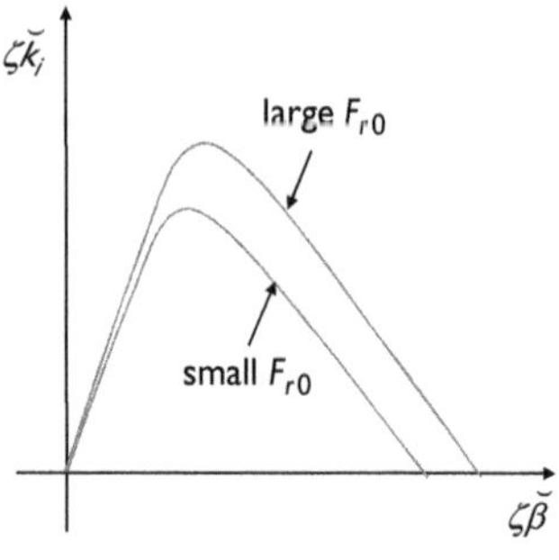

Figure 4.2 Relationship between normalized frequency $\zeta\breve{\beta}$ and the normalized growth rate $\zeta\breve{k}_i$.

between normalized frequency $\zeta\breve{\beta}$ and the normalized growth rate $\zeta\breve{k}_i$, as shown in FIG. 4.2, in which $\zeta = l/h_0$. This results in the maximum growth rate increases corresponding to the Froude number. Further, the frequency of the stability limit, e.g., $\breve{k}_i = 0$, increases as the Froude number increases. Note that the experimental growth rate obtained by Kimura et al. (1995) is smaller than the theoretical curve. This may be that they ignored viscous effects in the governing equations of Eqs. (4.1.2) and (4.1.3).

4.2 HORIZONTAL VORTEX

4.2.1 Kelvin Helmholtz Instability

Kelvin Helmholtz instability considers developing a small disturbance in the boundary of two layers with different densities, as shown in FIG. 4.3. Although it uses a vertical two-dimensional coordinate, it can be discussed in a horizontal two-dimensional coordinate, neglecting the gravity force. Supposing the irrotational flow field, a discontinuous velocity profile is given with constant velocities in two layers. Two uniform flows contact the boundary plane ($y = 0$) at the initial stage. This theory assumes that the boundary plane is distorted by the small fluctuations, and the deviation of the boundary height is expressed by $\eta(x, t)$.

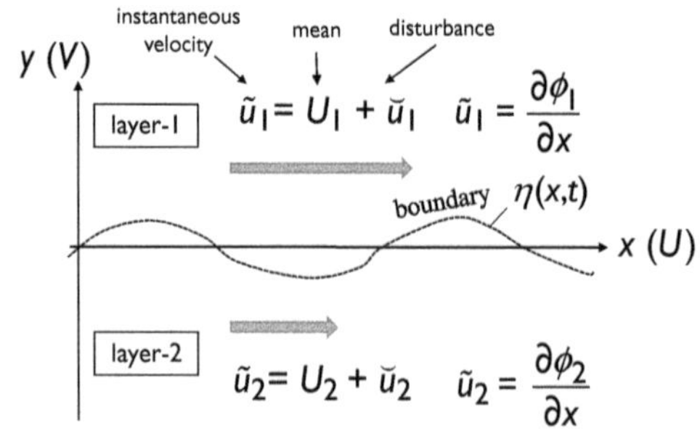

Figure 4.3 Coordinate system in two uniform flows with different densities.

Instantaneous streamwise velocities in the layers 1 and 2, $\tilde{u}_1$, $\tilde{u}_2$ are decomposed into the mean velocities U_1, U_2 and the deviations $\breve{u}_1$, $\breve{u}_2$, as indicated in Eq. (4.2.1).

$$\tilde{u}_1 = U_1 + \breve{u}_1 \tag{4.2.1a}$$

$$\tilde{u}_2 = U_2 + \breve{u}_2 \tag{4.2.1b}$$

Assuming that mean velocity in the vertical direction is zero, instantaneous vertical velocities are equal to corresponding deviations like Eq. (4.2.2):

$$\tilde{v}_1 = \breve{v}_1 \tag{4.2.2a}$$

$$\tilde{v}_2 = \breve{v}_2 \tag{4.2.2b}$$

Assuming the present flow is irrotational, the instantaneous velocities in each layer are expressed by Eq. (4.2.3):

$$\tilde{u}_1 = \frac{\partial \phi_1}{\partial x}, \quad \tilde{v}_1 = \frac{\partial \phi_1}{\partial y} \tag{4.2.3a}$$

$$\tilde{u}_2 = \frac{\partial \phi_2}{\partial x}, \quad \tilde{v}_2 = \frac{\partial \phi_2}{\partial y} \tag{4.2.3b}$$

in which ϕ_1, ϕ_2 are the velocity potentials in the layers 1 and 2. A deviation of the boundary elevation η and potentials of velocity deviation $\breve{\phi}_1$, $\breve{\phi}_2$ are given by the wave number k and wave celerity c.

$$\eta = \hat{\eta} e^{ik(x-ct)} \tag{4.2.4}$$

$$\breve{\phi}_1 = \hat{\phi}_1(y) e^{ik(x-ct)} \tag{4.2.5a}$$

$$\breve{\phi}_2 = \hat{\phi}_2(y) e^{ik(x-ct)} \tag{4.2.5b}$$

in which $\hat{\eta}$, $\hat{\phi}_1$, $\hat{\phi}_2$ are amplifications of the corresponding variables.

The wave celerity is decomposed into the real number and the imaginary number, i.e., $c = c_r + ic_i$. Then, the expression $e^{ik(x-ct)} = e^{c_i kt} e^{ik(x-c_r t)}$ is obtained. This implies that an initial disturbance develops in time when $c_i > 0$. The combination of the kinetic boundary condition and the unsteady

Bernoulli equation gives the following form (mathematical expansion is described in appendix 3):

$$c = c_r + ic_i = \frac{\rho_1 U_1 + \rho_2 U_2}{\rho_1 + \rho_2} \pm \left[\frac{g}{k} \frac{\rho_2 - \rho_1}{\rho_1 + \rho_2} - \rho_1 \rho_2 \left(\frac{U_1 - U_2}{\rho_1 + \rho_2} \right)^2 \right]^{1/2} \quad (4.2.6)$$

When the number inside the square root in Eq. (4.2.6) is positive, $c_i = 0$, it means neutrally stable. In other words, when $k \leq \frac{g}{\rho_1 \rho_2} \frac{\rho_2{}^2 - \rho_1{}^2}{(U_1 - U_2)^2}$, the present two-layer flow is neutrally stable. In contrast, $k > \frac{g}{\rho_1 \rho_2} \frac{\rho_2{}^2 - \rho_1{}^2}{(U_1 - U_2)^2}$ and $c_i > 0$, the flow is unstable in which the disturbance grows.

Considering no velocity difference between the two layers, i.e., $U_1 = U_2$, a content of the square root is $\frac{g}{k} \frac{\rho_2 - \rho_1}{\rho_1 + \rho_2}$. Therefore, when the fluid in the lower layer 2 is lighter than that in the upper layer 1 ($\rho_2 < \rho_1$), the content of the square root is negative, i.e., $c_i \neq 0$. Further, when $c_i > 0$, this flow is unstable. By contrast, in the opposite condition, ($\rho_2 > \rho_1$), it is neutrally stable. When the light fluid is over the heavy fluid, even if a parcel of the light fluid is instantaneously pushed down, it is put back by the buoyancy force. Thus, this case is stable.

Consider the condition of no density difference, ($\rho = \rho_1 = \rho_2$); Eq. (4.2.6) is rewritten by

$$c = c_r + ic_i = \frac{U_1 + U_2}{2} \pm \sqrt{-\left(\frac{U_1 - U_2}{2} \right)^2} = \frac{U_1 + U_2}{2} \pm i\frac{1}{2}|U_1 - U_2| \quad (4.2.7)$$

This suggests that when there is a velocity difference between two layers and $c_i > 0$, this situation is unstable and induces growth of the initial disturbance. Even if there is no inflection point in the velocity profile highlighted by the Rayleigh's inflection point theorem, the only velocity difference causes the instability.

Considering the y axis as the lateral axis in FIG. 4.3, this theory is applied to the horizontal two-dimensional field. Since the gravity force is removed in this coordinate system, Eq. (4.2.7) can be used under the constant density condition. A large-scale eddy with a vertical axis called a horizontal vortex is often observed in the confluence zone, behind the bridge peer, around the vegetation patch, as explained in the following section.

4.2.2 Horizontal Turbulence in Rivers

Horizontal fluid dynamics are more important than vertical ones in large-scale rivers, in which width is much larger than the water depth. Water flow

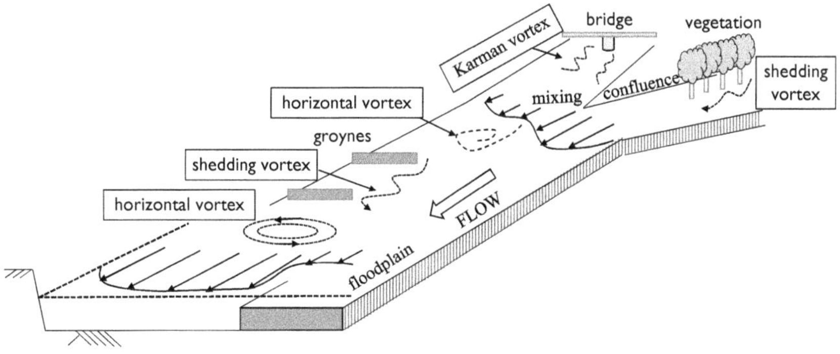

Figure 4.4 Vortices with vertical axis observed in natural rivers.

becomes locally unstable in the regions where the streamwise velocity varies significantly with the lateral direction. These regions are introduced as follows.

4.2.2.1 Confluence zone

Confluence means the point where two rivers meet. When there is quite a difference in velocities of meeting rivers, the inflection point is formed in the lateral profile of streamwise velocity downstream of the meeting point, as shown in FIG. 4.4. The mixing layer is developed by the velocity shear, and momentum exchange is promoted in the lateral direction. However, the mixing process disappears during the streamwise traveling, reducing the velocity difference by the mixing itself.

4.2.2.2 Compound channel

The compound channel is composed of a deep main channel and shallow floodplain, which is often observed in the rivers in flood. The over floodplain flow is slower than the main channel, because the low-depth floodplain flow is influenced significantly by the bottom roughness, such as vegetation. It results in formation of the mixing layer induced by the velocity difference. This mixing layer continues downstream unlike that induced by the confluence zone. Hence, large-scale horizontal vortices are generated periodically. They are found to transport a large amount of sediment from the main channel to over the floodplain.

4.2.2.3 Bridge peer wake

The flow separates behind the bridge peer and the deadwater zone is formed there. Shedding vortices are generated in a boundary between the mainstream and the deadwater zone. They grow up and to be the von Karman vortex street.

4.2.2.4 *Vegetation*

Emergent vegetation is one of the permeable obstacles that increases a friction against the mainstream. Particularly, a vegetation group on the river sides induces the velocity difference between the mainstream and the vegetation zone. It results in the production of horizontal mixing in the same manner as the compound open-channel flow. In contrast, an isolated vegetation generates the shedding vortices in the same manner as the bridge peer.

4.2.2.5 *Groynes*

A groyne is one of the artificial structures that navigates flow direction or prevents bank erosion. A large-scale horizontal circulation is generally observed. Furthermore, the mixing layer appears between the mainstream and the separation zone behind the groyne. Hence, the shedding vortex with vertical axis is produced periodically. Neighboring groynes aligned in the streamwise direction form a side cavity zone, as shown in FIG. 4.4. The long free-surface waves are known to occur in the lateral and streamwise directions there.

4.2.2.6 *Side walls*

A side wall generates the friction against the mainstream as well as the bottom. Hence, the streamwise velocity is significantly reduced and, as a result, in the horizontal boundary layer is developed. Although the turbulence structure has not been revealed more intensively than the vertical one, the horizontal bursting events are expected to occur near the side wall.

4.3 SURFACE VELOCITY DIVERGENCE

We can find fluctuation and turbulence in the free surface of rivers. These structures are detected as a local velocity shear or vorticity when the simultaneous multi-point measurements such as PIV are conducted. This section explains a surface velocity divergence, which is one of the key elements in the free-surface turbulence. The surface velocity divergence controls significantly the gas or heat exchanges between air and water. Therefore, it is crucial to reveal hydraulic properties of the surface velocity divergence for environmental management in the artificial channel, rivers and lakes with the surface stream.

4.3.1 Definition of Surface Velocity Divergence

Instantaneous surface velocity divergence at the air–water interface $\tilde{\beta}$ (where the tilde indicates instantaneous components) is defined as

$$\tilde{\beta} \equiv \frac{\partial \tilde{u}}{\partial x} + \frac{\partial \tilde{w}}{\partial z} = -\frac{\partial \tilde{v}}{\partial y} \tag{4.3.1}$$

Here, the coordinate system is relative to the water surface, and thus the instantaneous vertical velocity is always zero at the free surface. When the vertical velocity is assumed to be zero in the free surface, the left-hand side of Eq. (4.3.1), i.e., a velocity gradient $\partial\tilde{v}/\partial y$ means the occurrence of upward or downward currents. Hence, Eq. (4.3.1) implies that a positive divergence corresponds to the upward current and a negative divergence corresponds to the downward one, respectively.

FIG. 4.5 shows the instantaneous horizontal distribution of $\tilde{\beta}$ and the velocity vectors $(\tilde{u} - \tilde{u}_r, \ \ \tilde{w} - \tilde{w}_r)$ measured by the PIV method (see Sanjou et al. 2017), in which the free-surface velocity U_s = 12.3 cm/s and the water depth H = 8 cm. $(\tilde{u}_r, \ \ \tilde{w}_r)$ is a reference velocity vectors. These profiles were acquired as a time series at 0.033 s intervals. In the convergence zone $\tilde{\beta} < 0$, the surrounding momentum concentrates and a downward current is generated. In contrast, in the divergence zone $\tilde{\beta} > 0$, the momentum diverges as an upward current flows. A convergence zone marked C1 corresponding to negative divergence is traveling downstream during 0.033 seconds (FIG. 4.5 a and b). Surrounding fluids are found to come close to this zone, and the sign of instantaneous divergence implies that the downward currents are accompanied there. At t = 0.033 s, a new convergence zone marked C2 appears upstream in the measurement area. FIG. 4.5c focuses on C2 and the reference velocity components' corresponding location are subtracted, and this results in what we can observe around fluids approaching C2.

4.3.2 Hydrodynamic Property

The strength of the surface divergence is calculated by the root mean square of time series of instantaneous surface velocity divergence. This index is called surface divergence intensity $\beta' \equiv \sqrt{\overline{\beta^2}}$. The dependency of the divergence intensity on the bulk-mean velocity U_m is as shown in FIG. 4.6. This results in β' increases quadratic functionally corresponding to an increase of U_m in the same manner as the previous data. The measured data implies that β' varies with bulk-mean velocity and remains almost constant under the same bulk-mean velocity condition, irrespective of water depth. Although U_m has significant impacts on the production of the surface divergence, this is a complicated mechanism related to turbulence structure in the stream.

4.3.3 Scaling of Surface Velocity Divergence

What hydrodynamic parameters govern the surface velocity divergence? Here, let's conduct a dimensional analysis focusing on β'; additionally,

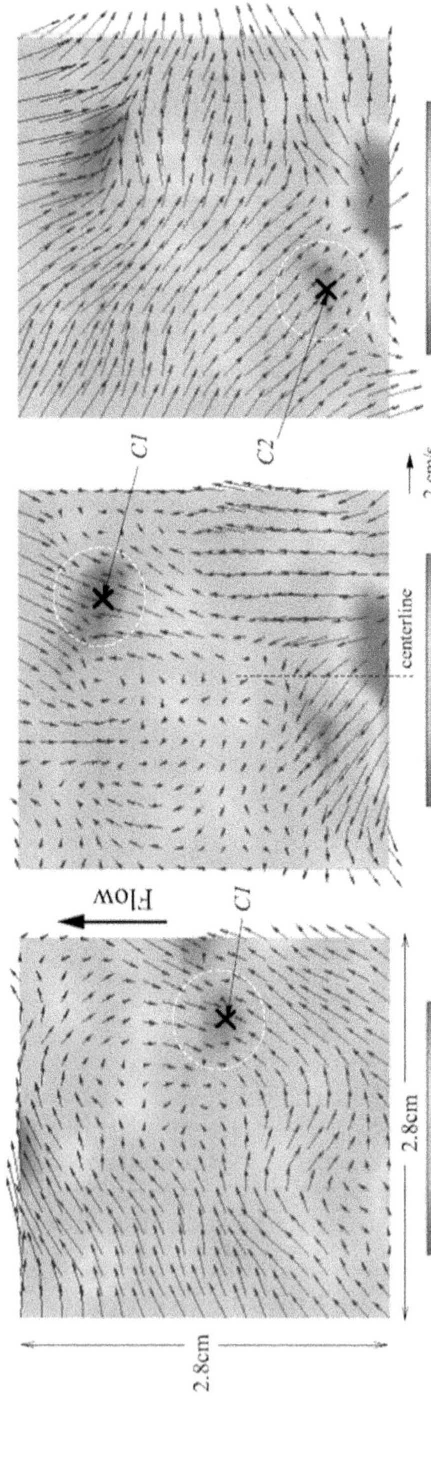

Figure 4.5 Time variation of instantaneous surface velocity divergence and horizontal velocity vectors (after Sanjou et al. 2017). The color contour plot is provided by the original paper.

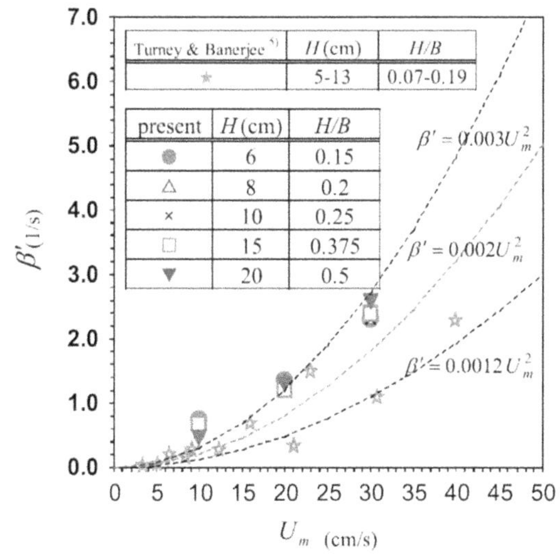

Figure 4.6 Relation between the surface divergence intensity and the bulk-mean velocity (after Sanjou et al. 2017).

water depth H, ρ and free-surface velocity U_s are also considered for dimensional analysis. The Π-theorem provides normalized numbers.

$$\Pi_{\beta'} = \nu\beta'/U_s^2, \quad \text{where } \nu \equiv \mu/\rho, \tag{4.3.2}$$

$$\Pi_H = U_s H/\nu = \mathrm{Re}_s, \tag{4.3.3}$$

in which Re_s is a U_s-based Reynolds number. Therefore, the surface divergence intensity normalized by the surface velocity and kinematic viscosity is expected to depend on the Reynolds number as follows:

$$\frac{\nu\beta'}{U_s^2} \sim func_1(\mathrm{Re}_s) \tag{4.3.4}$$

in which $func_1$ is a function of U_s-based Reynolds number.

FIG. 4.7 shows the relation between the normalized divergence $\Pi_{\beta'}$ and Reynolds number Re_s, in which, despite scattering, being comparatively large at a low Reynolds number the fitting function

$$\nu\beta'/U_s^2 = 1.3 \times 10^{-3}\,\mathrm{Re}_s^{-0.392} \cong 1.3 \times 10^{-3}\,\mathrm{Re}_s^{-2/5} \tag{4.3.5}$$

is observed.

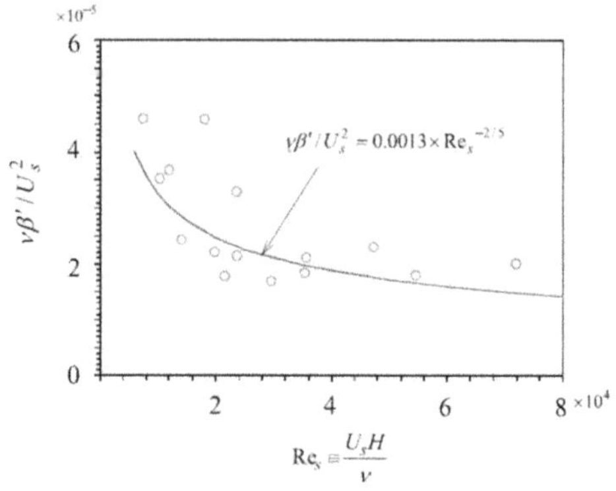

Figure 4.7 Relation between the normalized surface divergence intensity and Reynolds number (after Sanjou et al. 2017).

4.4 2-D DEPTH-AVERAGED GOVERNING EQUATIONS

Bending, meandering and gradually expanding and shrinking zone are often observed in rivers with a large variation in the lateral direction. They are hard to be considered by the one-dimensional analysis in hydraulics. In contrast, when the 3-D N-S equations are applied to a comparatively wide area in natural rivers, a calculation cost becomes excessive. The lateral scale is much larger than the vertical one in rivers. Hence, we focus on only horizontal velocities averaged in the vertical direction in order to save the calculation cost. In this section, we derive the depth-averaged continuity equation and the momentum equations.

4.4.1 Mathematical Preparations

FIG. 4.8 shows the coordinate system, in which x, y, z are streamwise, lateral and vertical directions, respectively. h, η are the water depth and the bottom elevation from the reference level. $H = h + \eta$ is the water level from the reference level. Here, η is known value. U, V are the depth-averaged velocity components, respectively. They are unknown variables. Since a following analysis uses an assumption of static pressure, the water depth h is an unknown variable instead of the pressure. When we consider the bottom slope, there is a gap between the depth direction and the vertical direction. However, we define conveniently the y direction is along the vertical axis, assuming small bottom slope. The vertical velocity is assumed to be zero. We need two mathematical preparations. First, the Leibniz rule is given by Eq. (4.4.1).

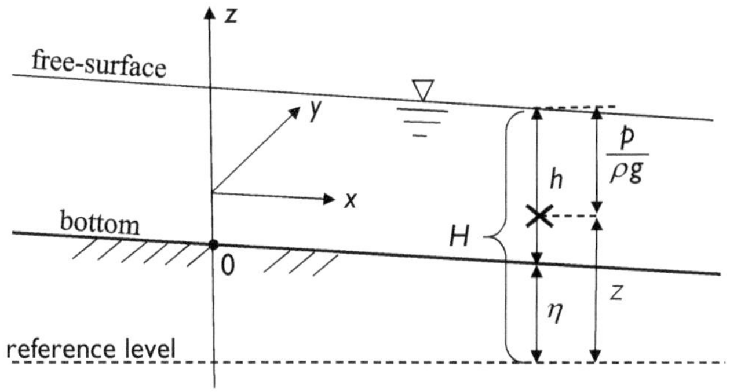

Figure 4.8 Coordinate system in depth-averaging technique.

$$\frac{d}{dx}\left(\int_{f(x)}^{g(x)} W(x,y)dy\right) = \int_{f(x)}^{g(x)} \frac{\partial W}{\partial x}dy + W(x,g(x))\frac{\partial g}{\partial x} - W(x,f(x))\frac{\partial f}{\partial x} \tag{4.4.1}$$

Here, let us apply Eq. (4.4.1) to the unsteady term $\frac{\partial U}{\partial t}$ that appears in the RANS equations given by Eq. (2.6.3):

$$\int_{\eta(x,y)}^{H(x,y)} \frac{\partial U}{\partial t}dz = \frac{\partial}{\partial t}\left(\int_{\eta}^{H} Udz\right) - U(x,y,H)\frac{\partial H}{\partial t} + U(x,y,\eta)\frac{\partial \eta}{\partial t} \tag{4.4.2}$$

Defining the depth-averaged velocity by $\bar{U} = \frac{1}{h}\int_{\eta}^{H} Udz$, Eq. (4.4.2) is re-written as

$$\int_{\eta}^{H} \frac{\partial U}{\partial t}dz = \frac{\partial}{\partial t}(h\bar{U}) - U\frac{\partial H}{\partial t} + U\frac{\partial \eta}{\partial t} \tag{4.4.3}$$

Second, a kinematic boundary condition is required to derive the depth-averaged governing equations. Applying this boundary condition to the free surface and the bottom, the following forms are obtained, respectively:

$$\frac{\partial H}{\partial t} + U(x, y, H)\frac{\partial H}{\partial x} + V(x, y, H)\frac{\partial H}{\partial y} = W(x, y, H) \tag{4.4.4}$$

$$\frac{\partial \eta}{\partial t} + U(x, y, \eta)\frac{\partial \eta}{\partial x} + V(x, y, \eta)\frac{\partial \eta}{\partial y} = W(x, y, \eta) \tag{4.4.5}$$

4.4.2 Derivation of Depth-Averaged Governing Equations

4.4.2.1 *Momentum equations*

Assuming that the external force is only gravity and the viscous force is the only component associated with the z- direction gradient of U, x component of RANS in Eq. (2.6.3) is expressed by

$$\frac{\partial U}{\partial t} + U\frac{\partial U}{\partial x} + V\frac{\partial U}{\partial y} + W\frac{\partial U}{\partial z} = F_x - \frac{1}{\rho}\frac{\partial P}{\partial x} + (\nu + \nu_t)\frac{\partial^2 U}{\partial z^2} \tag{4.4.6}$$

in which the static pressure approximation $P = \rho g(h - z)$ is used. The first and second terms in Eq. (4.4.6) are written by $F_x - \frac{1}{\rho}\frac{\partial P}{\partial x} = gi_0 - g\frac{\partial h}{\partial x}$. $i_0 = -\frac{\partial \eta}{\partial x}$ is the bottom slope. Furthermore, using the continuity equation $\frac{\partial U}{\partial x} + \frac{\partial V}{\partial y} + \frac{\partial W}{\partial z} = 0$, the convection term in Eq. (4.4.6) can be rewritten by $U\frac{\partial U}{\partial x} + V\frac{\partial U}{\partial y} + W\frac{\partial U}{\partial z} = \frac{\partial U^2}{\partial x} + \frac{\partial UV}{\partial y} + \frac{\partial UW}{\partial z}$. Therefore, Eq. (4.4.6) is re-formed as follows:

$$\frac{\partial U}{\partial t} + \frac{\partial U^2}{\partial x} + \frac{\partial UV}{\partial y} + \frac{\partial UW}{\partial z} = gi_o - g\frac{\partial h}{\partial x} + \frac{\partial}{\partial z}\left((\nu + \nu_t)\frac{\partial U}{\partial z}\right) \tag{4.4.7}$$

We integrate this from η to H in the z direction as follows:

$$\int_\eta^H \frac{\partial U}{\partial t}dz + \int_\eta^H \frac{\partial U^2}{\partial x}dz + \int_\eta^H \frac{\partial UV}{\partial y}dz + \int_\eta^H \frac{\partial UW}{\partial z}dz$$
$$= gi_o\int_\eta^H dz - g\int_\eta^H \frac{\partial h}{\partial x}dz + \int_\eta^H \frac{\partial}{\partial z}\left((\nu + \nu_t)\frac{\partial U}{\partial z}\right)dz \tag{4.4.8}$$

Referring to Eq. (4.4.2), Eq. (4.4.9) is derived by the Leibniz rule.

$$\begin{aligned}&\frac{\partial}{\partial t}\left(\int_{\eta}^{H} Udz\right)+\frac{\partial}{\partial x}\left(\int_{\eta}^{H} U^2dz\right)+\frac{\partial}{\partial y}\left(\int_{\eta}^{H} UVdz\right)\\&\quad - U|_H\frac{\partial H}{\partial t} - U^2|_H\frac{\partial H}{\partial x} - U|_H V|_H\frac{\partial H}{\partial y} + U|_H W|_H\\&\quad - U|_\eta\frac{\partial \eta}{\partial t} - U^2|_\eta\frac{\partial \eta}{\partial x} - U|_\eta V|_\eta \times \frac{\partial \eta}{\partial y} + U|_\eta W|_\eta\\&= ghi_o - gh\frac{\partial h}{\partial x} + \left[(\nu+\nu_t)\frac{\partial U}{\partial z}\right]_\eta^H\end{aligned} \tag{4.4.9}$$

Eqs. (4.4.4) and (4.4.5) give following forms:

$$W|_H = \frac{\partial H}{\partial t} + U|_H\frac{\partial H}{\partial x} + V|_H\frac{\partial H}{\partial y} \tag{4.4.10}$$

$$W|_\eta = \frac{\partial \eta}{\partial t} + U|_\eta\frac{\partial \eta}{\partial x} + V|_\eta\frac{\partial \eta}{\partial y} \tag{4.4.11}$$

Substituting them into the second and third lines in Eq. (4.4.9), these lines become zero. Let's move to the third term in Eq. (4.4.7). The following conditions hold at the free surface ($z = H$) and the bottom ($z = \eta$), respectively:

$$(\nu+\nu_t)\left.\frac{\partial U}{\partial z}\right|_H = 0\left(\because \left.\frac{\partial U}{\partial z}\right|_H \fallingdotseq 0\right) \tag{4.4.12}$$

$$(\nu+\nu_t)\left.\frac{\partial U}{\partial z}\right|_\eta \fallingdotseq \nu\left.\frac{\partial U}{\partial z}\right|_\eta = \frac{\tau_{0x}}{\rho} \tag{4.4.13}$$

Hence, Eq. (4.4.7) is rewritten by

$$\frac{\partial}{\partial t}\left(\int_{\eta}^{H} Udz\right)+\frac{\partial}{\partial x}\left(\int_{\eta}^{H} U^2dz\right)+\frac{\partial}{\partial y}\left(\int_{\eta}^{H} UVdz\right) = ghi_o - gh\frac{\partial h}{\partial x} - \frac{\tau_{0x}}{\rho} \tag{4.4.14}$$

Introducing the following depth-averaged variables, $\bar{U} \equiv \frac{1}{h}\int_{\eta}^{H} Udz$, $\overline{U^2} \equiv \frac{1}{h}\int_{\eta}^{H} U^2dz$, $\overline{UV} \equiv \frac{1}{h}\int_{\eta}^{H} UVdz$, the x component of the depth-averaged momentum equation is obtained. The y component of that is also derived in the same way. They are expressed by Eqs. (4.4.15) and (4.4.16), respectively:

$$\frac{\partial h\overline{U}}{\partial t} + \frac{\partial h\overline{U^2}}{\partial x} + \frac{\partial h\overline{U}\overline{V}}{\partial y} = ghi_o - gh\frac{\partial h}{\partial x} - \frac{\tau_{0x}}{\rho} \tag{4.4.15}$$

$$\frac{\partial h\overline{V}}{\partial t} + \frac{\partial h\overline{U}\overline{V}}{\partial x} + \frac{\partial h\overline{V^2}}{\partial y} = -gh\frac{\partial \eta}{\partial y} - gh\frac{\partial h}{\partial y} - \frac{\tau_{0y}}{\rho} \tag{4.4.16}$$

The bottom shear stress is found, often by following the empirical formulae using a friction coefficient C_f.

$$\tau_{0x} = \rho C_f \times \sqrt{U^2 + V^2} \times \overline{U}, \quad \tau_{0y} = \rho C_f \times \sqrt{U^2 + V^2} \times \overline{V} \tag{4.4.17}$$

4.4.2.2 Continuity equation

The time-averaged continuous equation is expressed by Eq. (4.4.18).

$$\frac{\partial U}{\partial x} + \frac{\partial V}{\partial y} + \frac{\partial W}{\partial z} = 0 \tag{4.4.18}$$

The integral of Eq. (4.4.18) is as follows:

$$\int_{\eta}^{H} \frac{\partial U}{\partial x}dz + \int_{\eta}^{H} \frac{\partial V}{\partial y}dz + \int_{\eta}^{H} \frac{\partial W}{\partial z}dz = 0 \tag{4.4.19}$$

Applying the Leibniz rule to Eq. (4.4.19), the following form is obtained:

$$\begin{aligned}&\frac{\partial}{\partial x}\left(\int_{\eta}^{H} Udz\right) + \frac{\partial}{\partial y}\left(\int_{\eta}^{H} Vdz\right)\\ &- \left(U|_H\frac{\partial H}{\partial x} + V|_H\frac{\partial H}{\partial y} - W|_H\right) + \left(U|_\eta\frac{\partial \eta}{\partial x} + V|_\eta\frac{\partial \eta}{\partial y} - W|_\eta\right) = 0\end{aligned} \tag{4.4.20}$$

Applying the kinetic boundary condition to the third and the fourth terms on the left-hand side of Eq.(4.4.20), becomes $\frac{\partial H}{\partial x}$ and $-\frac{\partial \eta}{\partial x}$, respectively. Hence, the depth-averaged continuity equation is given by Eq. (4.4.21):

$$\frac{\partial h}{\partial t} + \frac{\partial}{\partial x}\left(\int_{\eta}^{H} U dz\right) + \frac{\partial}{\partial y}\left(\int_{\eta}^{H} V dz\right) = 0 \leftrightarrow \frac{\partial h}{\partial t} + \frac{\partial h\overline{U}}{\partial x} + \frac{\partial h\overline{V}}{\partial y} = 0 \tag{4.4.21}$$

The unknown variables $h(x, y, t)$, $\overline{U}(x, y, t)$, $\overline{V}(x, y, t)$ can be solved numerically by differentiating Eqs.(4.4.15), (4.4.16) and (4.4.21) with proper boundary conditions.

Chapter 5

3-D Turbulence Structure

5.1 GENERATION MECHANISM OF SECONDARY CURRENTS

5.1.1 Introductory Remarks

Secondary currents are found to be produced in open-channel flows as well as closed channels such as confined duct flows. They influence significantly bottom deformation, sediment accumulation and velocity profile although they are much smaller than the mainstream velocity.

Prandtl classified the secondary currents into two categories, as shown in FIG. 5.1. One is the first kind of secondary currents that are induced by a centrifugal force in a bending section (FIG. 5.1(a)). By contrast, a second kind of secondary current is expected to be produced by the turbulence structure. They contribute to production of longitudinal streaks in a river bed or the velocity dip phenomena, in which the position of the maximum velocity drops below the free surface.

5.1.2 First Kind of Secondary Currents

Centrifugal force is produced in the skewed channels and it results in the occurrence of the first kind of secondary current. The centrifugal force exerts on the fluid passing in the bending region, which is proportional to the square of mainstream velocity. The mainstream velocity is larger in the free surface than in the bottom due to the bed shear stress. Hence, the centrifugal force is also more significant in the free surface compared to the bottom. Consequently, an outward flow is produced in the transverse direction near the free surface and an inward flow is produced near the bottom. Also, upward and downward currents are formed near the inside and the outside walls. A cross-sectional scaled circulation is observed in FIG. 5.1(a). Such secondary currents due to the centrifugal force are rarely seen in straight channels. The geometric condition is needed to produce this secondary current.

The velocity of secondary currents reaches around a 10% order of the mainstream velocity. They transport bottom sediments to the inside of the

DOI: 10.1201/9781003112198-5

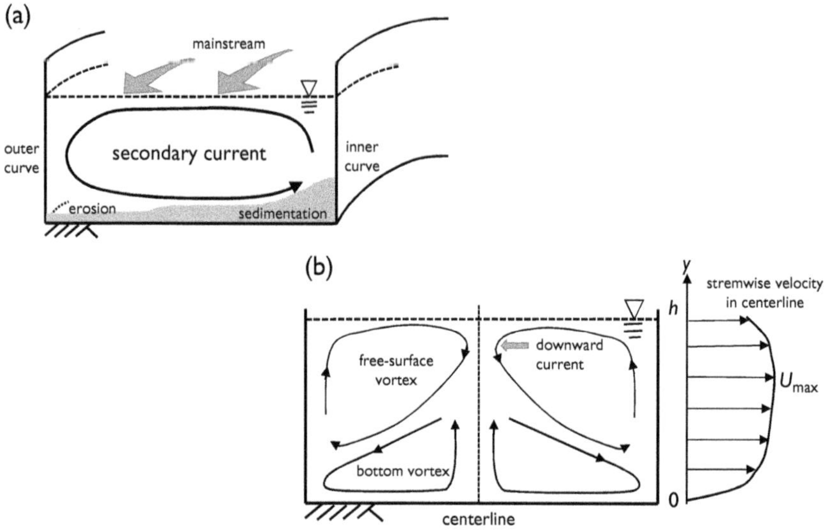

Figure 5.1 Sketch of secondary currents: (a) first kind of secondary currents (b) second kind of secondary currents.

Figure 5.2 Example of sand bar formed inside of bending river flow.

bend. The downward flow and the bottom friction promote erosion significantly on the bottom near the outside wall, and scoured sediment is conveyed laterally. Thus, a large amount sediment accumulates near the inside wall, which is related to the point sand bars built inside of the river meander, as shown in FIG. 5.2.

5.1.3 Second Kind of Secondary Currents

A second kind of secondary current is observed in non-circular turbulent flows. Einstein & Li (1956) proposed that turbulence anisotropy induces the secondary currents by using the following transport equation of mean streamwise vorticity Ω:

$$\frac{D\Omega}{\partial t} = \frac{\partial^2(\overline{v^2} - \overline{w^2})}{\partial y \partial z} - \left(\frac{\partial^2}{\partial z^2} - \frac{\partial^2}{\partial y^2}\right)\overline{vw} + \nu\nabla^2\Omega \tag{5.1.1}$$

in which $\Omega \equiv \frac{\partial W}{\partial y} - \frac{\partial V}{\partial z}$. The first term on the right-hand side of Eq. (5.1.1) means a difference of the normal components of Reynolds stress. Near the wall parallel to the y direction, w' is expected to be reduced compared to v'. In contrast, near the wall parallel to the z direction, the opposite tendency is observed. Distribution of these normal Reynolds stresses vary significantly near the corner, and this term does not vanish. The second term on the right-hand side of Eq. (5.1.1) is a second-order differential of shear components of Reynolds stress. When we consider that $\overline{vw}$ has a linear profile with a normal direction to each wall as well as the turbulence formed in the boundary layer, $\frac{\partial^2}{\partial z^2}\overline{vw}$ and $\frac{\partial^2}{\partial y^2}\overline{vw}$ vanish. The third term is comparatively small in a large Reynolds number condition. Therefore, the right-hand side of Eq. (5.1.1) is dominated by the first term, which means the existence of the anisotropy in turbulence. Because the right-hand side of Eq. (5.1.1) is not zero, as mentioned previously, and it suggests $\frac{D\Omega}{Dt} \neq 0$. It results in the space and time variation of the streamwise vorticity, i.e., the formation of secondary currents.

There are still unclear points and many researchers have been intensively studying (e.g., Bradshaw 1987, Perkins 1970, Tamburrino & Gulliver 1999, Gavrilakis 1992) although this idea is a basic concept of secondary currents' production.

Yang et al. (2012) have proposed another idea of secondary current production. Here, their theory is explained using a Cartesian coordinate system for simplicity, although they discussed a generalized coordinate system in consideration of a bottom configuration. Two-dimensional continuity equation for mean velocities and momentum equation of uniform flow are as follows:

$$\frac{\partial V}{\partial y} + \frac{\partial W}{\partial z} = 0 \tag{5.1.2}$$

$$\frac{\partial(\rho UV - \tau_{xy})}{\partial y} + \frac{\partial(\rho UW - \tau_{xz})}{\partial z} = \rho g S \tag{5.1.3}$$

in which $S = \sin\theta$ (θ is a bottom slope) and τ_{xy}, τ_{xz} are shear stresses.

The variations of momentum and stresses are smaller with z directions than those of the y direction near the bottom. Hence, Eq. (5.1.3) is rewritten as

$$\frac{\partial(\rho UV - \tau_{xy})}{\partial y} = \rho g S \tag{5.1.4}$$

$$\frac{\partial(\rho UW - \tau_{xz})}{\partial z} = 0 \tag{5.1.5}$$

After integral calculus of Eq. (5.1.5), a following form is obtained:

$$UW - \nu\frac{\partial U}{\partial z} + \overline{uw} = 0 \tag{5.1.6}$$

When we consider a laminar flow, the Reynolds stress is zero. Hence, Eq. (5.1.6) is rewritten as

$$W = \frac{\nu}{U}\frac{\partial U}{\partial z} \tag{5.1.7}$$

This suggests that the lateral gradient of mainstream distribution, $\frac{\partial U}{\partial z}$, generates a lateral velocity W. The direction of the lateral flow corresponds to the sign of $\frac{\partial U}{\partial z}$. That is to say, it flows with the spanwise direction from the side wall to the centerline.

When we consider the turbulent flows, neglecting the viscous effects, Eq. (5.1.6) is rewritten as

$$UW + \overline{uw} = 0 \tag{5.1.8}$$

Introducing a gradient diffusion hypothesis using the spanwise eddy viscosity ν_{tz},

$$-\overline{uw} = \nu_{tz}\frac{\partial U}{\partial z} \tag{5.1.9}$$

is obtained. Both Eqs. (5.1.8) and (5.1.9) lead to the following expression:

$$W = \frac{\nu_{tz}}{U}\frac{\partial U}{\partial z} \tag{5.1.10}$$

Eq. (5.1.10) suggests that the lateral velocity W becomes larger near the bottom since the streamwise velocity U is smaller there than the main flow region. This finding implies that the secondary currents originate at the bottom.

In the viscous sublayer, $U = \frac{y}{\nu}U_*^2 = \frac{\tau y}{\rho\nu}$ holds. Therefore, Eq. (5.1.11) is derived from Eq. (5.1.10).

$$W = \frac{\nu_{tz}}{\tau}\frac{\partial\tau}{\partial z} = \frac{\nu_{tz}}{\tau}\frac{d\tau}{dz} \tag{5.1.11}$$

Here, Webel & Schatzmann (1984) obtained the constant value of $\nu_{tz} = 0.177$ by experiments. Substituting this result into Eq. (5.11),

$$\frac{W}{U_*} = \frac{\nu_{tz}}{U_* h}\frac{h}{\tau}\frac{d\tau}{dz} \simeq 0.177\frac{h}{\tau_b}\frac{d\tau_b}{dz} \tag{5.1.12}$$

is obtained. Yang & Lim (1997, 1998) showed that the boundary layer shear stress varies linearly in the z direction and the simplified Eq. (5.1.12) to

$$\frac{W}{U_*} \simeq 0.177 \tag{5.1.13}$$

The velocity coefficient is generally known as $\frac{U}{U_*} = 15 \sim 20$ in open-channel flows. Hence, it gives

$$\frac{W}{U} \simeq 0.01 \tag{5.1.14}$$

This theoretical result is consistent with the velocity scale of secondary currents reported by laboratory experiments of Nezu & Nakagawa (1993).

Next, we focus on the main flow region. Eq. (5.1.14) is integrated with the y direction from y from h as follows:

$$\begin{aligned}
&\int_y^h \frac{\partial(\rho UV - \tau_{xy})}{\partial y}dy = \int_y^h \rho g S dy \\
&\leftrightarrow\ -UV + \frac{\tau_{xy}}{\rho} = ghS\left(1 - \frac{y}{h}\right) \\
&\leftrightarrow\ -UV - \overline{uv} = U_*^2\left(1 - \frac{y}{h}\right) \\
&\leftrightarrow\ \frac{-\overline{uv}}{U_*^2} = \left(1 - \frac{y}{h}\right) + \frac{UV}{U_*^2}
\end{aligned} \tag{5.1.15}$$

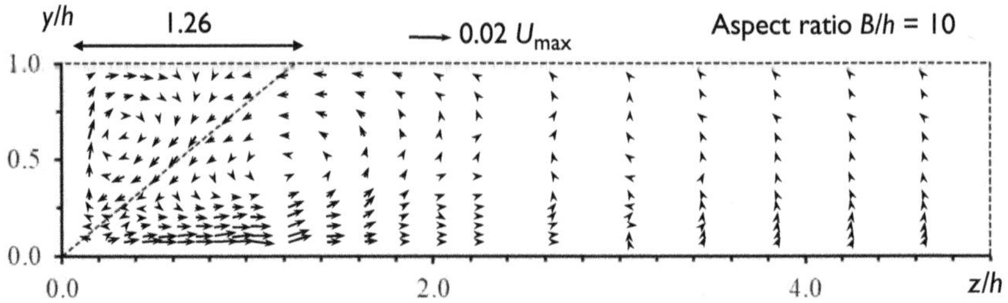

Figure 5.3 Division line of free-surface vortex and bottom vortex in cross-sectional plane with velocity data measured by Nezu & Nakagawa (1993) (created with reference to Yang et al. 2012).

in which $V = 0$ at $y = h$ and $\tau_{xy} = 0$ are assumed. Furthermore, the viscous effects are ignored in the main flow region. Eq. (5.1.15) shows that the Reynolds stress is closely related to the production of vertical velocity V. In summary, the lateral current is produced in the bottom region and the vertical current is produced in the main flow region. It results in secondary currents, as shown in FIG. 5.1(b). It should be noted that anisotropy of turbulence is not required in this theory.

Nezu & Nakagawa (1993) pointed out that the free surface and the bottom vortices have different sizes; the bottom vortex is larger than the free-surface one shown in FIG. 5.3. The velocity is much smaller in the bottom region than in the free-surface region. When the streamwise velocity U contributes more significantly to the production of the secondary currents than the velocity gradient, $\frac{\partial U}{\partial z}$, Eq. (5.1.10) supports the theoretical tendency.

Considering one point in a cross-sectional plane in an open-channel flow, distances from the side wall and the bottom wall are given by L_w and L_b, respectively. Yang & Lim (1997) and Yang & McCorquadale (2004) introduced the normalized distances and defined a division line between the regions of the free-surface and the bottom vortices, where influences by both walls are same. Yang & Lim (1997) derived a following differential equation of the gradient of the division line, $k = L_w/L_b$:

$$k^3 + (h/b)k - 2 = 0 \tag{5.1.16}$$

in which h is the water depth and b is the flume width. Assuming $h/b \simeq 0$ is limited to wide width flumes, a solution of Eq. (5.1.16) is obtained, i.e., $k = 1.26$.

FIG. 5.3 indicated the division line with $k = 1.26$ on the secondary currents' velocity vectors measured by Nezu & Nakagawa (1993). It is noted that two kinds of corner vortices, i.e., the free-surface vortex and the bottom vortex, could be classified clearly by the theoretical division line proposed by Yang et al. (2012).

The friction properties are quite different in the free surface and the bottom wall. This may induce the free surface and the bottom vortices that are different in size, unlike a rectangular duct flow. It remains unknown which vortex is formed earlier. But, the second kind of secondary currents are found to influence significantly mass and momentum transports and the distribution of the wall shear stress.

Such secondary flows are also promoting a deformation in the riverbed, e.g., formation of longitudinal sand streaks although they are only a few percent of the mainstream velocity. The downward second kind of secondary current erodes the movable bottom composed of gravel and sand, and there are accumulations in place where the upward currents occur as well as the first kind of secondary currents. There is an interaction between the velocity distribution and the bottom configuration, and thus they complicate hydrodynamic characteristics.

Influences of the second kind of secondary currents in the open channel are classified by the aspect ratio of the flume width/water depth, i.e., b/h. In the case of wide-width channels, $b/h > 5$, the secondary currents produced near the corners have little impact on the flow in the centerline of the channel. In contrast, in the case of the narrow-width channels, $b/h < 5$, the developed secondary currents induce downward currents near the free surface in the centerline. The streamwise momentum in the free-surface flow is conveyed downward, and it results in the elevation of the maximum streamwise velocity drops below the free surface. This phenomenon is called velocity dip.

5.2 THREE-DIMENSIONAL TURBULENCE IN A STRAIGHT CHANNEL

5.2.1 Secondary Cells

The open-channel flow is found to have a large-scale three-dimensional structure as well as the pipe flow and the boundary layer flow. A spiral current is induced by the secondary currents and it makes the bottom shear stress vary across the channel.

The previous studies represented several longitudinal secondary cells that appear and related upwelling and downwelling currents are produced in the whole depth region. It results in an undulating distribution of the bottom shear stress near the flume bed. Furthermore, we can observe lateral currents facing each other and separating from each other in the free surface. Such a three-dimensional structure has great impacts on not only the bottom shear stress but also sediment and gas transfers.

Tominaga et al. (1989) reported experimentally that the secondary current determines a cross-sectional distribution of the mean streamwise velocity. Nezu & Nakagawa (1993) explained that there are universal properties in three-dimensional turbulence of open-channel flow, irrespective of the Reynolds number.

Tamburrino & Gulliver (1999, 2007) revealed that a boil vortex street is formed in the free surface where the upwelling currents are formed, and streaks appear in the free surface where the downwelling currents are formed. The pattern of secondary currents was also found to fluctuate with the lateral direction. Furthermore, they investigated that detailed properties of the longitudinal secondary cell structure with a depth scale related to the second kind of the secondary currents. Particularly, the number of the secondary cell is formulated as follows:

(downwelling region) even integer close to $\frac{B/H}{2\sim 3}$

(upwelling region) odd integer close to $\frac{B/H}{2\sim 3}+1$,
in which B/H is the flume aspect ratio.

Wang & Cheng (2006) conducted velocity measurements in the open-channel flume in which the bottom roughness is alternated by roughness strips and roughness elements. They suggested that upwelling currents are generated in locally rougher bottom region as well as Nezu and Nakagawa. They proposed an experimental formula representing relation between maximum vertical velocity and amplification of lateral distribution of bottom height. Rodriguez & Garcia (2008) presented that an interaction between rough bottom and smooth wide wall influences significantly the secondary currents formation and the bottom shear stress profile. Albayrak & Lemmin (2011) conducted simultaneous velocity measurement of large-scale PIV (LS-PIV) in the free surface and acoustic Doppler velocity profiler (ADVP) under the free surface, and pointed out the lateral profiles of the mean currents, the bottom shear stress and Reynolds stress undulate due to the secondary currents. These highly advanced measurements supported the results of Tamburrino and Gulliver (1999).

Common knowledge was obtained in the previously mentioned studies, and they may be concluded as shown in FIG. 5.4.

5.2.2 Three-Dimensional Large-Scale Motions

Many researchers have reported the existence of three-dimensional large-scale motions (LSMs) related to streamwise rotations. Also, a larger fluid parcel called a very large-scale motion (VLSM) is found to be formed in various kinds of currents, such as boundary layer, closed channel and open-channel, etc.

Kim & Adrian (1999) pointed that the VLSMs is a longitudinal alignment of the LSMs which is may be a group of hairpin vortices. Although Cameron et al. (2017) have investigated experimentally the characteristics of the VLSM in a rough bed open-channel flow, a generation mechanism of the VLSM has not yet been revealed. Hutchins & Marusic (2007) called such very large

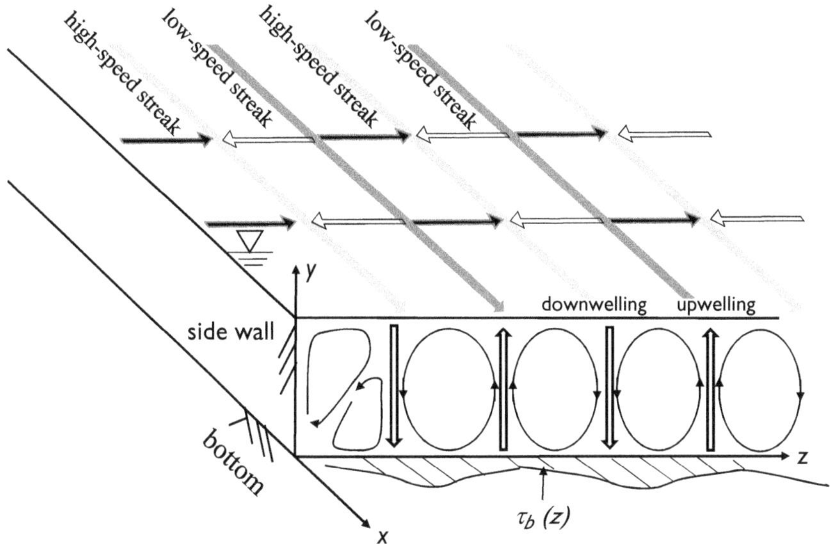

Figure 5.4 Phenomenological model of three-dimensional turbulence in open-channel flow suggested by previous works.

motions in the boundary layer super structure. Monty et al. (2009) compared properties of pre-multiplied spectra in the pipe flow and the boundary layer flow, and they concluded the similar characteristics in the VLSMs.

Grinvald & Nikora (1988) conducted a long duration velocity measurement in a river flow, and suggested that the horizontal turbulence induces a low-frequency motion with a large scale corresponding to the river width. Furthermore, they indicated that these large-scale motions influence significantly a depth scale motion and bottom-oriented turbulence structure. Cameron et al. (2017) proposed that streak motions with low and high momentums meander corresponding to the VLSMs with a peak spectrum in the low-frequency region. Adrian & Marusic (2012) pointed out that the meandering VLSMs and the secondary current formed in the open channel have an important relationship. That is to say, the VLSMs have a complex three-dimensional structure.

Cameron el al. (2017) carried out a long-duration PIV measurement in the open-channel flume, in which all three velocity components in the two-dimensional plane could be measured by the stereoscopic method. Their accurate measurements provided valuable time-series data of instantaneous velocity field, and allowed the two-dimensional spectrum and correlation analyses. To examine the length scale of LSMs and VLSMs in the open channel, they transferred the frequency spectrum to the wave number spectrum, introducing Taylor's frozen turbulence hypothesis (refer to section 3.8). FIG. 5.5 shows a summarized sketch of a part of their results, in which H is the water depth.

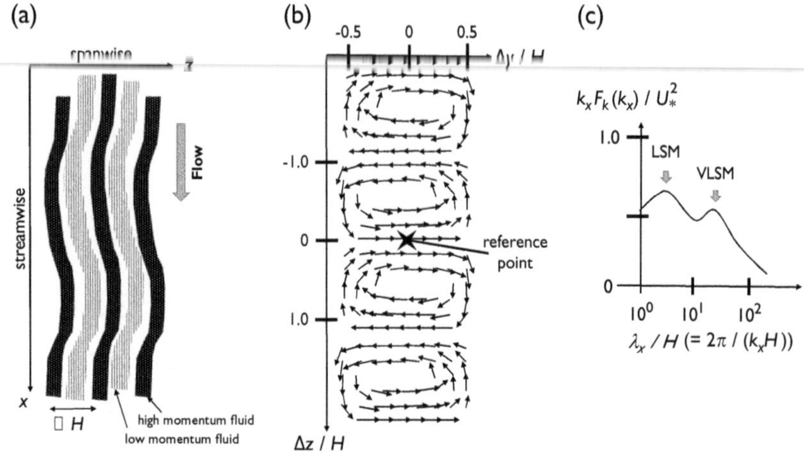

Figure 5.5 Proofs of existence of very large-scale motions in open-channel flows obtained in experiments by Cameron et al. (2017): (a) fluids alternating in spanwise direction and elongated in streamwise direction; (b) vector distribution of two-point correlation functions (R_{uv}, R_{uw}) (c) distribution of pre-multiplied spectrum with two peaks corresponding to LSM and VLSM.

FIG. 5.5(a) shows a general view of instantaneous streamwise momentum distribution in the horizontal plane. Low- and high-momentum fluids are elongated with the stramwise direction, and they meander along alternating lateral motions. They are similar to the superstructures observed in the boundary layer flows. The lateral spacing of neighboring streaks seems to be a depth order, i.e., $\Delta z \sim H$.

FIG. 5.5(b) shows a sketch of two-point correlation vectors (R_{uv}, R_{uw}) depicted in the cross-sectional plane, in which a reference point is chosen as $(y, z) = (0, 0)$. R_{uv} and R_{uw} are defined by following forms (refer to section 3.8.2).

$$R_{uv}(y, z, \Delta y, \Delta z) = \frac{\overline{u(y, z)v(y + \Delta y, z + \Delta z)}}{u'(y, z)v'(y + \Delta y, z + \Delta z)} \tag{5.2.1}$$

$$R_{uw}(y, z, \Delta y, \Delta z) = \frac{\overline{u(y, z)w(y + \Delta y, z + \Delta z)}}{u'(y, z)w'(y + \Delta y, z + \Delta z)} \tag{5.2.2}$$

in which a prime symbol means root mean square. Although these values decay farther away from the reference point, this sketch draws vectors with the same length due to visual effect. These vectors mean vertical and spanwise fluctuation (v, w) at the timing of $u < 0$. Of particular significance is that a pair of counter-rotational cells are aligned across the channel.

This means a close relationship among the three velocity components in the current forming the depth scale rotation like secondary currents.

FIG. 5.5(c) shows a general view of pre-multiplied wave number spectrum $k_x F_k(k_x)$, in which k_x is a wave number in the streamwise direction and λ_x is a corresponding wave length. It is very interesting that two peaks of the LSM and VLSM appear near the depth scale, i.e., $H \le \lambda_x \le 10H$. Cameron et al. (2017) reported that there is only a single peak of the LSM near the bottom wall.

5.3 THREE-DIMENSIONAL TURBULENCE IN A CURVED CHANNEL

5.3.1 Hydrodynamics in a Curved Channel

The hydrodynamics of the curved channel flow are influenced significantly by the balance of the transverse centrifugal force and the pressure difference due to the lateral free-surface gradient. It results in the production of the Prandtle's first kind of secondary currents. They are found to redistribute kinetic energy-forming complex three-dimensional flow structure (de Vriend 1981). That is, the production of the secondary current feeds back the streamwise momentum transport, and the cross-sectional distribution of the streamwise velocity is varied, including the maximum velocity position. Further, transport of dissolved matters such as gas, nutrient and sediment in the curved channel is also different from that observed in the straight channel.

Blanckaert & Graf (2004) evaluated experimentally the momentum convection term and turbulent transport term. They pointed out that momentum transport due to the streamwise circulation influences the distribution of the streamwise velocity. Abhari et al. (2010) conducted numerical simulation and laboratory experiment in the curved channel, and discussed the secondary currents strength along the bend. Gholami et al. (2014) showed that two secondary currents having opposite rotation directions are formed in the sharp bend channel, and they remain downstream of the bend exit. Han et al. (2011) predicted numerically that the curved vane could control generation of secondary current. The vanes are expected to decrease the strength of the secondary currents and related bottom shear stress.

5.3.2 Evolution of Streamwise Velocity

The streamwise momentum equation in cylindrical coordinates (FIG. 5.6) is introduced by Batchelor (1970) and Blanckaert & Graf (2004) as follows:

$$\frac{\partial U_s}{\partial s} = -\left[\left(\underline{\frac{1}{1+n/R}U_s\frac{\partial U_s}{\partial s}} + \underline{U_n\frac{\partial U_s}{\partial n} + U_z\frac{\partial U_s}{\partial z} + \frac{1}{1+n/R}\frac{U_s U_n}{R}}\right)\right]$$

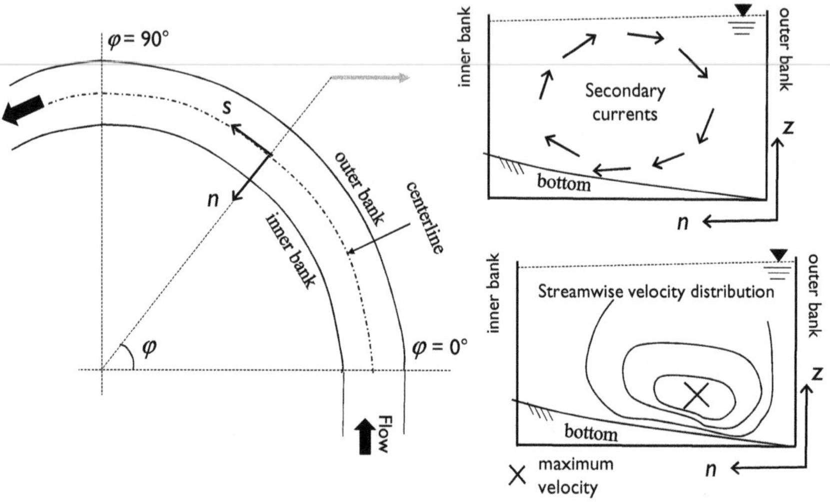

Figure 5.6 Cylindrical coordinate and general cross-sectional views of the secondary current and streamwise velocity distribution in the curved channel.

$$-\frac{1}{1+n/R}g\frac{\partial z}{\partial s}-\left(\frac{1}{1+n/R}\frac{\partial \overline{u_s^2}}{\partial s}+\frac{\partial \overline{u_s u_z}}{\partial z}+\frac{\partial \overline{u_s u_n}}{\partial n}+\frac{2}{1+n/R}\frac{\overline{u_s u_n}}{R}\right) \tag{5.3.1}$$

in which, the subscripts, n, s, z mean the lateral, streamwise and vertical directions.

Blanckaert & Graf (2004) decomposed the mean velocity with the j direction, U_j into the depth averaged velocity $\langle U_j \rangle$ and the deviation u_j^* like

$$U_j = \langle U_j \rangle + u_j^* \tag{5.3.2}$$

Using Eq. (5.3.2), Eq. (5.3.1) is rewritten by

$$\frac{\partial U_s}{\partial t} =$$

$$-\left[\underbrace{\left(\frac{1}{1+n/R}U_s\frac{\partial U_s}{\partial s}\right.}_{\text{i}}\underbrace{+\langle U_n\rangle\frac{\partial U_s}{\partial n}+\left.\frac{1}{1+n/R}\frac{U_s\langle U_n\rangle}{R}\right)}_{\text{ii}}+\underbrace{\left(u_n^*\frac{\partial U_s}{\partial n}+\frac{1}{1+n/R}\frac{U_s u_n^*}{R}+U_z\frac{\partial U_s}{\partial z}\right)}_{\text{iii}}\right]$$

$$\underbrace{-\frac{1}{1+n/R}g\frac{\partial z}{\partial s}}_{\text{iv}}-\left(\underbrace{\frac{1}{1+n/R}\frac{\partial \overline{u_s^2}}{\partial s}}_{\text{v}}+\underbrace{\frac{\partial \overline{u_s u_z}}{\partial z}+\frac{\partial \overline{u_s u_n}}{\partial n}+\frac{2}{1+n/R}\frac{\overline{u_s u_n}}{R}}_{\text{vi}}\right) \tag{5.3.3}$$

i. non-uniformity
ii. convective momentum transport by streamwise flow
iii. convective momentum transport by secondary currents
iv. gravity
v. non-uniformity
vi. turbulent momentum transport

Blanckaert & Graf (2004) concluded that the convective momentum transport by the streamwise circulation, i.e., secondary currents, changes the cross-sectional distribution of the sreamwise velocity. The convective momentum term is comparable with the gravity term in (5.3.3). Particularly, this momentum transport causes an increase of the streamwise velocity in outward direction. Further, the position of the maximum streamwise velocity is shifted toward the bottom (FIG. 5.6), although the maximum velocity appears near the free-surface in the straight open-channel flow. The turbulent transport term is smaller than the convective term, and may have minor role to form the three-dimensional current structure.

5.3.3 Evolution of Streamwise Velocity

FIG. 5.7(a) shows a sketch of distributions of depth-averaged streamwise velocity in 90° bend channel. In the inlet of $\varphi = 0°$, the maximum point of the velocity is shifted toward the inner curve region, where the streamwise gradient of the pressure is larger compared to the outer curve. This is

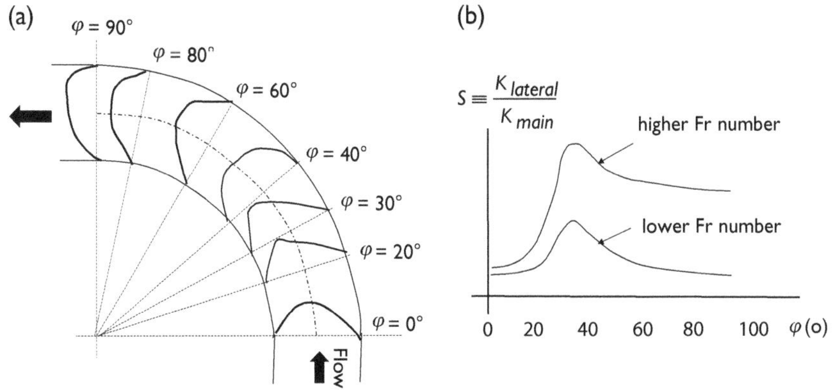

Figure 5.7 Streamwise evolutions of (a) depth-averaged stremwise velocity and (b) strength of secondary currents in 90° bend channel. They are drawn referring to Abhari et al. (2010).

understood from the fact that the outer wall plays a role as a kind of obstacle to the stream passing the inlet section. Thereafter, a generation of secondary currents pushes it back to the channel centerline. Ultimately, it is shifted to the outer region. Abhari et al. (2010) concluded that the maximum velocity is closest to the inner wall at the section of $\varphi = 30°$, and this property was reported in numerical simulation by Line et al. (1999).

5.3.4 Strength of Secondary Currents

Abhari et al. (2010) defined a strength of secondary currents S by a ratio of kinetic energy in lateral flow to total kinetic energy as follows:

$$S \equiv \frac{E_{lateral}}{E_{total}} \tag{5.3.4}$$

in which $E_{lateral} \equiv \frac{\langle U_n \rangle^2}{2}$ and $E_{total} \equiv \frac{\langle U_s \rangle^2 + \langle U_n \rangle^2 + \langle U_z \rangle^2}{2}$.

FIG. 5.7(b) shows a sketch of streamwise variation of secondary currents strength based on Abhari et al.'s result. The strength is almost constant up to $\varphi = 10°$ near the bend inlet. It increases downstream and reaches a maximum value at the section of $\varphi = 30°$. It decreases rapidly up to about $\varphi = 60°$. Thereafter, it decreases gradually and becomes almost constant near the bend exit of $\varphi = 90°$. This tendency is observed, irrespective of the Froude number in the bend inlet. Increasing the Froude number induces stronger secondary currents in the bend, keeping the streamwise evolution form, as shown in FIG. 5.7(b). The larger Froude number corresponds to the mainstream includes larger kinetic energy in the inlet, and the kinetic energy can be redistributed much more to the secondary currents. The result of Abhari et al. gives $S = 0.22$, 0.54 and 1.0 for $\mathrm{Fr} = 0.09$, 0.18 and 0.27, respectively.

5.4 THREE-DIMENSIONAL FLOW IN A FLOOD

5.4.1 Meandering Flow in a Flood

The meandering flow is composed of repetitions of multiple bends, to which the knowledge of the single-bend channel in the previous section can be applied. Some previous studies pointed out that sedimentation and scours are promoted significantly in rectangular meandering channels. For example, Whiting & Dietrich (1993) explained the sediment transport process associated with the development of bed deformation in the bend. Silva et al. (2006) conducted laboratory experiments by varying channel sinuosity, and investigated the effects of sinuosity on the streamline behaviors and bed-deformation process. These bed-deformation properties have great influences on the mean flow and turbulence.

By contrast, after inundation over the floodplain, the complex flow consists of the meandering main-channel currents and the straight overbank current. Straight compound channel has been studied by many researchers, and the hydrodynamics of it is noted in detailed in section 7.3. Here, the meandering compound channel is focused. It is one of high interest in river engineering and flood disaster fields. There exist complicated distributions of mean velocity and turbulence components in these flows. In particular, cross-sectional secondary currents develop significantly associated with an interaction between meandering inbank main-channel flow and straight overbank floodplain flow. These currents are much larger compared with those observed in straight compound channels, and they promote mass and momentum exchanges between them, as well as sediment transport and bed erosion. Further, organized motions may be generated in the same manner as typical horizontal vortices in straight compound channels. It is, therefore, very important to reveal the mean flow, turbulence structure and coherent motion in such flows.

Schroder et al. (1991) have tried to conduct 3-D LDA measurements in large-scale meandering compound flume in order to reveal longitudinal variation of secondary flow pattern. Sellin et al. (1993) have examined the relation between the mean-flow properties and the meandering sinuosity. Shiono & Muto (1998) have conducted turbulence measurements in meandering compound open-channel flow by using LDA. They revealed the distributions of secondary currents and turbulence in continuous bends. Wormleaton et al. (2005) have measured the velocity profiles, depth variations and sediment transport under the overbank condition in large-scale meandering channels. Wormleaton & Marriot (2007) also reproduced the mean flow profiles in the separation and recirculation zone by using a 3-D $k - \varepsilon$ model.

Sanjou & Nezu (2009) focused on a 3-D structure of horizontal coherent motion and its advection property in the channel-transition zone from the straight to meandering compound open-channel flows by using a multi-layer scanning PIV and comparing them with those of continuous meandering channel flows. As a result, a strong relation between the horizontal vortices and secondary currents was obtained, and then a phenomenological flow model was proposed.

5.4.2 Phenomenological Model of Compound Meandering Flow

The natural rivers generally consist of straight flow zone and meandering flow zone. FIG. 5.8 shows the horizontal topographic patterns observed in compound open-channel flows. FIG. 5.8(a) indicates the previously mentioned straight-compound open-channel flows. The streamwise velocity becomes larger in the main channel than over the floodplains and, thus, the shear instability causes some coherent horizontal vortices, together with an

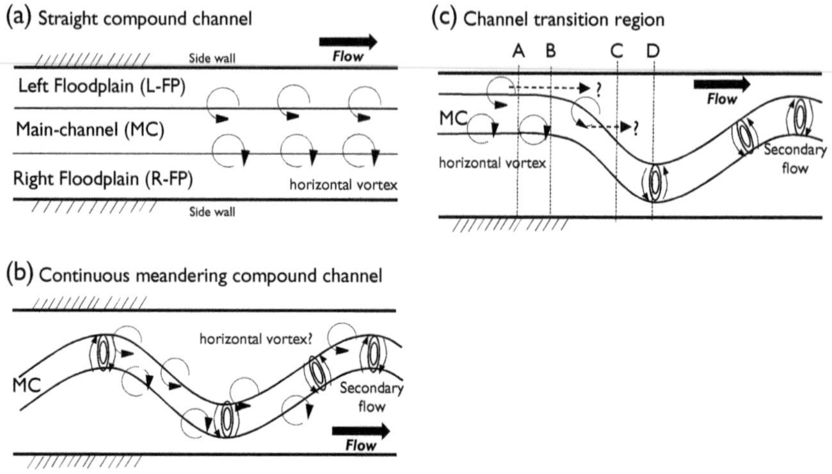

Figure 5.8 Topographic variation of compound channels.

apparent shear stress, near the junction between the main channel and the floodplain.

FIG. 5.8(b) indicates the continuous meandering compound open-channel flow, in which typical secondary currents are generated by the shear stress between the inbank and overbank layers, and become quite different from those observed in the previously mentioned straight-compound channels (FIG. 5.8(a)). That is, the former is generated by the vortex-stretching terms (Prandtl's first kind secondary currents), whereas the latter is generated by the anisotropy of turbulence (second kind ones), as reviewed by Bradshaw (1987).

FIG. 5.8(c) is the channel-transition zone from the straight to meandering compound open-channel flows, in which the structure of horizontal vortex shown in FIG. 5.8(a) should vary significantly in the downstream direction because of the addition of a centrifugal force, i.e., vortex-stretching effect. Sanjou & Nezu (2009) highlighted such a channel-transition zone from the straight to meandering compound open-channel flow. They investigated the developing process of secondary currents and turbulence structure in the curved main channel by PIV measurements.

Sanjou & Nezu (2009) proposed the phenomenological model shown in FIG. 5.9. The horizontal vortices typically generated in the straight-compound channel keep their formation even in the first meandering zone near the channel-transition section, and then they disappear on the way to cross over the main-channel due to the spatial variation of streamwise velocity. At the same time, a part of mass and momentum accompanied with horizontal vortex are transported downward toward the main-channel bed by the secondary currents, which is peculiar to the meandering compound open-channel flow. Further, the spiral-like secondary currents will also

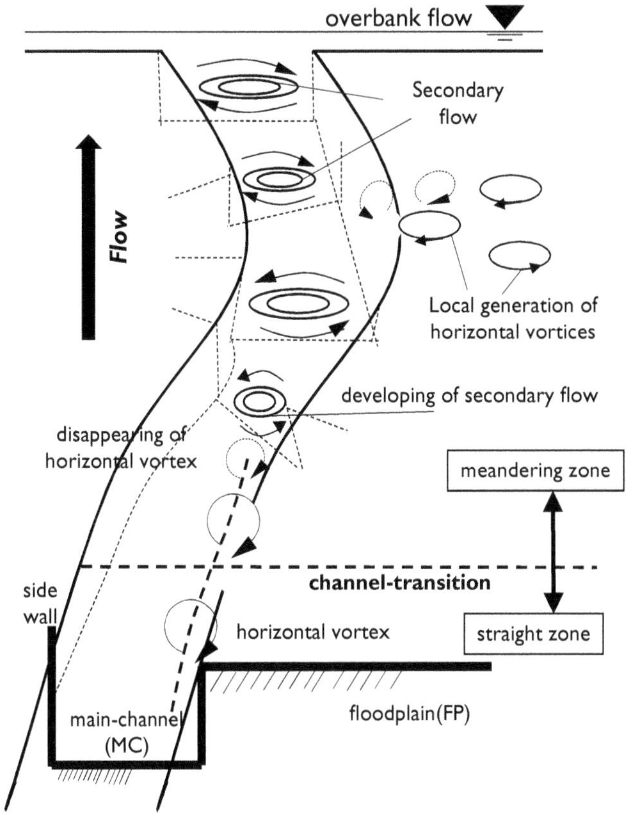

Figure 5.9 Phenomenological model of turbulence structure in meandering compound channel proposed by Sanjou & Nezu (2009).

transport the mass and momentum in the streamwise direction. The mean velocity distribution is quite different between the straight channel zone and the continuous meandering one. Several horizontal vortices are generated locally in a wide region including over the floodplains. Their rotational direction corresponds to the velocity shear at the vortex core. These vortices may be inclined to the channel bed because the convection velocity varies in the vertical direction, and they disappear with a short life span. The cross-sectional secondary currents induce out the effective circulation motion in the main-channel zone. It is therefore inferred that these combinations between the horizontal vortex and the secondary currents should promote the significant mass and momentum exchanges between near the main-channel bed and near the free surface.

Chapter 6

Turbulence Transport

6.1 INTRODUCTORY REMARKS

Turbulence promotes significant mass transports such as heat, dissolved gas and suspended sediments. This chapter explains the turbulent diffusion equation related to the Euler method. Also, particle analysis about the diffusion process in the uniform flow based on the Lagrange specification was highlighted by Taylor (1922) and Richardson (1926).

The velocity shear induced by the bottom friction yields a difference in diffusion speed in the flow direction between the free surface and the bottom layers. In such a case, a convection diffusion called dispersion should be considered.

6.2 TURBULENT DIFFUSION EQUATION

We consider one-dimensional mass transport in the unidirectional laminar flow, as shown in FIG. 6.1(a). A streamwise profile of the concentration and the flow velocity are defined by $\tilde{c}(x)$ $(=C(x))$ and $\tilde{u}$ $(=U)$, respectively. A streamwise transport flux without the flow velocity is given by $F_x = -D_x \frac{\partial \tilde{c}}{\partial x}$ considering Fick's law, in which D_x is the streamwise diffusion coefficient. When the diffusion in the unidirectional flow is considered, the flux is expressed by $F_x = \tilde{u}\tilde{c} - D_x \frac{\partial \tilde{c}}{\partial x}$, including the convection contribution, $\tilde{u}\tilde{c}$. The inflow flux, F_x, and outflow flux, $F_x + \frac{\partial F_x}{\partial x}\Delta x$, pass the sides of control volume, as shown in FIG. 6.1(a). The difference of results in a time variation of mass within the control volume. Considering a phenomena with time interval Δt, the following relation is obtained:

$$\frac{\partial \tilde{c}\Delta x \Delta y}{\partial t}\Delta t = \left\{F_x - \left(F_x + \frac{\partial F_x}{\partial x}\Delta x\right)\right\}\Delta y \Delta t \quad \leftrightarrow \quad \frac{\partial \tilde{c}}{\partial t} + \frac{\partial \tilde{u}\tilde{c}}{\partial x} = D_x \frac{\partial^2 \tilde{c}}{\partial x^2} \tag{6.2.1}$$

DOI: 10.1201/9781003112198-6

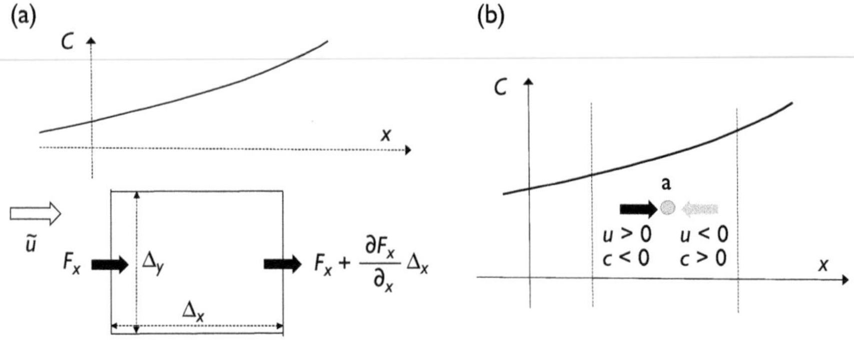

Figure 6.1 Mass flux in diffusion equation: (a) concentration transport passing two-dimensional control volume; (b) directions in instantaneous turbulence transports.

This is a one-dimensional diffusion equation. Expanding to three dimensions,

$$\frac{\partial \tilde{c}}{\partial t} + \frac{\partial \tilde{u}\tilde{c}}{\partial x} + \frac{\partial \tilde{v}\tilde{c}}{\partial y} + \frac{\partial \tilde{w}\tilde{c}}{\partial z} = D_x \frac{\partial^2 \tilde{c}}{\partial x^2} + D_y \frac{\partial^2 \tilde{c}}{\partial y^2} + D_z \frac{\partial^2 \tilde{c}}{\partial z^2} \tag{6.2.2}$$

is derived, in which subscripts y and z mean vertical and spanwise directions, respectively.

The velocity and concentration vary in time for the turbulent flows. Hence, the Reynolds decomposition is applied to them as follows:

$$\tilde{u} = U + u \quad \text{and} \quad \tilde{c} = C + c \tag{6.2.3}$$

Averaging in time after substitution of Eq. (6.2.3) into Eq. (6.2.1) results in $\frac{\partial \overline{(C+c)}}{\partial t} + \frac{\partial \overline{(U+u)(C+c)}}{\partial x} = D_x \frac{\partial^2 \overline{(C+c)}}{\partial x^2}$. Since $\overline{c} = \overline{u} = \overline{Uc} = \overline{uC} = 0$ holds, a following form is derived:

$$\frac{\partial C}{\partial t} + \frac{\partial UC}{\partial x} = D_x \frac{\partial^2 C}{\partial x^2} - \frac{\partial \overline{uc}}{\partial x} \tag{6.2.4}$$

A one-dimensional zero mean flow with turbulence, i.e., $u \neq 0$, is considered, as shown in FIG. 6.1(b) for modeling of the second term on the right-hand side in Eq. (6.2.4). How the concentration varies in time at the point a for a positive gradient of the concentration, i.e., $\frac{\partial C}{\partial x} > 0$, depends on a direction of turbulent component u. A lower concentration parcel is transported from the upstream point to point a for the case of $u > 0$ and, thus, the concentration reduces instantly, i.e., $c < 0$. In contrast, in the case of $u < 0$, the concentration increases instantly, i.e., $c > 0$. Therefore, correlation of turbulence components of velocity and concentration is positive, i.e., $uc > 0$, irrespective of the sign of u. This fact suggests that mass transport induced by the turbulence is controlled by the concentration gradient, $\frac{\partial C}{\partial x}$. Thus, the second term on the right-hand side in Eq. (6.2.4) is expressed by using the turbulent diffusion coefficient K_x:

$$-\overline{uc} = K_x \frac{\partial C}{\partial x} \tag{6.2.5}$$

Since it is generally known that the turbulent diffusion is much larger than the molecular diffusion, i.e., $K_x \gg D_x$, Eq. (6.2.4) is rewritten by

$$\frac{\partial C}{\partial t} + \frac{\partial UC}{\partial x} = K_x \frac{\partial^2 C}{\partial x^2} \tag{6.2.6}$$

Expansion to three dimensions gives the following form:

$$\frac{\partial C}{\partial t} + \frac{\partial UC}{\partial x} + \frac{\partial VC}{\partial y} + \frac{\partial WC}{\partial z} = K_x \frac{\partial^2 C}{\partial x^2} + K_y \frac{\partial^2 C}{\partial y^2} + K_z \frac{\partial^2 C}{\partial z^2} \tag{6.2.7}$$

This is called the turbulent diffusion equation.

6.3 PARTICLE TRACKING ANALYSIS

6.3.1 One-Particle Analysis

The Lagrangian specification allows us to perceive intuitively the vertical diffusion of particles in the flow (Taylor 1922). We consider a particle diffusion problem with a one-point source in unidirectional flow, as shown in FIG. 6.2. The width of vertical spread and diffusion coefficient tracking a particle motion released at the source are theoretically derived. Here, a unidirectional two-dimensional turbulent flow ($U \neq 0,\ V = 0$) is focused, in which u and v are streamwise and vertical turbulent component, respectively. Assuming that a particle follows perfectly the fluid motion, instantaneous speed of the particle in the y direction is given by v and the vertical traveling distance for the time interval dt is vdt.

The vertically spread width at $t = t$, $Y(t)$, is given by an integral of the traveling distance, $v(t)dt$, as follows:

$$Y(t) = \int_0^t v(t')dt' \quad \leftrightarrow \quad v(t) = \frac{dY}{dt} \tag{6.3.1}$$

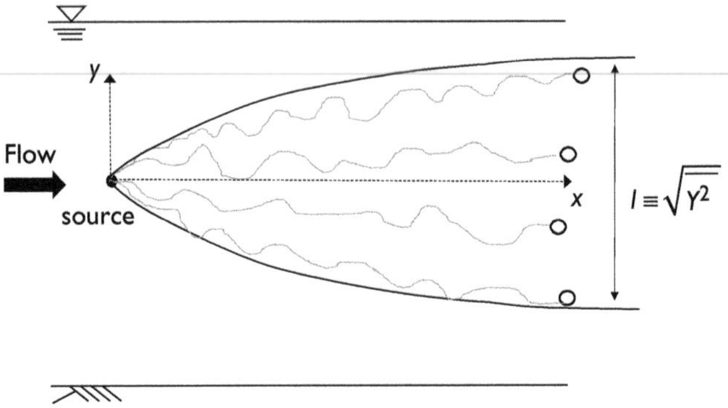

Figure 6.2 Particle tracer lines based on Lagrangian specification.

in which $t = 0$ is an initially release time. The Lagrange correlation function is defined as

$$R_l(\tau) = \frac{\overline{v(t)v(t+\tau)}}{v'^2} \tag{6.3.2}$$

This is an ensemble-averaged correlation of velocity signals with the time-interval τ using velocity data, in which an over bar means the ensemble average operator. Here, there are some properties of $R_l(\tau)$, e.g., $R_l(0) = 1$, $R_l(\infty) = 0$ and $R_l(\tau) = R_l(-\tau)$.

The time differential of square of $Y(t)$ is calculated as follows:

$$\begin{aligned} \frac{1}{2}\frac{d\overline{Y^2}}{dt} &= \overline{Y\frac{dY}{dt}} = \overline{\int_0^t v(t')dt' \times v(t)} = \overline{\int_0^t v(t)v(t')dt'} \\ &= \int_{-t}^{0} v'^2 R_l(\tau)d\tau = \int_0^t v'^2 R_l(\tau)d\tau = v'^2 \int_0^t R_l(\tau)d\tau \end{aligned} \tag{6.3.3}$$

in which $t' - t = \tau$. The integral of Eq. (6.3.3) leads to

$$\overline{Y^2} = 2v'^2 \int_0^t \left[\int_0^\eta R_l(\tau)d\tau \right] d\eta \tag{6.3.4}$$

The turbulent diffusion coefficients with x, y and z are given by K_x, K_y, and K_z, respectively. In this problem, we assume isotropic turbulent diffusion, i.e., $K = K_x = K_y = K_z$. Let us consider a distribution of time mean concentration $C(y)$ in a unidirectional flow, including turbulence, as shown in FIG. 6.3. The mass flux with the y direction is given by

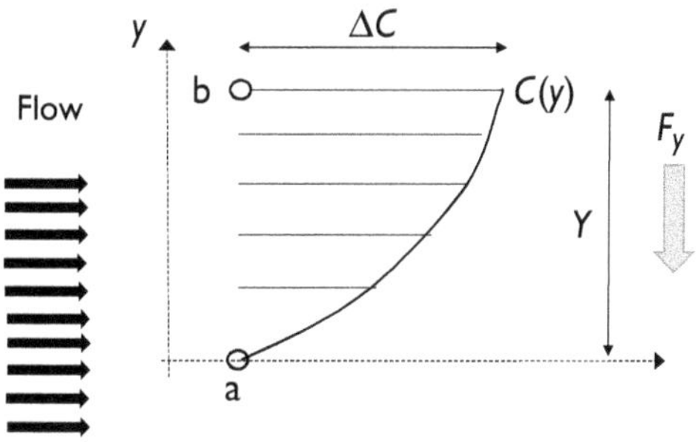

Figure 6.3 Vertical mass flux in the unidirectional flow.

$$F = -K\frac{dC}{dy} \tag{6.3.5}$$

Then, a concentration difference between point a on the x-axis and point b, which is Y apart from the point, is expressed by $\Delta C = Y\frac{dC}{dy}$. The concentration difference, ΔC, is transported with the y direction by the fluid having a velocity v. Hence, the ensemble-averaged turbulence transport flux is as follows:

$$F = -\overline{v\Delta C} = -\overline{vY}\frac{dC}{dy} \tag{6.3.6}$$

A combination of Eqs. (6.3.5) and (6.3.6) leads to

$$-K\frac{dC}{dy} = -\overline{vY}\frac{dC}{dy} \quad \leftrightarrow \quad K = \overline{vY} \tag{6.3.7}$$

Eq. (6.3.1) gives a relation, $vY = Y\frac{dY}{dt} = \frac{1}{2}\frac{dY^2}{dt}$. Further, Eq. (6.3.3) leads to

$$K = \overline{vY} = \frac{1}{2}\frac{d\overline{Y^2}}{dt} = v'^2 \int_0^t R_l(\tau)d\tau \tag{6.3.8}$$

Here, the following time scale is introduced:

$$T_* \equiv \int_0^\infty R_l(\tau)d\tau \tag{6.3.9}$$

We calculate the vertical spreading width $l \equiv \sqrt{\overline{Y^2}}$ and the turbulent diffusion coefficient for the limits $t \to 0$ and $t \to \infty$, respectively.

a. Case of $t \to 0$

Since the Lagrangian correlation function becomes $R_l = 1$, Eq. (6.3.8) obtained a time-depenent turbulent coefficient $K = v'^2 t$. Furthermore, $\frac{1}{2}\frac{d\overline{Y^2}}{dt} = v'^2 t \leftrightarrow \overline{Y^2} = v'^2 t^2$. It results in

$$\frac{1}{2}\frac{d\overline{Y^2}}{dt} = v'^2 t \quad \leftrightarrow \quad l = \sqrt{\overline{Y^2}} = v't \tag{6.3.10}$$

b. Case of $t \to \infty$

Eq. (6.3.9) leads a relation $\int_0^\infty R_l(\tau)d\tau = T_*$. Hence, the constant turbulent diffusion coefficient $K = v'^2 T_*$ is obtained due to Eq. (6.3.8). Further, $\frac{1}{2}\frac{d\overline{Y^2}}{dt} = v'^2 T_* \leftrightarrow \overline{Y^2} = 2v'^2 T_* t$ is led and, thus, the spreading width is as follows:

$$l = \sqrt{\overline{Y^2}} = \sqrt{2v'^2 T_*}\sqrt{t} \tag{6.3.11}$$

6.3.2 Two-Particle Analysis

Richardson (1926) considers two-particle tracking released at the same point as shown in FIG. 6.4. The relative distance between two particles Y increases with time. This is because vortices with a smaller scale than Y diffuses the fluid around the particles. In contrast, the two particles are moved in the same way by a vortex with a larger scale than Y and, therefore, such a larger vortex has no influence on the diffusion, i.e., development of Y. The two-particle analysis provides the time-dependent Y and the turbulent diffusion coefficient K in the following way.

a. Case of large diffusion time, t

The vortex scale becomes large, corresponding to a long diffusion time interval. Here, we assume the vortex in the inertial subrange of the frequency domain controls the diffusion phenomenon. Hence, the relative distance Y, diffusion interval t and the turbulent energy dissipation rate ε decide the diffusion process of the two particles. Since the vortex of inertial subrange is focused on which is much larger than the Kolmogorov scale, there is no influence of the kinetic viscosity.

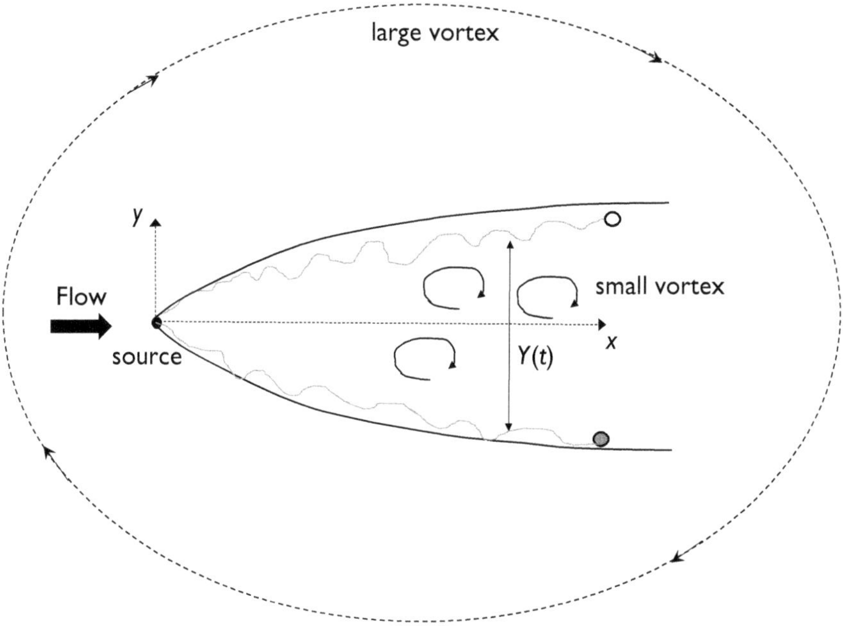

Figure 6.4 Influences of various scale vortices on relative diffusion of two particles.

The dimensional analysis leads the following relation:

$$Y_r \propto \varepsilon^{1/2} t^{3/2} \tag{6.3.12}$$

Further, Eq. (6.3.8) gives

$$K = \overline{vY_r} = \overline{\frac{dY_r}{dt} Y_r} = \frac{1}{2}\frac{d\overline{Y_r^2}}{dt} \tag{6.3.13}$$

Hence, the following equation is obtained using $l_r \equiv \sqrt{\overline{Y_r^2}}$:

$$K = \frac{1}{2}\frac{d\overline{Y_r^2}}{dt} \propto \varepsilon^{1/3} \sqrt{\overline{Y_r^2}}^{4/3} = \varepsilon^{1/3} l_r^{4/3} \tag{6.3.14}$$

This is called Richardson's 4/3 power law, and it obeys the field observation data.

Eq. (6.3.12) leads a relation $\frac{1}{2}\frac{d\overline{Y_r^2}}{dt} \propto \varepsilon t^2$. It results in

$$l_r \equiv \sqrt{\overline{Y_r^2}} \propto \varepsilon^{1/2} t^{3/2} \tag{6.3.15}$$

b. Case of small diffusion time, t

Since an initial relative distance $Y_r(0)$ influences the diffusion during a short diffusion time, $Y_r(0)$, t and ε should be control parameters. Therefore, the dimensional analysis obtains the following relation:

$$K = \frac{1}{2}\frac{d\overline{Y_r^2}}{dt} \propto \{\varepsilon Y_r(0)\}^{2/3} t \qquad (6.3.16)$$

The integral of Eq. (6.3.16) leads to

$$l_r \equiv \sqrt{\overline{Y_r^2}} \propto \{\varepsilon Y_r(0)\}^{1/3} t \qquad (6.3.17)$$

Substitution of Eq. (6.3.17) into Eq. (6.3.16) gives the following form:

$$K \propto \{\varepsilon Y_r(0)\}^{1/3} l_r \qquad (6.3.18)$$

6.3.3 Eddy-Dependent Trajectories

A tracer parcel with an initial length scale corresponding to the relative distance between the two particles is conveyed downstream and diffused in the unidirectional turbulent flow. The characteristics are explained by both the one-particle analysis and two-particle analysis. A gravity center of the tracer parcel is diffused laterally, as discussed in the one-particle analysis in section 6.3.1. The envelop just after the release, i.e., $t \simeq 0$, is expressed by $l \propto t$ (see Eq. (6.3.10)). After a sufficient time interval, a large vortex compared to the tracer parcel makes meandering trajectories. The lateral amplitude, $l \propto t^{1/2}$, is due to Eq. (6.3.11), and it suggests that increasing the rate of the amplitude reduces in time.

The size of the tracer parcel, l_r, is diffused by small vortices corresponding to the two-particle analysis. This is expressed by $l_r \propto t$ at $t \simeq 0$ (see Eq. (6.3.17). For the larger diffusion time, Eq. (6.3.15) provides $l_r \propto t^{3/2}$. Since the turbulent flow includes various vortices with different length scales, the tracer trajectories are explained by a combination of the one- and two-particle analysis shown in FIG. 6.5.

6.4 DISPERSION

The streamwise velocity near the bottom of the open channel reduces significantly due to the friction, as shown in FIG. 6.6. This velocity profile influences the diffusion process. Dissolved matters released at the point source in the two-dimensional open channel diffuse vertically during the

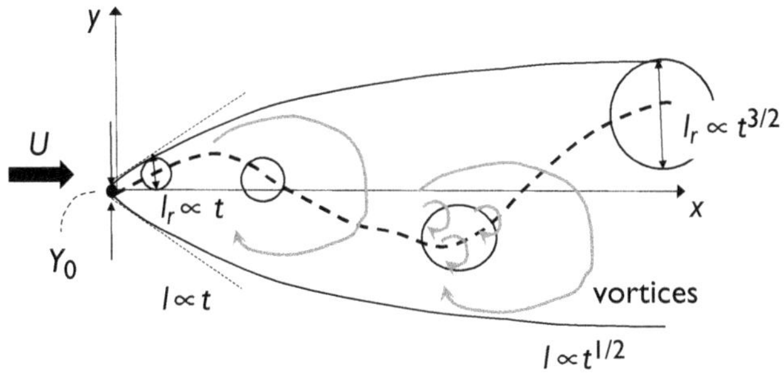

Figure 6.5 Meandering trajectory of tracer parcel in the unidirectional turbulent flow.

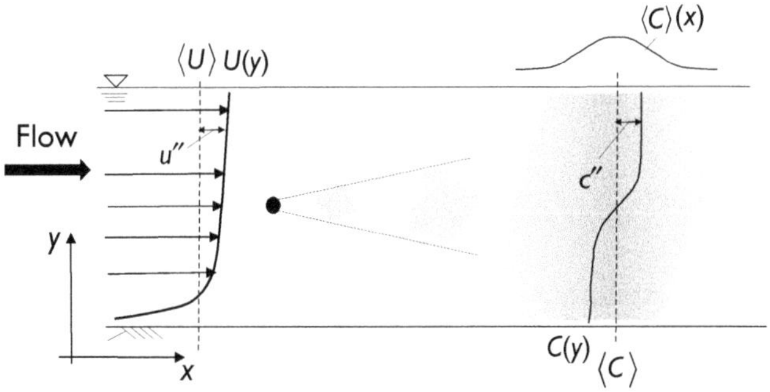

Figure 6.6 Dispersion of contaminant in open-channel flow.

streamwise travel. It is noted that the matters in the free-surface travel downstream with higher convection speed compared to those near the bottom. It results in a concentration map of the matters is distorted like a parallelogram. Here, the depth-dependent concentration $C(y)$ is introduced at a streamwise location x. The corresponding depth-averaged concentration is defined by $\langle C \rangle$. $\langle \rangle$ means the depth-averaging operator. Therefore, $\langle C \rangle$ means the depth-averaged velocity. Deviation from the depth-averaged value for the concentration and the velocity are defined by c'' and u'', respectively. Hence, the velocity and concentration are expressed as follows:

$$U = \langle U \rangle + u'', \quad C = \langle C \rangle + c'' \tag{6.4.1}$$

A one-dimensional turbulent diffusion equation is expressed using a streamwise turbulent diffusion coefficient, K_x, as follows:

$$\frac{\partial C}{\partial t} + \frac{\partial UC}{\partial x} = K_x \frac{\partial^2 C}{\partial x^2} \tag{6.4.2}$$

This is rewritten by the depth-averaged values and the deviations as follows:

$$\frac{\partial(\langle C\rangle + c'')}{\partial t} + \frac{\partial(\langle U\rangle + u'')(\langle C\rangle + c'')}{\partial x} = K_x \frac{\partial^2(\langle C\rangle + c'')}{\partial x^2} \tag{6.4.3}$$

Assuming that K_x is constant, Eq. (6.4.3) leads to

$$\begin{aligned} &\frac{\partial\langle\langle C\rangle + c''\rangle}{\partial t} + \frac{\partial\langle(\langle U\rangle + u'')(\langle C\rangle + c'')\rangle}{\partial x} = K_x \frac{\partial^2\langle\langle C\rangle + c''\rangle}{\partial x^2} \\ &\leftrightarrow \frac{\partial\langle C\rangle}{\partial t} + \frac{\partial\langle U\rangle\langle C\rangle}{\partial x} = K_x \frac{\partial^2\langle C\rangle}{\partial x^2} - \frac{\partial\langle u''c''\rangle}{\partial x} \end{aligned} \tag{6.4.4}$$

in which $\langle c''\rangle = \langle u''\rangle = \langle v''\rangle = \langle Uc''\rangle = \langle Cu''\rangle = 0$ are used.

An analogy with Fick's law obtained is the following form:

$$-\langle u''c''\rangle = E_x \frac{\partial\langle C\rangle}{\partial x} \tag{6.4.5}$$

in which E_x is a streamwise dispersion coefficient. Therefore, a following advection-dispersion equation is obtained. This equation is often used for the prediction of propagation of contaminants in rivers. Generally speaking, the dispersion coefficient is much larger than the turbulent diffusion coefficient, $K_x \ll E_x$.

6.5 PECLET NUMBER

A convection time scale in uniform flow with characteristic velocity U and length L is expressed by $T_{con} = L/U$. In contrast, the dimensional analysis defines a diffusion time scale as $T_{dif} = L^2/D$, in which D is a diffusion coefficient. The ratio of the convection velocity to the diffusion velocity is the Peclet number, Pe, which is as follows:

$$\text{Pe} \equiv \frac{\textit{convection speed}}{\textit{diffusion speed}} = \frac{1/T_{con}}{1/T_{dif}} = \frac{LU}{D} \tag{6.5.1}$$

This is a product of the Reynolds number and Schmit number, i.e., $\text{Pe} \equiv \text{Re}\cdot\text{Sc}$. When the Peclet number is larger than one, i.e., $\text{Pe} > 1$, the convection is more important than the diffusion. In contrast, when the Peclet number is smaller than one, i.e., $\text{Pe} < 1$, then the contaminant diffusion is more important than the convection.

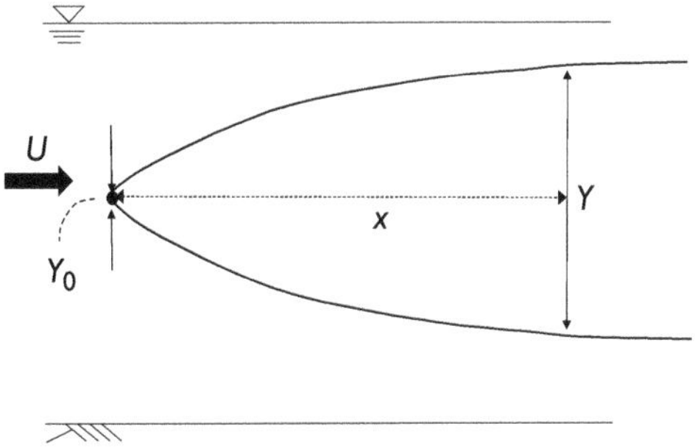

Figure 6.7 The definition of diffusion width and traveling distance.

The diffusion width with a point source in uniform flow could be expressed by the Peclet number, in which Y_0 is an initial width at the source, as shown in FIG. 6.7. The Peclet number at the source is defined by $\mathrm{Pe} = \frac{Y_0 U}{D}$. Hence, the following relation is obtained:

$$\mathrm{Pe} \sim \frac{Y_0 Ut}{Y^2} = \frac{Y_0 x}{Y^2} \quad \rightarrow \quad Y \sim \left(\frac{Y_0 x}{\mathrm{Pe}} \right)^{1/2} \tag{6.5.2}$$

Eq. (6.5.2) implies that the development of the diffusion width, Y, is smaller at the fixed downstream position x for the larger Peclet number since the convection governs the mass transport. In contrast, the diffusion width develops more significantly for the smaller Peclet number since the mass transport is controlled by the diffusion.

Chapter 7

Applications to River Flow

7.1 INTRODUCTORY REMARKS

This chapter explains the turbulent dynamics of river flow while introducing previous studies by taking up characteristic phenomena. Rivers have various flow fields with different spatiotemporal scales. Here, we pay attention to the vegetation flow, compound open channel, deadwater zone, sediment transport, debris flow, unsteady open-channel flow, accelerated flow due to bed formation and flood protection forest. Each topic is too deep to be written in this book. Hence, interested readers are encouraged to research and read the relevant literature.

7.2 VEGETATION FLOW

7.2.1 Research Background

It is very important in a river environment to maintain the vegetation zone, which is a significant part of an aquatic ecosystem. In a submerged, vegetated open-channel flow observed in rivers in a flood, a velocity difference becomes remarkable within a canopy layer and an over-canopy layer due to a drag force of plants. This property usually causes a shear instability followed by generations of coherent turbulence motions, e.g., sweeps and ejections. As such, mean flow properties and turbulence structures have significant relations with increase in bed friction and mass and momentum transports near the vegetated zone; a lot of researchers have studied this intensively.

Brunet et al. (1994) have analyzed single-point velocity statistics in a wind tunnel with waving wheat models, and they reported that the turbulence near the canopy edge exhibits a striking similarity to that of a plane mixing layer. Raupach et al. (1996) have revealed turbulence energy budget and coherent structure, and they have pointed out the analogy between the vegetated canopy flow and the mixing layer. Poggi et al. (2004) have classified the whole flow depth region into three layers, i.e., the first is the lower layer in which sweep and ejection events are dominant, the second is the middle layer in

DOI: 10.1201/9781003112198-7

which K-H waves are generated by the inflectional instability and the third is the upper layer that is similar to the boundary layer. They proposed that the length scale and first-order closure models are able to reproduce the mean velocity and Reynolds stress profiles. Nepf & Vivoni (2000) have revealed the effect of water depth on the turbulence structure. They also showed that there are two distinct flow zones within a canopy: the lower canopy, called the "longitudinal exchange zone" and the upper canopy, called "the vertical exchange zone". The extent of each zone is set by the depth of submergence and canopy density. Wilson et al. (2003) have examined the effects of two forms of flexible vegetation on the turbulence structure. The additional surface area of the fronds increases the momentum absorbing area of the plants. Carollo et al. (2005) have noticed a relation between the flow resistance law and the vegetation density. Ghisalberti & Nepf (2006) have reported that the waving motion of the flexible canopy decreases the momentum transport through the shear layer relative to a rigid canopy on the basis of turbulence measurements with 3-D ADV system. Nikora (2010) discussed the intensive interaction between fluid motion and aquatic organisms. Nikora's study plays an important role for connecting fluid mechanics and biomechanics. Developments of high-quality devices allow us to conduct outdoor turbulence measurements. Sukhodolov & Sukhodolova (2010) obtained cross-sectional distributions of mean velocity components, turbulence intensities and Reynolds stress tensor in a natural stream using an ADV system, and they provided an interesting comparison of non-vegetation and vegetation sections. Okamoto et al. (2012) revealed organized turbulence motions induce vertical mass and momentum transfer near the top edge of the plants.

The flexibility of vegetation stems is also one of the key issues to understand how flow–plant interaction occurs. In particular, spanwise vortices produced near the plant edge by the flow instability induce coherent waving motion of the submerged plants. This interesting phenomena is called "monami" e.g., Nepf (2012). The turbulence structure, including mass and momentum, transfers observed in flow with flexible plants are found to be different from those in the rigid stem condition (Ghisalberti & Nepf 2006, Ghisalberti & Nepf 2009, Okamoto & Nezu 2009).

7.2.2 Emergent Vegetation

We can often see a lot of vegetation taking root in the river bed. They are taller than the water depth during the low-discharge stage. Such an emergent vegetation influences significantly the whole depth zone. The vegetation patch with a large allocation density behaves like a porous material, and both the drag and friction loss increases. It results in the water depth also becoming large. Further, shedding vortices with a vertical axis occur behind an individual plant stem. Hence, a stem-scale horizontal turbulence is closely related to the kinetic energy consumption and mass transport, as shown in FIG. 7.1(a).

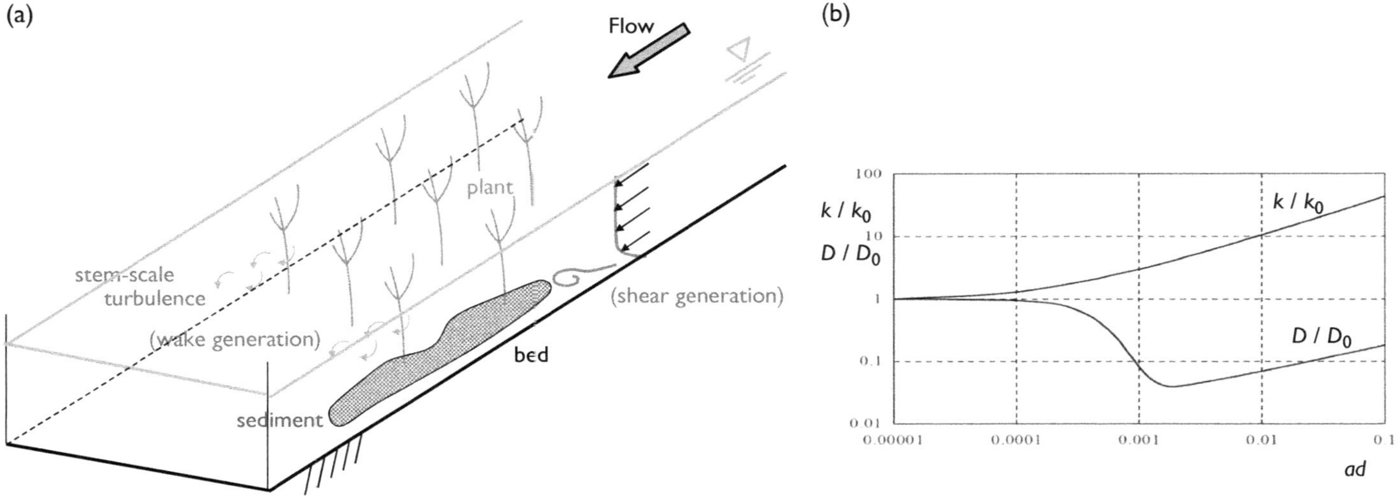

Figure 7.1 Flow passing emergent vegetation: (a) turbulence structure in emergent vegetation flow; (b) variation of normalized turbulent kinetic energy and diffusion coefficient in the emergent vegetated flow drawn, referring to Nepf (1999).

Nepf (1999) proposed the physical model of relation among drag, turbulence and diffusion. Nepf also carried out the turbulence measurement in the laboratory flume with emergent vegetation models, and explained the wake formed behind the plant stem contributes to the turbulence generation. This wake generation, G_w, is much stronger than the bottom shear generation, G_s, which is the main turbulence source in the standard open-channel turbulence without vegetation. This is because the velocity shear is reduced due to the deceleration through the vegetation resistance. The hydrodynamic characteristics in the vegetated flow is influenced significantly by the vegetation allocation density. Nepf defined the vegetation density per meter by $a \equiv d/S^2$, in which d is the stem diameter and S is the mean spacing between stems.

Nepf's physical model expressed the normalized diffusion coefficient D by combination of the turbulence diffusion and mechanical diffusion:

$$\frac{D}{Ud} = \alpha\,(C_D ad)^{1/3} + \left(\frac{\beta^2}{2}\right)ad \tag{7.2.1}$$

in which α and β are scale factors.

The first term and the second term on the right-hand side refer to the turbulence diffusion and the mechanical one, respectively. Of particular significance is that the first term is introduced by the assumption that the wake generation equals the turbulence energy dissipation, i.e., $G_w = \varepsilon$. Then, turbulent kinetic energy, k, could be connected with the drag coefficient, C_D. The mechanical diffusion is induced by the physical collision of the stream with the plants. The fluid particle also moves in the lateral direction for collision avoidance during streamwise traveling. Assuming the variance of the lateral position is expressed by the Gaussian probability function, the second term is obtained.

A fraction of mean energy partitioned to turbulence depends on the ratio of the form drag to the viscous drag. Therefore, the turbulence generation is controlled by the plant shape, flexibility and stem Reynolds number.

The turbulence structure in the vegetated flow is quite different from that in the flow without vegetation. Here, let's consider the normalized turbulence energy, k/k_o, and the diffusion coefficient, D/D_o, in which a subscript o means the open channel without vegetation. FIG. 7.1(b) shows outlines of them, depending on ad under the same U. As ad increases, k/k_o becomes larger since the stem wake is promoted. In contrast, the channel width or depth scale eddies collide with the plant stems and, as a result, the vortex scale is reduced. Furthermore, U also is decelerated. Hence, the vortex's length scale becomes shorter, and then D/D_o decreases. Particularly, the characteristic length of turbulence is reduced to the stem diameter with vegetation density greater than 1%. This is much smaller than that in an open-channel flow without vegetation, which is the water-depth scale.

7.2.3 Submergent Vegetation

When the water depth increases over the vegetation height, a two-layer flow appears in the open-channel with submergent vegetation. The streamwise velocity is reduced in the canopy layer within the bottom and, in contrast, the velocity is accelerated over the canopy. Such a flow situation is often observed in the deep-depth rivers with the larger discharge rate. A larger-scale vortex is formed due to the shear instability near the vegetation height. Therefore, turbulence transport is promoted and friction loss is significantly increased. Coherent turbulence events are also different from those of the open channel with the flat bottom.

Poggi et al. (2004) and Ghisalberti & Nepf (2006) have pointed out that whole depth region could be classified into several layers on the basis of the vertical profiles of mean streamwise velocity and Reynolds stress. Nezu & Sanjou (2008) classified the vegetated open-channel flow into the three kinds of layers, as shown in FIG. 7.2(a). The lower layer within the canopy is termed "wake zone" and this layer was defined by Neph & Vivoni(2000) and Ghisalberti & Nepf (2006) with a penetration height of momentum transfer, h_p. This height is an upper limit of the wake zone and the same as an elevation of 10% of the maximum Reynolds stress. In the wake zone, horizontal momentum transfers are dominated by stem wakes and vertical transport is comparably smaller than the horizontal one. A middle layer includes the vegetation edge, termed the "mixing layer zone", which is characterized by a lower limit of h_p and an upper limit of $h_{\log}$. The velocity profile has one inflection point near the canopy edge and velocity fluctuation evolves to form large-scale coherent motions of the sweeps and ejections.

A free-surface layer is termed the "logarithmic zone", in which the velocity profile obeys well with the following logarithmic law of roughness boundary layer:

$$\frac{U}{U_*} = \frac{1}{\kappa} \log \frac{y - d}{y_0} \tag{7.2.2}$$

in which d and y_0 are the displacement thickness and roughness height, respectively. The von Karman constant, κ, was used as the standard value in the open-channel flows, i.e., $\kappa = 0.412$.

Hydrodynamic properties in vegetated canopy rivers, in which velocity distributions are largely changed, in not only vertical but also spanwise directions. In particular, it is necessary to reveal such dispersive properties of velocity components, since the velocity dispersion has significant relations with turbulent transport and generation of turbulence due to the vegetation wakes.

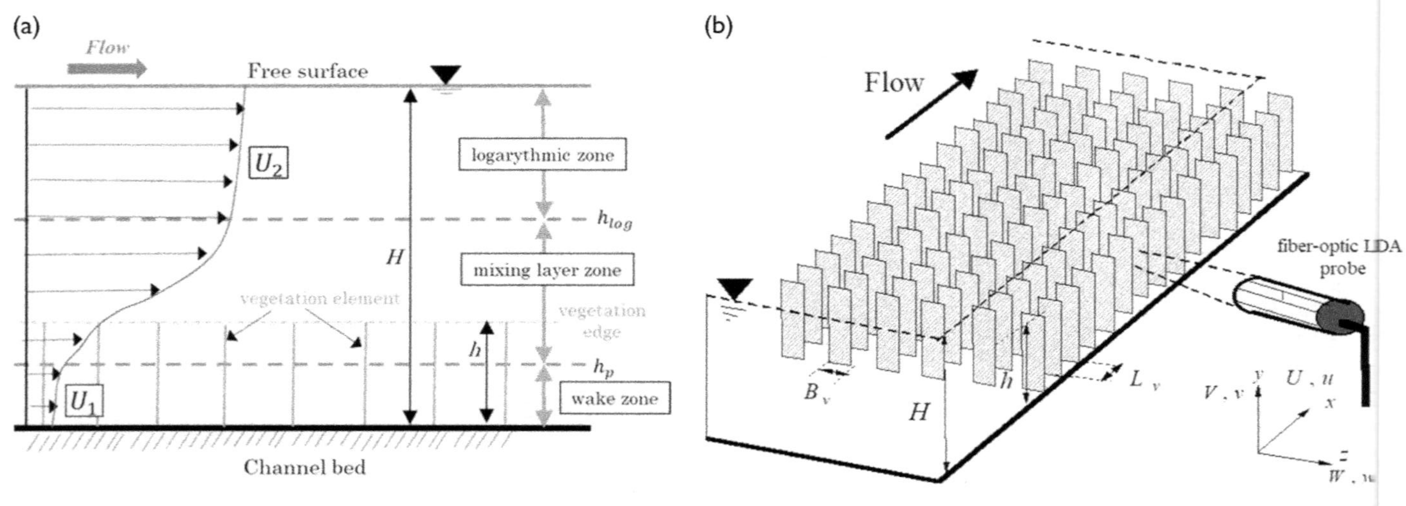

Figure 7.2 Flow passing submerged vegetation: (a) vertical structure of mean flow profile in vegetated open-channel flow; (b) experimental setup for LDA measurements in open-channel flume with model plants.

Turbulence measurements varying the population density systematically contribute to reveal the turbulent structure and dispersive effects (Nezu & Sanjou 2008). FIG. 7.2(b) shows a typical example of experimental setup and the coordinate system. The experiments were conducted in a 10 m long and 40 cm wide tilting flume in the environmental hydraulics laboratory in Kyoto University. The elements of the vegetation model were composed of non-flexible strip plates. These vegetation elements were attached vertically on the flume bed. H is the water depth and h is the vegetation height. B_v and L_v are neighboring vegetation spacings in the spanwise and streamwise directions, respectively. x, y and z are the streamwise, vertical and spanwise coordinates. The vertical origin, $y = 0$, is chosen as the channel bed. The time-averaged velocity components in each direction are defined as U, V and W and the turbulent fluctuations are defined as u, v and w, respectively. In order to measure two components of instantaneous velocity, i.e., $\tilde{u} \equiv U + u$ and $\tilde{v} \equiv V + v$, within and above the canopy vegetations, a two-component fiber-optic laser Doppler anemometer (LDA) (DANTEC made) system was used. This probe was located at 7 m downstream from the channel entrance. The four laser beams of 2W Ar-ion laser were projected into within the canopy layer as well as over the canopy vegetation. We could successfully measure the velocity profiles in the whole flow region from the bottom bed to the water surface, including within the canopy vegetation, because the laser beams were not blocked by any vegetation strips.

The combinations of B_v and L_v were selected as three patterns to examine the effects of the vegetation density, as shown in FIG. 7.3. The vegetation density λ is defined as follows:

$$\lambda \equiv \sum_i A_i / S \tag{7.2.3}$$

A_i is the frontal square of the vegetation strip and S is the referred bed area of $40 \times 40\,\mathrm{cm}^2$. The largest vegetation density is $\lambda = 1.56$ (caseA), the middle density is $\lambda = 0.77$ (caseB) and the sparsest density is $\lambda = 0.38$ (caseC). In order to examine the lateral variations of velocity components and their dispersive properties, LDA measurements were conducted along eight vertical traversing lines between the neighboring plants, $0 \le z/B_v \le 1.0$. To reveal the effects of vegetation on turbulent structure, it is of essential importance to consider the horizontal space averaging in the time-averaging N-S equations (RANS), as pointed out by Raupach et al. (1991) and Finnigan (2000). A hydrodynamic value, Φ, is defined as follows:

$$\Phi \equiv \langle \Phi \rangle + \Phi'' \tag{7.2.4}$$

in which the angle bracket < > denotes the horizontal space averaging, and the double prime denotes the deviation from the space-averaged quantity.

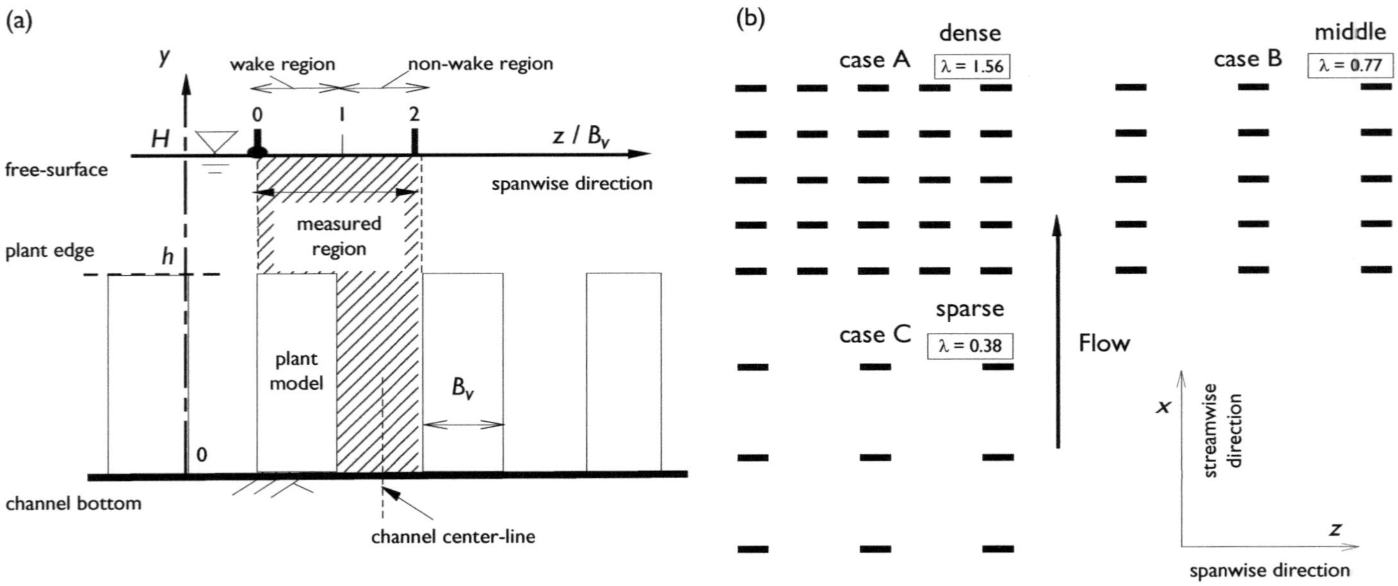

Figure 7.3 Model plant: (a) cross-sectional geometry of model plant made of acrylic plate and (b) horizontal view of allocation patterns.

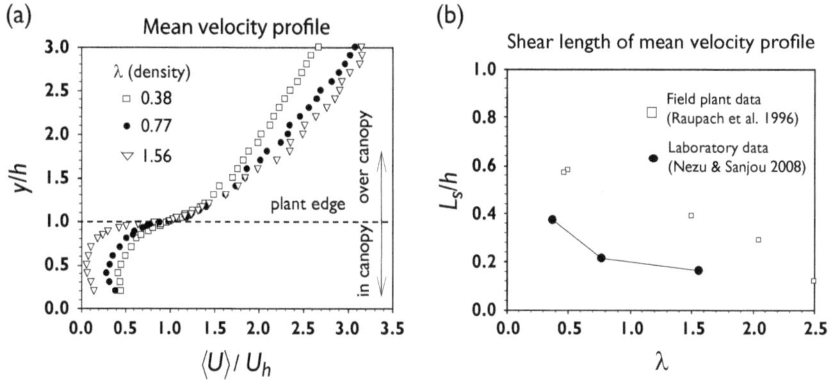

Figure 7.4 Time-mean velocity properties: (a) vertical profile of horizontal-averaged streamwise velocity compared among different population densities; (b) dependency of the shear length scale on vegetation density.

FIG. 7.4(a) shows the vertical profiles of space-averaged streamwise velocity $\langle U(y) \rangle$, which is normalized by the vegetation-edge velocity, $U_h \equiv \langle U(y) \rangle$. This figure reveals that the typical drag effects of vegetation decelerate the streamwise velocity more largely in the canopy layer of $y/h \leq 1$ as the vegetation density becomes larger.

Consequently, a significant inflection point appears near the canopy edge $y/h = 1$, the feature of which is consistent with that of plant canopy flows, as reviewed by Finnigan (2000). Furthermore, a second peak, i.e., the counter-gradient velocity profile, is observed near the bottom bed, which may be caused by three-dimensional dynamics of wakes. In order to evaluate the inflectional velocity strength reasonably, the shear length scale, L_s, is defined as

$$L_s \equiv \frac{U(h)}{(\partial U/\partial y)_{y=h}} \tag{7.2.5}$$

in the same manner as used by Raupach et al. (1996). They pointed out that L_s has a significant relation with the wavelength of coherent eddies observed in mixing layers and canopy flows. FIG. 7.4(b) shows the relation between L_s and λ, in which previous field plant data of Raupach et al. (1996) are also plotted for a comparison. L_s is in good correlation with vegetation density, and it is found obviously that the present flume value of L_s becomes smaller with an increase of λ, which is consistent with plant data reviewed by Raupach et al. (1996).

7.2.4 Flexible Vegetation

Unsteady motion of submerged vegetation is classified into four patterns, as shown in FIG. 7.5 (Carollo et al. 2005, Okamoto et al. 2016). They depend

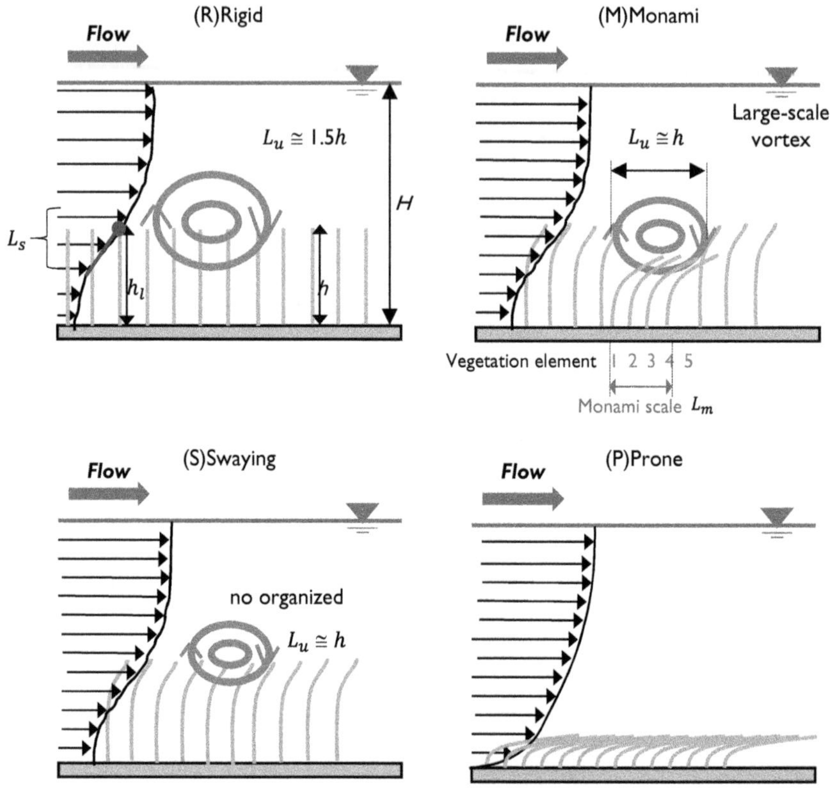

Figure 7.5 Four types of unsteady, submerged, flexible vegetation.

on the flow condition and the flexural rigidity of the plant stem. The first type is rigid (R), in which the plant is erect without deformation. The second is swaying (S), in which the plant is gently waving. The third is "monami" (M), in which plants are deflected with organized motion. The forth is prone (P), in which the plant bends over and keeps a prone position.

Turbulence structure depends significantly on these motion patterns, as shown in FIG. 7.6. Here, plant size is assumed to be constant in four motion patterns. The elevation of vegetation edge is lower in the swaying and prone type than in the erect type. This shifts the inflection point of mean velocity profile, $U(y)$, toward the bottom (FIG. 7.6(a)). Okamoto et al. (2016) reported organized motions due to the shear instability that occurs periodically, even in the flow with flexible submerged vegetation. The length scale of them is reduced compared to the flow with rigid vegetation, because the vertical height in the vegetated canopy layer decreases corresponding to the lower inflection point. It results in a smaller Reynolds stress distribution, as shown in FIG. 7.6(b), and it makes the vertical momentum exchange weaker within the moving vegetation.

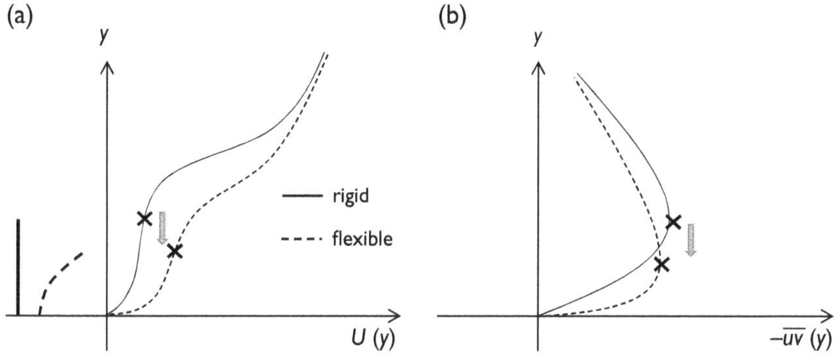

Figure 7.6 Comparison of vertical profiles in rigid and flexible vegetation: (a) mean velocity and (b) Reynolds stress.

A comparison of four patterns previously mentioned under a hydraulic condition with constant depth and bulk-mean velocity shows a clear difference in turbulence structure, as shown in FIG. 7.6. The height of submerged vegetation falling prone becomes shorter than that of rigid ones, and it results in a drop of the inflection point in mean velocity profile, as shown in FIG. 7.6(a).

Okamoto & Nezu (2009) reported generation of the coherent vortex due to the shear instability in the flexible vegetation as well as in the rigid vegetation. It is noted that a length scale of the vortex decreases with a reduction of Reynolds stress, because of shortened elevation of the inflectional point. This fact implies the vertical mass and momentum transfer is weakened in the flexible vegetation.

7.2.5 Summary of Vegetation Flow

The main findings for the vegetated flows are as follows. The turbulence structure including the velocity profile depends significantly on the vegetation height and allocation density. Particularly, there is quite a difference between the emergent and submerged condition. In the former, the turbulence intensity is more remarkable due to the stem-oriented wake generation than in the flow without vegetation. In contrast, the bottom-shear stress decreases due to the reduction of streamwise velocity in the whole depth region by the vegetation drag. Further, the turbulence diffusion is also reduced compared to the no-vegetation flow. In the latter, the flow is divided into the over-canopy layer and the within-canopy layer. The large velocity difference of them induces the large-scale spanwise vortices in the vegetation height. They play important roles in an increase of the current energy loss, mass and momentum exchanges between the two layers.

As previously mentioned, the aquatic vegetation influences decisively the river environment, the bottom friction and sediment transport. Hence, it is important for the river management to understand hydrodynamics and turbulence structure.

7.3 COMPOUND OPEN-CHANNEL FLOWS

7.3.1 Characteristics in Compound Channels

Compound open-channel flow appears in rivers in a flood. It is one of the significant topics in fundamental hydraulics as well as river engineering because it has various functions, such as preventing water disaster, valuable open spaces for aquatic lives and ecosystems. A river stream is observed within only the main channel in the ordinary stage, i.e., single channel. Thus the open spaces are used for public uses, e.g., parks, athletic stadium, open-air theater and so on. The main channel is designed as lmuch narrower than the whole width between both side banks. Therefore, it can keep a minimum water level required for aquatic environments and ship navigation.

Most amounts of discharge excess a volume capacity of the main channel results in the floodplain flows shown in FIG. 7.7. The compound channel consists of the deep main-channel flow and the shallow floodplain flow. It is generally known that the mainstream velocity is larger in the main channel than in the floodplain. A typical lateral gradient in the mainstream velocity is formed like a hyperbolic tangent curve, and shear instability produces a large-scale horizontal vortex with a vertical axis.

Such a two-stage flow causes a significant inflection instability near the junction region between the main channel and floodplains and, consequently, coherent horizontal vortices are generated and transported downstream in a complicated fashion. These vortices enhance mass and momentum exchanges significantly between the high-speed main-channel and low-speed floodplain flows and also influence the flow resistance and sediment transport significantly, as suggested by, e.g., Bridge (2003). Therefore, it is very important

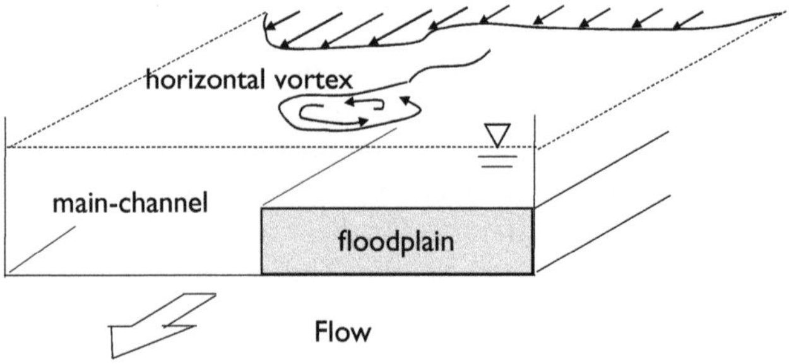

Figure 7.7 Compound open-channel flow.

in a river environment and hydraulic engineering to study turbulence structure and coherent vortices in such compound open-channel flows.

7.3.2 Previous Works

In compound open-channel flows that consist of a main channel and floodplains, Sellin (1964) found that there were some large-scale horizontal in the laboratory experiment. Such a coherent vortex influences significantly transverse transport of scalar variables such as suspended sediment. So, it is very important in hydraulic engineering and the river environment to investigate hydrodynamic characteristics of coherent horizontal vortices in compound open-channel flows, and really there are a lot of experimental and numerical studies on such coherent eddies.

Several highly accurate turbulence measurements have been conducted with a two-component laser Doppler anemometer (LDA) and PIV, and we can use very valuable database of observed turbulent structures. In the typical compound open-channel flows, the inflection point instability near the junction region between the main channel and floodplains appears and, thus, coherent horizontal vortices are generated and advected downstream, as pointed out by Sanjou & Nezu (2009). These vortices enhance mass and momentum exchanges significantly between the high-speed main-channel flow and the low-speed floodplain flow. Thus, such compound open-channel flows have been studied intensively in river environments and hydraulic engineering. Shiono & Knight (1991) proposed a depth-averaged analytical solution including the effects of secondary currents, and evaluated the lateral profiles of the apparent shear stress in large-scale compound flume. Further, Sellin et al. (1993) have investigated the friction law and 3-D velocity structure in a skewed two-stage flow with a large-scale experimental flume. A possibility of measuring turbulent structures in compound open-channel flows has been opened up relatively recently with the development of non-intrusive measurement devices such as LDA and PIV. Tominaga & Nezu (1991) have conducted accurate turbulence measurements in compound open-channel flows with a two-component LDA, and they revealed the cross-sectional distributions of primary velocity and secondary currents. The flow-visualization measurements such as PIV can reveal the horizontal vortex associated with shear instability near the junction region. Nezu et al. (1999) have conducted some LDA and PIV measurements and found that the coherent structure depends on the floodplain depth ratio. Many numerical simulations about compound open-channel flows have been also conducted. For example, Naot et al. (1993) have calculated steady compound open-channel flows by using an algebraic stress model (ASM) and reproduced the main features in compound open channels. Nadaoka & Yagi (1998) and Bousmar & Zech (2001) have simulated a structure of horizontal vortices by the use of a depth-averaged LES model, and discussed the stability of such a coherent structure numerically.

7.3.3 Cross-Sectional Features

FIG. 7.8 and FIG. 7.9 show sketches of secondary currents and streamwise velocity distribution comparing with and without one-line vegetation near the junction, in which $\langle U \rangle(z)$ is the depth-averaged velocity. In the non-vegetation case, there exist significant differences in streamwise velocities between the main channel and floodplain. Consequently, the spanwise gradient of mean velocity is formed significantly in the junction zone. Typical secondary currents are also observed there; that is, the strong inclined upflows are generated from the junction edge toward the free surface. The maximum magnitude of secondary currents attains up to a few percent of the streamwise velocity (Tominaga & Nezu 1991). In contrast, in the emergent vegetation case, the streamwise velocity is decelerated locally in the junction zone, and then larger-scale longitudinal vortices are recognized on both sides of the inclined upflow, as compared to those observed in the non-vegetation case (Sanjou et al. 2010). It is therefore inferred that the emergent vegetation has significant roles on turbulence structure and the associated mass/momentum transfers.

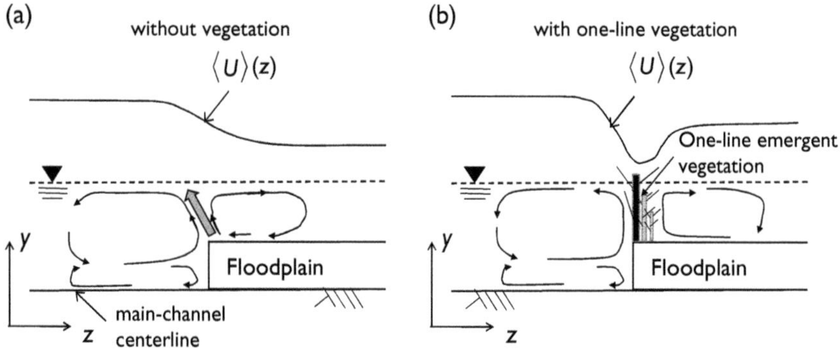

Figure 7.8 Distributions of secondary currents: (a) without vegetation; (b) with one-line vegetation near the junction.

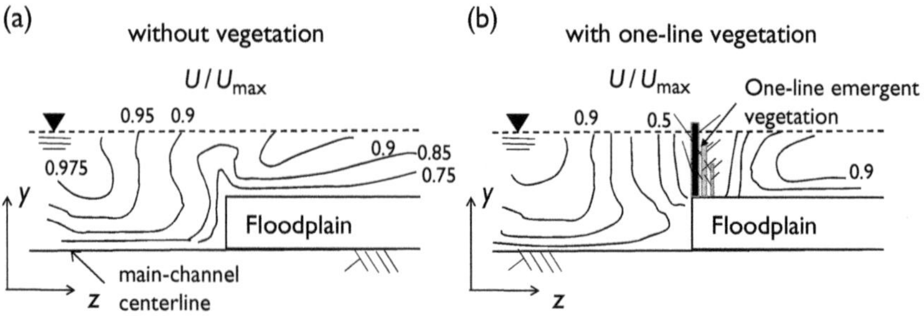

Figure 7.9 Distributions of streamwise velocity: (a)without vegetation; (b)with one-line vegetation.

7.3.4 Spanwise Profiles of Streamwise Velocity and Bed Shear Stress

FIG. 7.10(a) and FIG. 7.10(b) show the examples of measured spanwise profiles of the mean velocity $U(z)$ with H/D = 1.44 and 2.0, respectively. They are obtained by the PIV experiments in environmental hydraulics laboratory of Kyoto University. These values are normalized by the bulk mean velocity, U_m. U is larger significantly in the main channel than on the floodplain at all elevations in such shallow flows. It is therefore recognized that only one inflectional point exits near the junction between the main channel and floodplain in the whole depth region for a shallow-depth case.

At the lowest elevation point (y/D = 1.06) in the deep depth (H/D = 2.0), the streamwise velocity is larger in the main channel than on the floodplain. In contrast, at the free-surface elevation point (y/D = 1.62), the floodplain velocity is almost as large as the main-channel velocity, and consequently there is a local decrease of primary velocity $U(z)$ near the junction between the main channel and floodplain. This velocity dip is caused by secondary currents that transport low-speed momentum from the junction edge toward the free surface. It should be noticed that the velocity distribution $U(z)$ has two inflectional points at the main channel and floodplain sides, respectively. From these results, it is strongly inferred that the horizontal vortices have a complicated 3-D structure that combines with secondary currents.

FIG. 7.11 shows the sketch of bed shear stress profile normalized by the average value, referring to Tominaga & Nezu (1991). The increase of the bed shear stress on the floodplain is observed near the junction. This feature may be the result of the spanwise momentum transfer from the main channel, and it is more remarkable in the shallower floodplain case. In the deeper case, the monotonic variation is observed near the junction like the distribution of depth-averaged velocity, whereas a negative peak appears in

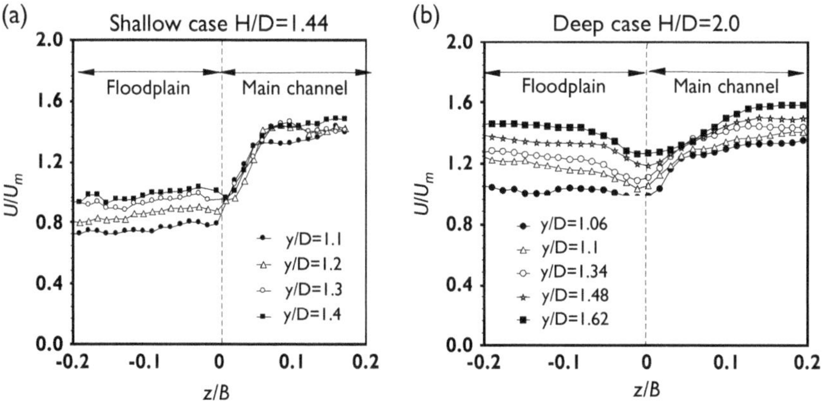

Figure 7.10 Spanwise profiles of streamwise velocity: (a) shallow depth; (b) deep depth.

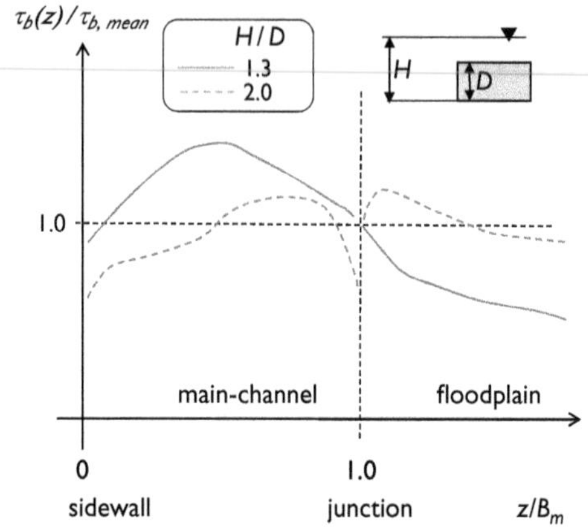

Figure 7.11 Spanwise profiles of bed shear stress: comparison between different relative depths.

the main channel near the junction in the shallower case, because both the bottom and junction wall frictions act on the currents.

7.3.5 Interaction between Main Channel and Floodplain

RANS in the streamwise direction in the uniform flow is expressed by a following form:

$$V\frac{\partial U}{\partial y} + W\frac{\partial U}{\partial z} = gI_e + \frac{\partial(-\overline{uv})}{\partial y} + \frac{\partial(-\overline{uw})}{\partial z} + \nu\left(\frac{\partial^2 U}{\partial y^2} + \frac{\partial^2 U}{\partial z^2}\right) \quad (7.3.1)$$

in which I_e is the energy slope in the streamwise direction. After integral in the depth direction of Eq. (7.3.1), assuming $V = 0$ and $-\overline{uv} = 0$ at the bottom wall and the free surface, a following form is given:

$$\tau_b/\rho = gI_eH' + \frac{d}{dz}H'(T - J) \quad (7.3.2)$$

in which $H' = H$ in the main channel, and $H' = H - D$ in the floodplain. $\tau_b/\rho = (-\overline{uv} + \nu\partial U/\partial y)_{y=0}$ is the streamwise bottom shear stress. $J = \frac{1}{H}\int_{y1}^{H'} UWdy$ and $T = \frac{1}{H}\int_{y1}^{H'} -\overline{uw}dy$ represent additional resistance components due to secondary currents and Reynolds stress, respectively.

FIG. 7.12(a) compares spanwise profiles of J and T for the case $H/D = 2.0$ between the LDA data by Tominaga & Nezu (1991) and the ADV data by Sanjou et al. (2010). The contribution of secondary currents, J, has a positive peak near the junction. In contrast, the contribution of

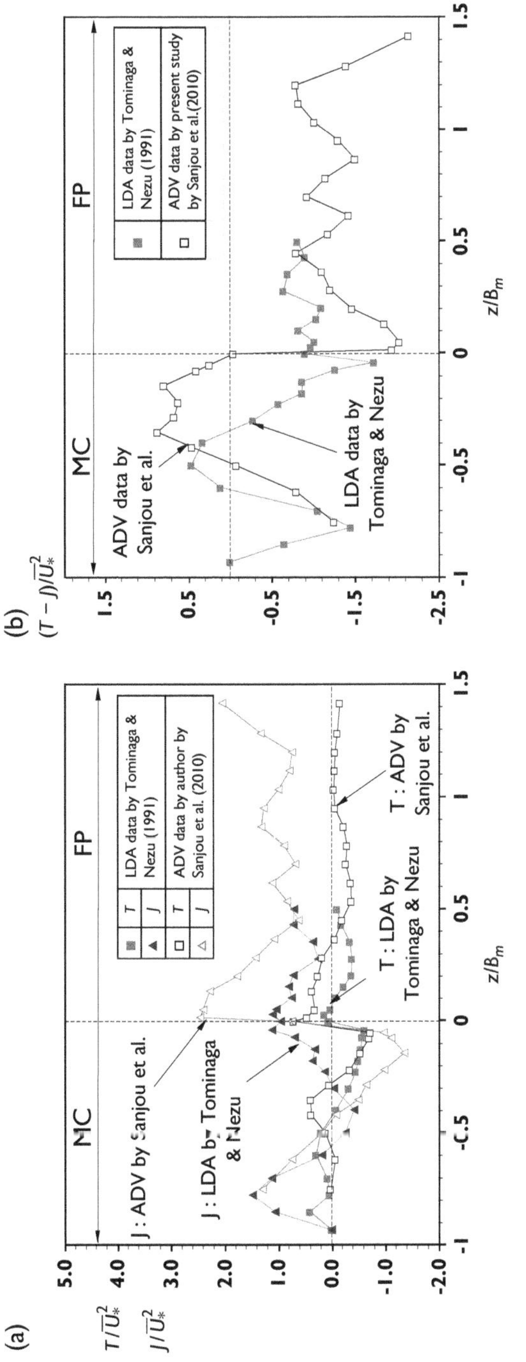

Figure 7.12 Spanwise profiles of bed shear stress: comparison between different relative depths: (a) *T* and *J*; (b) *T*-*J*.

turbulence, T, has negative and positive peaks in the main channel and in the floodplain near the junction. They correspond to counter rotating horizontal vortices. FIG. 7.12(b) shows the distribution of $(T - J)$. The spanwise gradient of $(T - J)$ is the additional bottom shear stress. There is a negative peak in the junction between the main channel and the floodplain, which implies the friction of fluid reduces significantly. The spanwise gradient of $(T - J)$ is positive, increasing the bottom shear stress.

7.3.6 Shiono and Knight Method (SKM)

Shiono & Knight (1991) derived a mathematical model called SKM. This gives the following theoretical solution of the depth-averaged velocity profile $\langle U \rangle (z)$:

$$\langle U \rangle = \{A_1 \exp(\gamma z) + A_2 \exp(-\gamma z) + gS_0 H'(1-\beta)/(f/8)\}^{1/2} \tag{7.3.3}$$

in which $\gamma = \left(\frac{2}{\lambda}\right)^{1/2}\left(\frac{f}{8}\right)^{1/4}/H'$, $\beta = \frac{\Gamma}{\rho g S_0 H'}$, $\Gamma = \frac{\partial}{\partial z}H'\langle \rho UW \rangle$ and $\langle -\overline{uw} \rangle = \lambda U_* H' \frac{\partial \langle U \rangle}{\partial z}$. $\langle \rangle$, f and S_0 mean the depth-averaging operation, friction coefficient and energy slope, respectively.

f, Γ and λ are the different constants in the main channel and floodplains, respectively. A_1 and A_2 are given using a boundary condition at the side walls and the junction between the main channel and floodplain. Shiono et al. (2009) have replaced $\frac{f}{8}$ in Eq. (7.3.3) by $\left(\frac{f}{8} + \frac{1}{2}\frac{N}{L}C_d S_0 H'\right)$ in order to consider drag force by vegetation. N is the number of plants and L is the spacing between the plants.

FIG. 7.13 shows the comparison between the SKM and the ADV data of Sanjou et al. (2010). In the non-vegetation case, one inflection point appears at the junction between the main channel and floodplain. In contrast, in the one-line vegetation case like FIG. 7.8 (right), the velocity is decelerated locally behind a tree. It should be noted that the SKM agrees well with the results of ADV quantitatively, and thus this implies that SKM can be applied to compound open-channel flows with emergent vegetation.

7.3.7 Horizontal Vortex

Sellin (1964) found that there were some large-scale horizontal vortices in the laboratory experiment. Such a coherent vortex influences significantly transverse transport of scalar variables, such as suspended sediment.

Nezu et al. (2003) developed a new dual-layer PIV system, in which two different height laser-light sheets were projected to a measurement region, and consequently they could analyze the velocity distributions at different elevations simultaneously (FIG. 7.14). 3-D coherent properties of horizontal

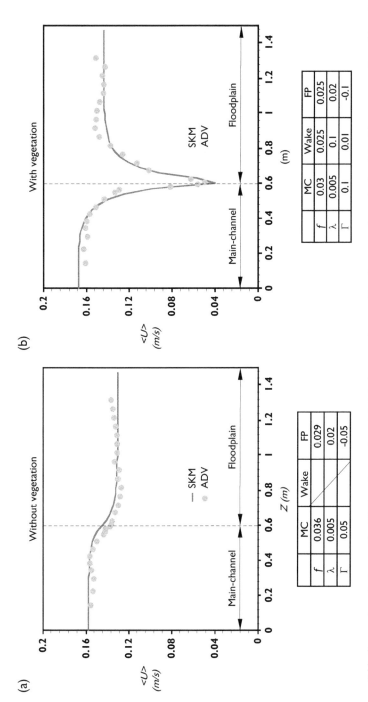

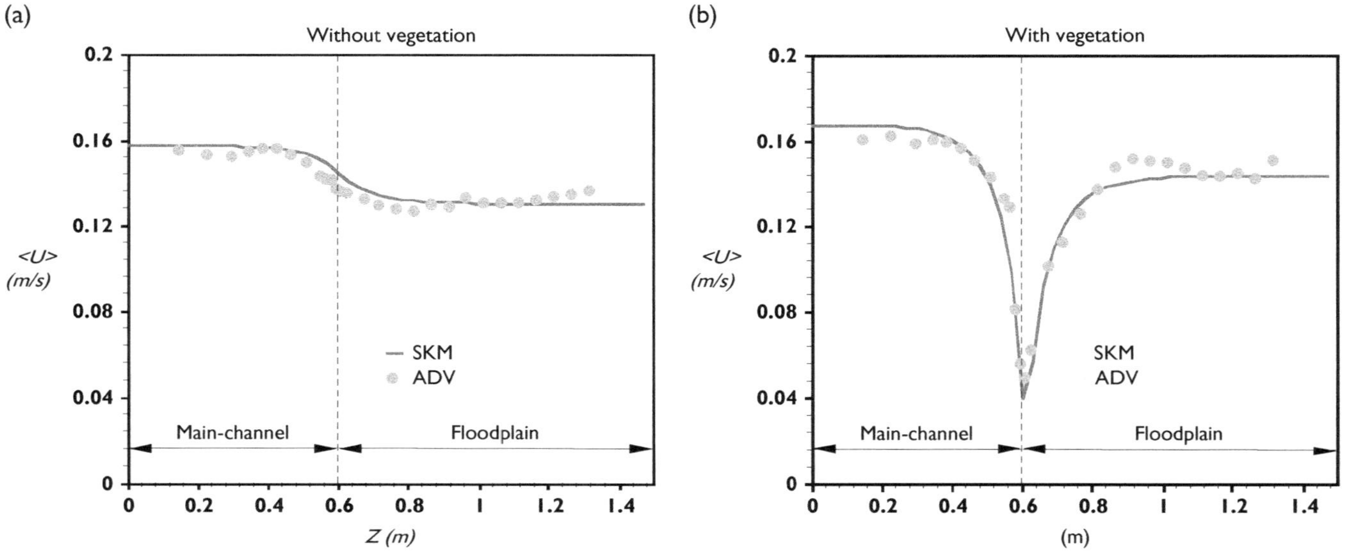

(a)

	MC	Wake	FP
f	0.036		0.029
λ	0.005		0.02
Γ	0.05		-0.05

(b)

	MC	Wake	FP
f	0.03	0.025	0.025
λ	0.005	0.1	0.02
Γ	0.1	0.01	-0.1

Figure 7.13 Comparison between the SKM and experimental data in spanwise profile of streamwise velocity: (a) without vegetation; (b) with vegetation.

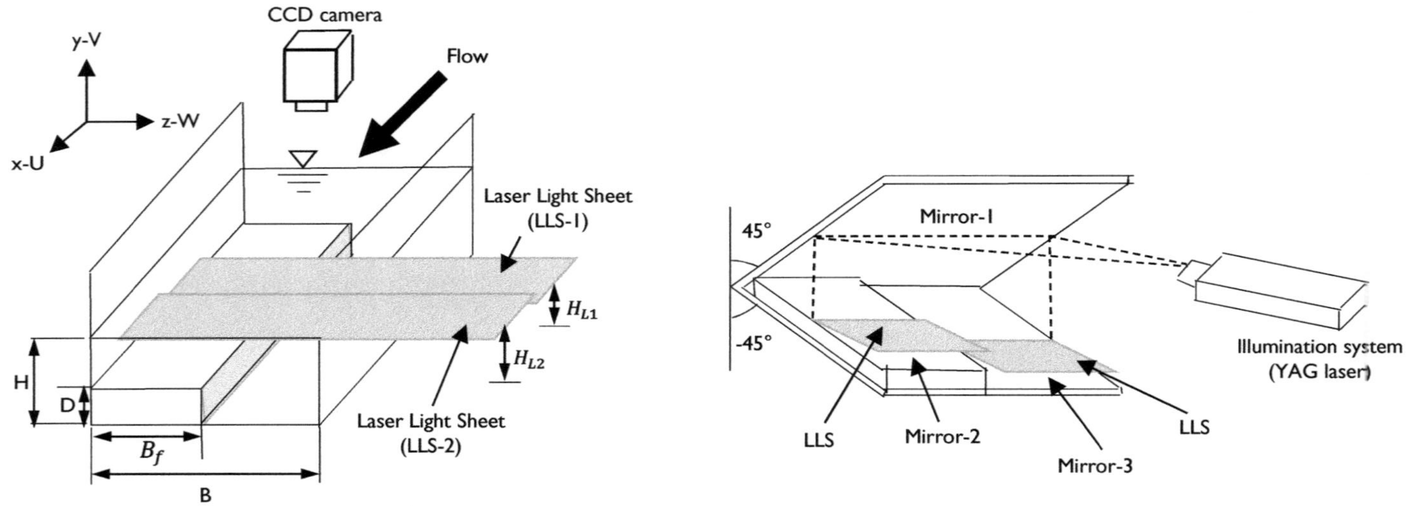

Figure 7.14 Dual-layer PIV system for detection of three-dimensional structure of horizontal vortex.

vortices, their convection mechanism and transverse momentum exchanges between the main channel and floodplain were investigated by using the dual-layer PIV system.

The velocity correlations between the lower and higher elevations, i.e., at $y = H_{L1}$ and H_{L2}, are discussed on the basis of the results of dual-layer PIV measurements. FIG. 7.15 shows the distributions of instantaneous velocities and the contour lines of the vorticity ω at two laser-light sheets LLS-1 and LLS-2 in the shallow-depth case (y/D = 1.1 and 1.2), in which the time t progressed from t = 0.0 s, 0.33 s and 0.66 s. The instantaneous vorticity is defined by

$$\omega_z = \frac{\partial \tilde{u}}{\partial z}\frac{\partial \tilde{w}}{\partial x} \tag{7.3.4}$$

In FIG. 7.15(a), a coherent horizontal vortex (A) observed on the upstream-side LLS-1 (y/D = 1.1) at t = 0.0 s is transported in the streamwise direction, and it appears on the downstream side LLS-2 (y/D = 1.2) at t = 0.33 s and 0.66 s. This noticeable finding is also recognized by the time variations of vorticity distributions. From this result, it is found that the coherent vortices observed on LLS-1 at t = 0.0 s and LLS-2 at t = 0.66 s are the parts of the same vertical vortex tube. Therefore, it is considered that the coherent horizontal vortices might have a 3-D tube structure.

FIG. 7.15(b) shows the instantaneous velocity vectors of the elevations y/D = 1.06 and 1.2 near the floodplain bed in the deep-depth case. At t = 0.0 s, a single eddy (A) exists near the junction between the main channel and floodplain at the lower elevation LLS-1 (y/D = 1.06), whereas twin eddies are observed at the higher elevation LLS-2 (y/D = 1.2); that is, the counter-rotating eddies (B) and (C) exist in the main channel and floodplain, respectively. These horizontal eddies (A), (B) and (C) are transported downstream with time. The other vortices (D) and (E) convected from the upstream region appear at 0.135 s and 0.27 s, respectively. FIG. 7.15(b) also shows the results of the elevations y/D = 1.48 and 1.62 near the free surface. At both layers of LLS-1 and LLS-2, the counter-rotating horizontal eddies are transported with time, and we can see the other eddies (a), (b), (c) and (d) convected from the upstream. It is, therefore, suggested from these results that there may be a transition from the single-eddy to twin-eddy structure in the vertical direction.

FIG. 7.16 shows a schematized 3-D structure of horizontal vortex inferred from the experimental results. In the shallow compound open-channel flows ($H/D \le 1.5$), a horizontal single eddy is generated near the junction region between the main channel and floodplain due to the single inflectional point of $U(z)$, as shown in FIG. 7.16(a). These single eddies have a significant correlation with the vertical direction and 3-D structure like a vortex tube. This vortex tube is inclined to the main stream near the

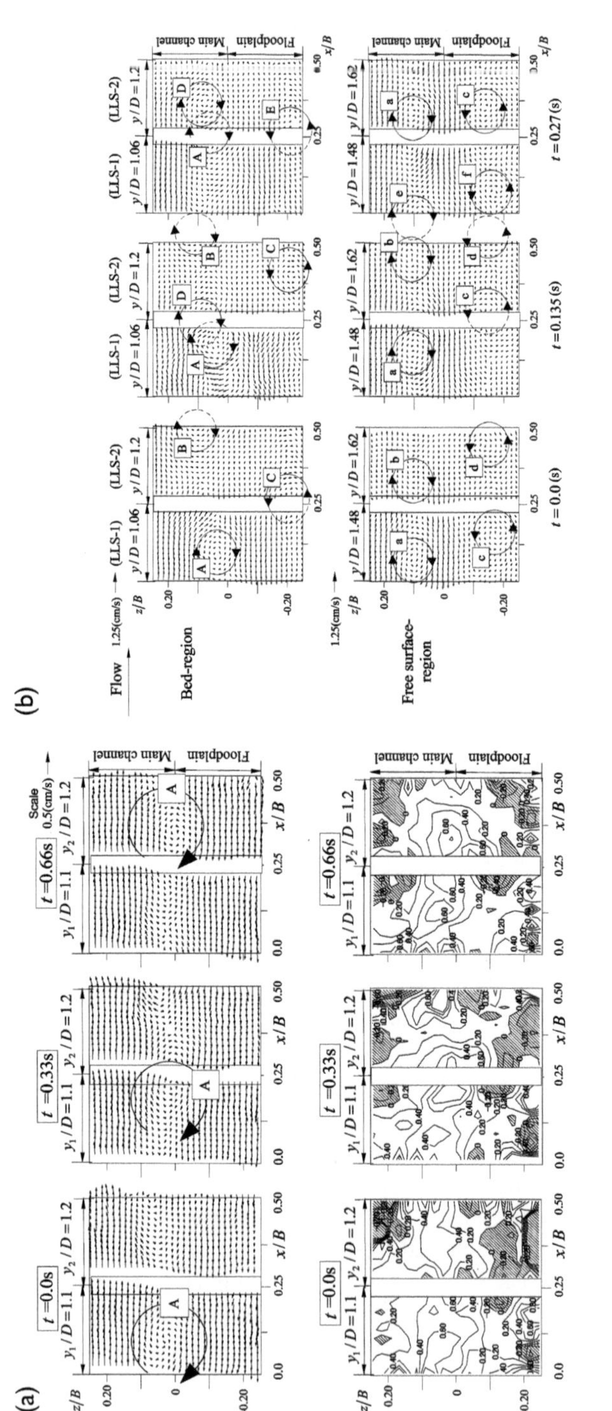

Figure 7.15 Instantaneous velocity vectors and their vorticity: (a) shallow-depth case ($H/D = 1.44$); (b) deep-depth case ($H/D = 2.0$).

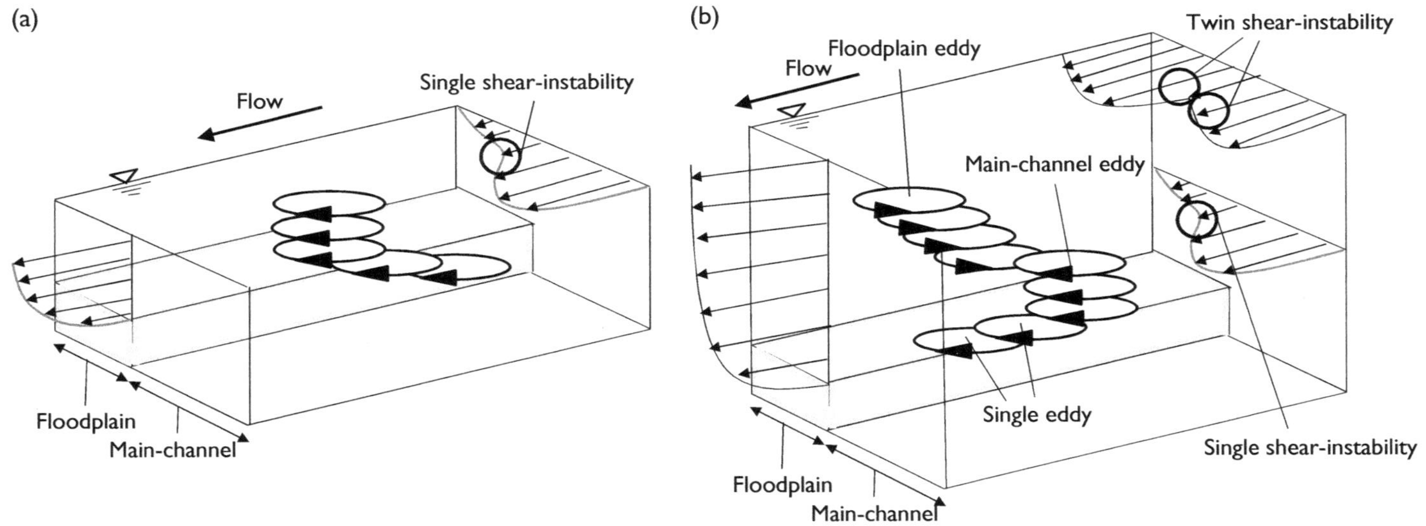

Figure 7.16 Schematized model for 3-D structure of compound channel flow: (a) shallow-depth case; (b) deep-depth case.

bed because the convection velocity is larger near the free surface than near the bed. In contrast, in the deep-depth case ($H/D > 1.5$), the spanwise profiles of $U(z)$ have two inflectional points in the free-surface region, as shown in FIG. 7.16(b). Consequently, the single-vortex street and two-vortex streets (twin eddies) were observed in the floodplain-bed and free-surface regions, respectively.

Nezu & Sanjou (2011) conducted Lagrangian tracking of horizontal vortex by using the moving camera sytem. The well-controlled carrier-mounting CMOS camera can move with a constant motor speed (variable one, max. 50 cm/s), as shown in FIG. 7.17(a). The digital images in the horizontal plane at the free surface were taken by the high-speed CMOS camera. This measurement considers reasonable trajectory lines and the development properties of horizontal vortices. The images on the free surface were recorded on the hard disk of the computer while the carrier moved from x = 3 m to 7 m. These measurements were conducted more than 30 times. FIG. 7.17(b) shows some examples of instantaneous velocity vectors for a typical horizontal vortex, which is indicated at every 1.0 m, i.e., at x = 3.0, 4.0, 5.0, 6.0 and 7.0 m. At x = 3.0 m, a counterclockwise horizontal vortex is formed near the junction zone in the main channel. The rotation direction corresponds well to the shear instability due to the difference between main-channel velocity and floodplain one. This horizontal vortex is convected downstream and still observed in each location at x = 4.0, 5.0, 6.0 and 7.0 m. The spatial and time variations of length scale of vortex seem to be very small and, it is, thus, inferred that the horizontal vortex is already developed at x = 3.0 m. The same feature was recognized in the other samples.

7.4 DEADWATER ZONE

7.4.1 Research Background

Deadwater zones in natural rivers occur in groyne fields and embayments (side cavity). A mixing layer is typically formed in the cavity opening or behind the groyne due to the velocity difference between the main channel and the deadwater zone. K-H instability induces significantly a large-scale horizontal circulation in the mixing layer, which often traps pollutants and nutrient salts. It results in a sudden drop in the water quality of the deadwater zone.

Many researches highlighted the mean velocity profile and development of the mixing layer. For example, van Prooijen and Uijttewaal (2002) focused on shallow mixing layers in which bed friction stabilizes the coherent structure. They proposed a theoretical model derived from depth-averaged shallow water equations and examined the predicted mean flow field, particularly the streamwise development of the velocity difference between

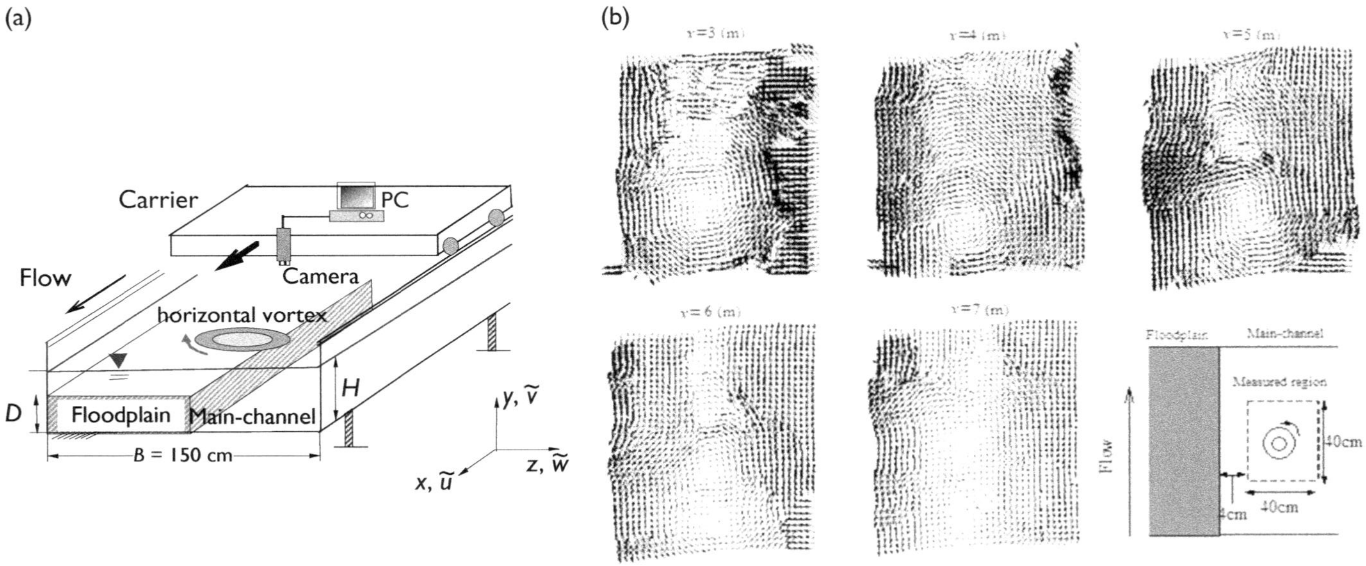

Figure 7.17 Lagragian tracking of horizontal vortex: (a) measurement setup and (b) example of instantaneous velocity vectors taken by the moving camera every 1 m traveling distance.

the high- and low-velocity sides, the development of the mixing layer width and the horizontal distribution of the streamwise velocity component by comparing these predictions with the measured data. Furthermore, their linear stability analysis based on the mean flow field allowed the determination of the spatial evolution of the energy densities and the characteristic length scales.

Therefore, many previous works have studied these issues from various perspectives.

7.4.2 Types of Local Deadwater Zones in Rivers

Local deadwater zones in rivers are observed behind isolated spur dikes, within groyne fields and in the embayment attached to the main stream, as shown in FIG. 7.18. A similar mixing layer due to shear instability should be considered for practical problems concerning sedimentation, mass transport and water preservation in these locations.

7.4.2.1 Isolated Groynes

A depth-averaged two-dimensional simulation around an isolated groyne by Tingsanchali & Maheswaran (1990) predicted characteristic lengths of the flow, such as the maximum width of the recirculating zone and the reattachment distance. Three-dimensional computational results conducted by Ouillon & Dartus (1997) provided distributions of velocity vectors, dynamic pressure, and bottom shear stress around an isolated groyne. It is noted that a point of maximum shear stress appears at the upstream corner of the groyne tip, which corresponds to the local scour zone observed on bridge piles. Also, they reported that the bottom shear stress in the region downstream from the groyne is twice that in the upstream region. Their calculated profiles of the bed shear stress are consistent with the bottom shear stress configuration observed in laboratory experiments. The horizontal flow field around a spur dike could be classified into characteristic zones, including flow separation and thalweg alignment, and these length

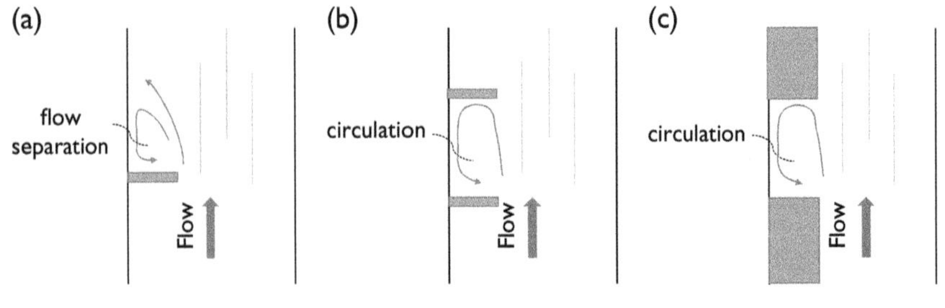

Figure 7.18 Local deadwater zones in river: (a) isolated groyne; (b) consecutive groynes; (c) embayment (side cavity).

scales and their related hydrodynamic characteristics were investigated (Ettema & Muste 2004). The results imply that single-spur dikes produce shedding vortices accompanied by surrounding complex three-dimensional behavior. Turbulence measurements around a spur dike model in a gravel bed flume with acoustic Doppler velocimetry (ADV) by Duan (2009) provided three-dimensional distributions of velocity components and Reynolds stress. The measured data revealed a secondary current structure and the relation between the local scoring and the shear stress. Sharma & Mohapatra (2012) measured the velocity components around a spur dike in a meandering trapezoidal flume with a rigid bottom; in their study, the dependencies of the separation zone and the streamwise velocity distribution on the dike position were examined.

7.4.2.2 Consecutive groynes

Consecutive spur dikes and groyne fields yield also deadwater zone. Horizontal vortices, related three-dimensional structures, and sediment processes are significantly influenced by cavity geometry and groyne rank number. Particle image velocimetry (PIV) measurements implied that the gyre structure within the cavity also depends on the mounting angle on the side wall of the channel (Weitbrecht et al. 2011). Uijttewaal et al. (2001) studied the dependence of the mass transfer coefficient on groyne geometry, water depth, groyne rank number and bottom elevation, and that transfer speed is significantly affected by aspect ratio. In a small groyne field where the aspect ratio is close to 1.0, i.e., a square horizontal shape, the mass transfer is dominated by the mixing layer formed along the cavity opening, whereas in a large groyne field, coherent shedding vortices are generated from the tip of the higher-ranking groyne. Weitbrecht et al. (2008) provided experimentally the relation between the exchange rate and the bed configuration of the cavity, and they suggested that there is a linear relation between the mass transfer coefficient and the morphometric groyne field parameter normalized by water depth, width and groyne length. Yossef and Vriend (2010) analyzed the time variation of bed form migration using the PIV technique, and these experiments yielded important results. The mixing layer dynamics induce variation in bed form celerity, and the time series of the instantaneous celerity vector field revealed a sediment transport process toward the cavity zone. There are significant differences in the development processes of the mixing layer and turbulence structures under emergent and submerged conditions (Yossef & Vriend 2011).

Sukhodolov et al. (2004) conducted velocity measurements with ADV in the Elbe River in Germany, and their analysis of the collected data provided mean horizontal velocity vectors, vertical profiles of the mean velocity, turbulent kinetic energy, and spectral properties. It is noteworthy that mean velocity profiles are consistent with the conventional power law, except near the free surface where the effects of wind-induced waves could be not ignored and upwelling occurs. Ercan & Younis (2009) predicted mean velocity components, turbulent kinetic energy, and bed shear stress for the reach of the Sacramento River and calculated the maximum erosion rate of the river bank.

7.4.2.3 Embayment

The side embayment in a main stream produces the mixing layer and cavity gyre in the same manner as the groyne field. Lateral intake zones in natural rivers also induce horizontal vortices that accompany the mainstream penetration to the intake channel (Neary & Odgaard 1993). There are some theoretical works focusing on two-dimensional simple rectangular cavities. For example, Mizumura & Yamasaka (2002) introduced a streamwise function within square- and rectangular-shaped cavities under limited conditions, i.e., two-dimensional assumption, steady flow, ignoring inertia, and rigid free-surface assumption. Hill (2014) assumed the horizontal gyre in the cavity to be in rigid body circular motion and gave the linear velocity distribution. Considering the shear stress and torque on the cavity opening and three walls, they introduced a rotational velocity profile normalized by the mainstream velocity on the basis of balancing resistance torques acting on the horizontal gyre.

Nezu & Onitsuka (2002) indicated that the geometry of the cavity opening also has a remarkable impact on the hydrodynamics in the cavity. In their experiments, a splitter plate was situated in a part of the opening between the main channel and the side cavity to vary the length of the cavity opening. Such an aperture ratio was found to influence the coherent separation vortices and the formation of the horizontal gyre. Flow patterns of different square sizes and the development of the mixing layer were discussed by Booij (2004). The effects of inserting a slotted plate in the junction between the main channel and the side cavity are significant. The results suggest that the slotted plate prevents the formation of the cavity-scale horizontal gyre, whereas small-scale vortices were produced in the small space of the slot zone.

7.4.3 Mean Currents in a Rectangular Embayment

FIG. 7.19(a) shows the horizontal distributions of mean velocity vectors reported by Sanjou & Nezu (2017); the contour in these corresponds to the value of the streamwise velocity. Large-scale circulations are observed in sizes comparable to the cavity area, irrespective of main channel streamwise velocity. Particularly, when $L_e/B_e = 1$, the circulation occupies the entire field of the cavity. B_e and L_e are lateral and longitudinal lengths of the rectangular embayment. In contrast, a counter-rotating circulation is produced in the upstream side of the cavity when $L_e/B_e = 2$, although the primary circulation occupies the majority of the deadwater zone. A bulge in

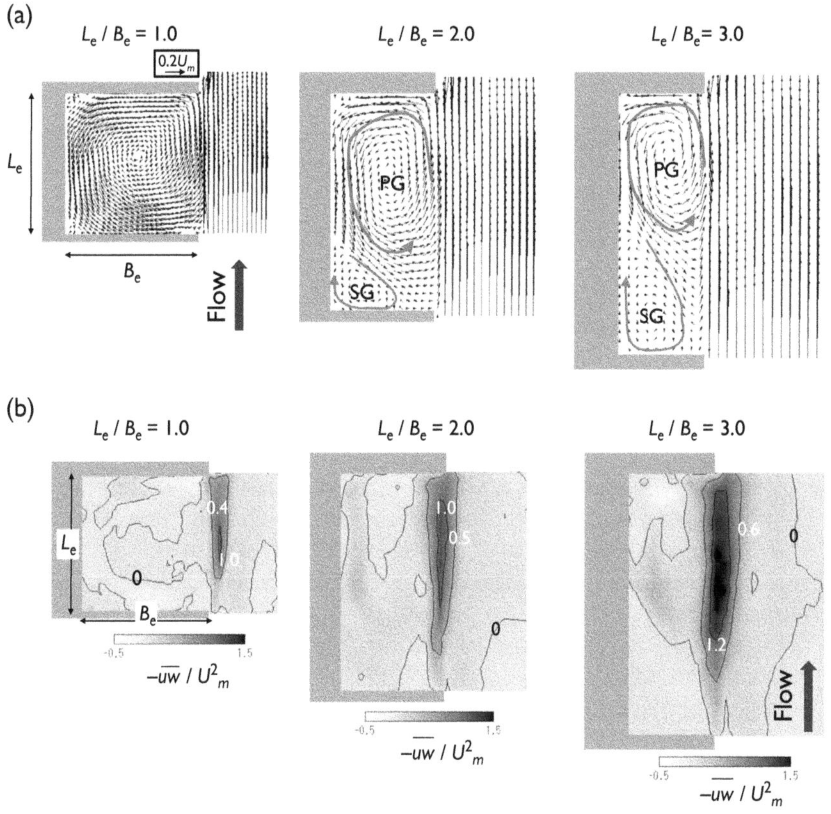

Figure 7.19 Horizontal distributions of hydrodynamic values: (a) mean horizontal velocities; (b) horizontal Reynolds stress normalized by bulk-mean velocity.

U situated at the upstream inside corner also supports the generation of a small horizontal vortex. Uijttewaal et al. (2001) previously noted the existence of this kind of secondary circulation, and they asserted it plays an essential role in local stagnation and sedimentation despite having a smaller length scale than that of the primary circulation. Both the primary gyre (PG) and the secondary gyre (SG) are formed in the case of $L_e/B_e = 3$ in the same fashion as with $L_e/B_e = 2$. The occupation area of the SG when $L_e/B_e = 3$ is larger than that when $L_e/B_e = 2$. The mainstream momentum flows in through the downstream side of the junction and flows out through the midsection of the junction after making one circulation in the downstream area of the side cavity. This fact suggests that the primary circulation is induced by intensive inflow of the mainstream, and furthermore, the width of the side cavity has a significant impact on the mean flow structure, e.g., the production of the SG, the length scale of the horizontal gyre and the position of the outflow on the cavity opening toward the mainstream.

7.4.4 Reynolds Stress Distribution in a Rectangular Embayment

FIG. 7.19(b) shows the horizontal distributions of the normalized Reynolds stress $-\overline{uw}/U_m^2$ (Sanjou & Nezu 2017). U_m is the bulk mean velocity in the main channel. These results suggest that large Reynolds stress values are locally observed in the boundary zone, i.e., the cavity opening. Although the mixing layer developed downstream for all cases, the peak value is the smallest when L_e/B_e = 1. The spanwise scale of the cavity, B_e, may influence the formation of the mixing layer. The outflow toward the main channel increases as the aspect ratio, L_e/B_e, increases, as mentioned previously, and this implies that such spanwise velocity is closely related to the development of the mixing layer and turbulence.

The statistically significant turbulence values are the Reynolds stresses observed in the main channel and the deadwater zone boundary, and it increases the flow resistance and promotes the formation of a mixing layer. However, it is not easy to accurately measure the Reynolds stress in natural rivers. Reynolds stress uses the momentum equation and the mean streamwise velocity profile, which is comparably easy to obtain in natural environments.

Momentum vectors related to a control volume ABCD situated in the horizontal boundary zone as shown in FIG. 7.20(a). The sides AD and BC correspond to the open end of the cavity and the outer edge of the boundary layer, respectively. U_{mc} is the mean velocity in the main channel. M_{AB} and M_{CD} indicate the inflow and outflow momenta per unit time of sides AB and CD. They are given by

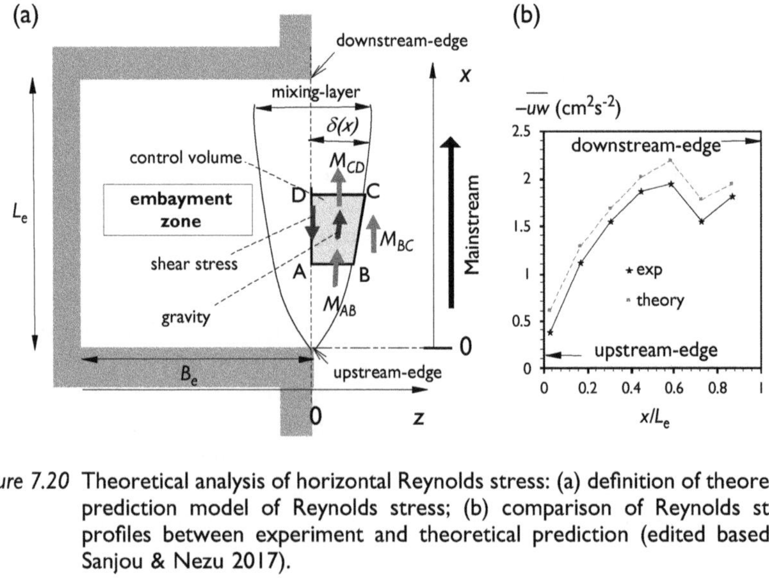

Figure 7.20 Theoretical analysis of horizontal Reynolds stress: (a) definition of theoretical prediction model of Reynolds stress; (b) comparison of Reynolds stress profiles between experiment and theoretical prediction (edited based on Sanjou & Nezu 2017).

$$M_{AB} = \int_0^\delta \rho U(z)^2 dz, \tag{7.4.1}$$

and

$$M_{CD} = \int_0^\delta \rho U(z)^2 dz + \frac{d}{dx}\left(\int_0^\delta \rho U(z)^2 dz\right) dx. \tag{7.4.2}$$

Ms_{AB} and Ms_{CD} indicate the inflow and outflow masses per unit time of sides AB and CD. They are given by

$$Ms_{AB} = \int_0^\delta \rho U(z) dz, \tag{7.4.3}$$

and

$$Ms_{CD} = \int_0^\delta \rho U(z) dz + \frac{d}{dx}\left(\int_0^\delta \rho U(z) dz\right) dx. \tag{7.4.4}$$

The difference between these mass flows $|Ms_{AB} - Ms_{CD}|$ is equal to the inflow mass, Ms_{BC}, through side BC, and the corresponding momentum, M_{BC}, can be expressed as

$$M_{BC} = U_{mc} \frac{d}{dx} \int_0^\delta \rho U(z) dz. \tag{7.4.5}$$

When the pressure gradient, $\partial P/\partial x$, and shear stress of side BC can be ignored, the primary forces acting on the control volume are the shear stress on side AD and gravity.

Thus, substitution of Eqs. (7.4.1), (7.4.2), and (7.4.3) in the momentum equation yields

$$\frac{d}{dx} \int U(z)^2 dz - U_{mc} \frac{d}{dx} \int U(z) dz = \delta g \sin\theta - \frac{\tau}{\rho} \tag{7.4.6}$$

As we target developed turbulent flows, i.e., $\tau/\rho \cong -\overline{uw}$, the Reynolds stress profiles in the main channel and the cavity junction can be predicted given the spanwise distribution of the mean velocity $U(z)$.

FIG. 7.20(b) shows an example of Reynolds stress predicted by this theoretical model, and the measured data by PIV is also included in this figure for comparison (Sanjou & Nezu 2017). The Reynolds stress tends to increase for larger aspect ratios over the whole region. The profile of the Reynolds stress corresponds with the development of the mixing layer. It should be noted that the width of the mixing layer continues developing within the whole region in the junction and, in contrast, the variation in the Reynolds stress is almost constant in the midsection. This is because an increase in shear stress inducing turbulence production is not expected due to the small velocity gradient, $\partial U/\partial z$, in the midsection of the cavity opening.

7.4.5 Three-Dimensional Structure in a Rectangular Embayment

FIG. 7.21 shows the three-dimensional phenomenological flow model proposed by Sanjou et al. (2012), in which there are mean horizontal gyres and instantaneous momentum transfer with high probability. Within the cavity with an aspect ratio of $L_e/B_e = 3.0$, two kinds of horizontal gyres are generated with counter-rotations, and their vertical variation is comparatively small compared with the horizontal structure. Contrary to the time-averaged gyres, there exist complicated three-dimensional properties for the

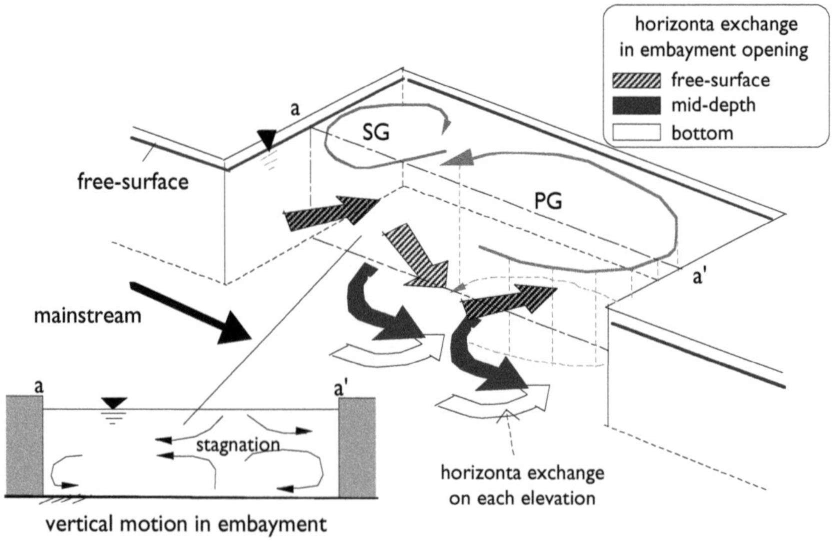

Figure 7.21 Three-dimensional phenomenological flow model proposed by Sanjou et al. (2012).

lateral momentum transfer through the junction between the mainstream and the side cavity. The vertical motions are noted significantly within the cavity. A combination of upward and downward currents produces a stagnation zone, and spanwise vortices are formed in both the upstream and downstream edges.

7.4.6 Free-Surface Oscillation

The deadwater zone, such as a side cavity, induces gravity waves and oscillations. They are known to generally increase drag force or momentum transfer (McGregor & White 1970). Kimura and Hosoda (1997) carried out both experimental and two-dimensional numerical simulation of the rectangular side cavity, in which a longitudinal seiche is formed. It results in that the seiche and shear instability influence the temporal velocity fluctuation observed in the cavity opening. Furthermore, small fluctuations in the velocity component initially are induced by the shear instability, and they are amplified to a large-scale vortex during one seiche cycle.

Transverse oscillations in an open channel with axisymmmetric cavities on both sides were measured by Meile et al. (2011). They asserted that the momentum transfer from the mainstream to the embayment has an impact on the sloshing phenomenon, and the free-surface oscillation is influenced by the impingement of a shedding vortex at the downstream edge. They emphasized that the Strouhal number is the key parameter that promotes wave excitation, and their measured data proved the existence of critical values of the Strouhal number where peak excitations of oscillations are clearly observed. They suggest the critical Strouhal number should be avoided for practical river management, e.g., reduction of water oscillations in cavity zones, such as the harbor, groyne field and embayment. The relation between the inherent instability of the mixing layer in a cavity opening and streamwise-oriented gravity standing waves was investigated by Wolfinger et al. (2012), which resulted in the occurrence of highly organized oscillations. They measured not only horizontal velocity components but also the pressure at the impingement wall of the cavity. Their results indicate the frequency of the separated turbulent layer in the cavity opening is consistent with the frequency of the gravity wave and there exists a threshold inflow velocity where the amplitude of the spectral peak of the pressure fluctuation reaches a maximum value. Furthermore, they concluded the coupling between the mixing layer and the gravity standing waves substantially enhances the increase in Reynolds stress, turbulence intensity and transverse velocity through the mixing layer. The laboratory experiments with a PIV system investigated the effects of standing waves on streamlines, turbulence structures and mass exchange in shallow flows past a single cavity (Tuna et al. 2013). The streamline near the bottom was found to be deflected by the standing wave. Their phase-averaged analysis provided detailed information about the development of the undulating vortex layer over time. Furthermore, they suggested that the exchange velocity evaluated by the integration of the transverse velocity has a strong correlation with the Reynolds stress and has a peak value at same position as does the Reynolds stress. The mass exchange velocity indicated a 40% enhancement due to the gravity standing wave.

7.5 SEDIMENT TRANSPORT

7.5.1 Introductory Remarks

Flows containing sand particles are much more complex than clear water flows. It has long been known that the friction factor decreases in the sediment-laden flows (e.g., Nomicos 1956), and it is closely related to the turbulence structure. This section deals with incipient motion of the particle, forces exerting on sediment particle and related to critical shear stress, unidirectional bed form, roles of bursting in particle movement and flow/particle interaction.

7.5.2 Particle Movements by Fluid

7.5.2.1 Forces exerted on sediment particles

FIG. 7.22 shows the forces exerted on a particle in the unidirectional shear flow like a straight river. F_D and F_L are the fluid forces parallel and perpendicular to the bottom surface, and called the drag and lift forces, respectively. F_G is the gravity force, F_R is the reaction force from the bottom and F_{vdW} is the van der Waals force sticking to the bottom surface. Further, the friction force F_{fr} acts against the streamwise flow. Whether the sediment moves or not depends on the balance among these forces, particularly the relation between the drag force and the friction force in the horizontal channel.

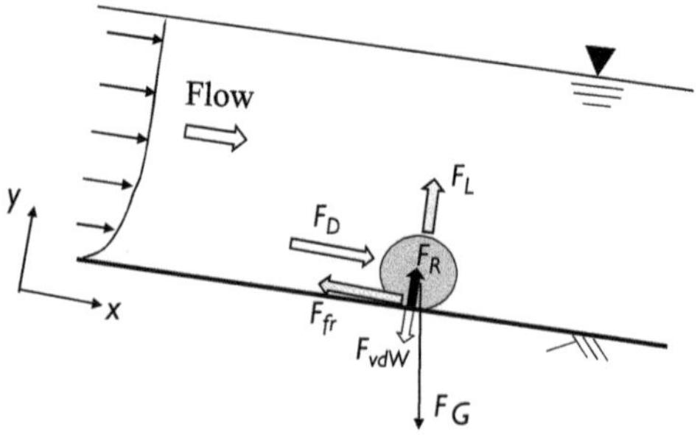

Figure 7.22 Forces exerted on a sediment particle exposed to the unidirectional flow.

The evaluation of the sediment movement has been studied intensively for a long time. Here, the Shields diagram and Hjulstrom curve are explained briefly.

The Shields diagram describes the critical bed shear stress to move the particle situated on the channel bottom. As reviewed by Guo (2020), it is widely used for the bottom stability analysis, bed load estimation and prediction of local erosion, etc. Shields (1936) obtained a relation between the initial sediment motion and the flow strength.

A dimensional analysis leads to

$$\frac{\tau_c}{(\rho_s - \rho)d} = f\left(\frac{U_{*c}d}{\nu}\right) \tag{7.5.1}$$

in which τ_c is the critical bed shear stress to move the particle and ρ and ρ_s are density of the fluid and sediment particle, respectively. d is the particle diameter and ν is the kinetic viscosity. The left-hand side of Eq. (7.5.1) is a non-dimensional parameter that is called a dimensionless critical bed shear stress, τ_{*c}. Substitution of $\tau_c \equiv \rho U_{*c}{}^2$ into Eq. (7.5.1) gives

$$\frac{U_{*c}{}^2}{(s - 1)d} = f\left(\frac{U_{*c}d}{\nu}\right) \tag{7.5.2}$$

in which $s = \rho_s/\rho$. The laboratory tests provide an outline of function in Eq. (7.5.2), as shown in FIG. 7.23(a). The vertical axis is the dimensionless critical bed stress or the critical Shields parameter θ_c. This is the ratio of fluid force to the submerged particle weight. That is to say, it expresses how the drag is excellent against the bottom friction force. The abscissa is the critical grain Reynolds stress. The viscous length is given by ν/U_{*c}. Hence, it means the ratio of the particle diameter to the viscous length. The region over this curve means the occurrence of sediment transport and that under

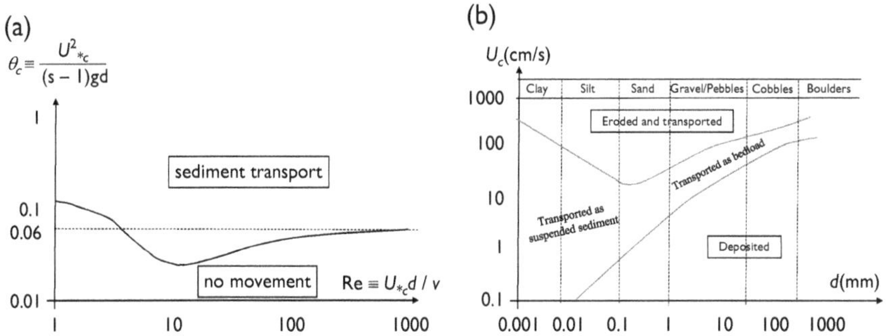

Figure 7.23 Charts of sediment movement: (a) Shields diagram and (b) Hjulstrom curve.

the curve means the deposition or no-motion in the unidirectional flow. The dimensionless bottom friction force is close to the constant in a larger Reynolds number condition.

FIG. 7.23(b) is a sketch of a Hjulstrom curve (Hjulstrom 1935) that illustrates the critical velocity, U_c, corresponding to the bottom erosion or deposition. An upper downward-convex curve means the fluid velocity at which the bottom-situated particle starts to move and erode. In the region where the particle size is small, the smaller the particle size, the larger the crucial flow velocity due to the cohesion. A lower curve means the fluid velocity at which the moving particles in the flow start to be deposited. It is easy to use because it is not necessary to consider the friction velocity. This is a great advantage in practice. The higher velocity can carry the larger particle without deposition. However, the drawback is that it does no take water depth into account. Even at the same velocity, the water depth and the bottom shear stress may change depending on the bed slope.

Neither the Shields diagram nor the Hjulstrom curve is applicable to the situation where particles of various sizes are mixed. Since the small particles are hidden behind the large particles, it becomes difficult for the fluid force to act on the small particle in such a case. Egiazaroff (1965) proposed a formula of the critical bed shear stress over the non-uniform bottom composed of mixed-sized particles as follows. (7.5.3) was derived by the log-law profile of the mean velocity, assuming that the roughness height is equal to a mean size of mixed particles, d_m.

$$\frac{U_{*ci}^2}{(s-1)gd_i} = \frac{0.1}{\{\log_{10} 19(d_i/d_m)\}^2} \tag{7.5.3}$$

in which d_i is a remarked particle size and U_{*ci} is a critical bottom friction velocity initiating the particle movement with a diameter d_i. (7.5.3) suggests that the normalized friction velocity about the particle i decreases with the relative size, d_i/d_m, and it is supported by many experimental results (Egiazaroff 1965).

7.5.2.2 *Particle movement*

There are three ways in which sediment particles are moved by the flow, as shown in FIG. 7.24.

Rolling/Sliding

Water flow causes the particles to roll or slide along the bottom surface. They are carried until the flow becomes too slow to push them, they are stopped by other particle, or they are trapped in a ditch, scoring hole or vegetation stem. Such motions occur depending on the particle characteristics and flow

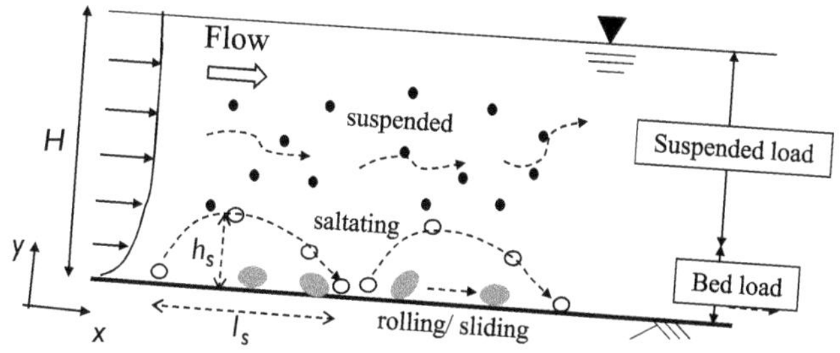

Figure 7.24 Different particle motions in open-channel turbulent flow.

condition. The particle is moved by the fluid force, but the size is comparatively too large to be lifted off the bed surface. Smaller particles tend to be in the following saltation or suspended modes.

Saltation

This is a hopping and bouncing movement of particles along the bed surface. While hopping, they strike other sediment particles or each other, causing them to roll and bounce. The saltation is observed near the bottom and classified to a bed load together with the previously mentioned rolling/ sliding motion. As reviewed by Ali and Dey (2019), the particle saltation depends on the drag force, F_D, and the lift force, F_L. These forces lift the bed particles off the bed surface to a certain elevation. Then, decrease of the lifting force, it returns to the bottom by the gravity force. The lifting force is related to the bursting entrainment, as explained in a following subsection.

Previous experimental studies estimated a saltation height that could be a function of a transport stage $T_* \equiv (\theta/\theta_c) - 1$. Here, θ and θ_c are Shields parameter and critical one, as mentioned. Zeehan and Subhasish concluded the relative saltation height, h_s/d, and saltation length, l_s/d, increase, as the T_* increases. For example, Auel et al. (2017) showed the following experimental formulae: $h_s/d = 0.7T_*^{0.3}$ and $l_s/d = 2.3T_*^{0.8}$.

Suspension

When sediment particles lifted up away from the bottom are supported by the turbulent flow and vortex motion, they are suspended for long-distance traveling. The Stokes equation implies that the terminal settling velocity decreases as the particle size becomes smaller. Further, the turbulent vortex contributing to the particle lift has a finite volume. Hence, the finer sediments are suspended more easily, and they still do not return to the bottom after quite a long time. The concentration flux of the suspended sediment is generally discussed, considering the turbulent diffusion process, as explained in section 6.2.

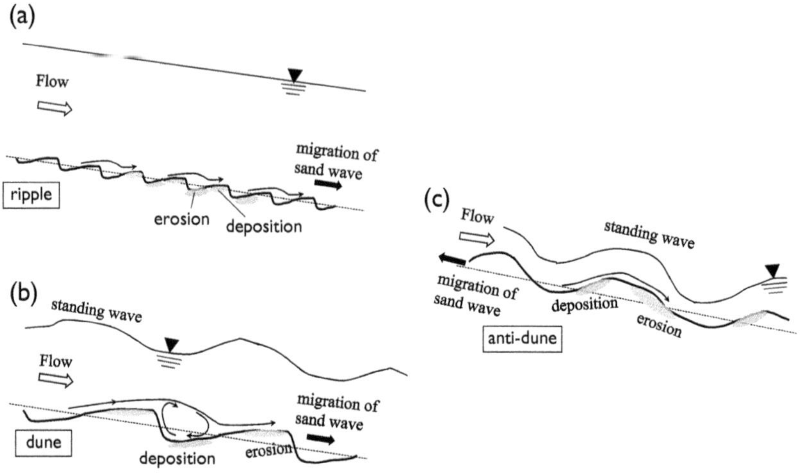

Figure 7.25 Examples of unidirectional bed forms: (a) ripple; (b) dune; (c) anti-dune.

7.5.2.3 Unidirectional bed forms

The bed form in the open-channel flow changes as the Froude number increases, as shown in FIG. 7.25. The hydraulic condition where the Froude number is less than 0.8 is called the lower regime, and we can see a ripple and dune. When the flow velocity increases and the Froude number exceeds 0.8, the bed form becomes flat. Furthermore, when the Froude number exceeds 1.0, the anti-dune occurs. The condition when the Froude number is larger than 0.8 is called the upper regime. Ferraro & Dey (2015) explained the basic mechanism of the bed form in the river flow and typical research so far.

A ripple is the smallest riverbed morphology, depending on the near-bed flow and the riverbed material. It is formed when the particle Reynold number, $\mathrm{Re}_d \equiv \frac{U_* d}{\nu}$, is 20 or less (Yalin 1964). The wavelength, λ, is determined by the particle size. Yalin (1964) proposed the formula, $\lambda = 1000d$. There are various planar shapes, from three-dimensional irregular ones to two-dimensional regular arrangements. Regarding the longitudinal cross-section shape, the slope on the stoss side is gentle and the slope on the lee side is large. The erosion occurs on the stoss side, and the particles are deposited on the lee side. It results in the sand waves moving downstream.

A dune is larger than a ripple, and the wavelength and wave height are closely related to the flow depth. The free-surface standing wave has an opposite phase to the bed sand wave. Similar to the ripple, the slope on the stoss side is gentle, and the slope on the lee side is steep. Also, the sand waves propagate downstream. As is well known, the wavelength of the dune is about five times the flow depth, i.e., $\lambda = 5h$.

An anti-dune is a sand wave that is in phase with the free-surface standing wave. It occurs when the Froude number is greater than 1. The anti-dune is known to migrate upstream. The standing free-surface wave and the sand wave are in phase. However, anti-dunes can be stationary or can move downstream, as observed in some experiments (e.g., Kennedy 1963). Particularly, the conceptual subdivision of the bed forms in the supercritical flow is indicated by Cartigny et al. (2014).

7.5.2.4 Turbulent transport

Transport of bed load

Sumer et al. (2003) studied the relation between the bed load discharge and the turbulence level by control of the external turbulence. They generated three kinds of external turbulence. The first is a horizontal single pipe placed laterally in the flow. The second is a series of grids shorter than the depth, covering a finite length of the channel. The third is a series of grids with varying clearance from the bottom, which cover the entire free surface. The experimental data showed a larger bottom turbulence increases more of the sediment transport. The sediment discharge normalized by the undisturbed one was found to obey a single curve, irrespective of the kinds of external turbulence under the fixed Shield parameter condition.

Transport of suspended sediment

The suspended sediments are transported downstream by the turbulence while constantly receiving the lifting force. On the other hand, the presence of suspended sediment changes the generation and transport of the turbulent energy. In this way, the suspended sediment and turbulence interfere with each other. For low-concentration two-dimensional sediment-laden flow, the following momentum-conservation, concentration and energy transport equations hold (Barenblatt 1996, Nikora & Going 2002). The coordinate system is shown in FIG. 7.22.

$$-\overline{uv} = U_*^2(1 - y/h) \approx U_*^2 \tag{7.5.4}$$

$$\overline{cv} = V_0 C \tag{7.5.5}$$

$$\overline{uv}\frac{dU}{dy}(1 - \mathrm{Ko}) + \varepsilon = 0 \tag{7.5.6}$$

in which h is the flow depth, V_0 is the settling velocity, $\sigma = (\rho_s - \rho)/\rho$, ρ_s is the sediment density, ρ is the fluid density, ε is the turbulent energy dissipation and $\mathrm{Ko} = -\sigma g\overline{cv}/\left(\overline{uv}\frac{dU}{dy}\right)$ is the Kolmogorov number.

Eqs. (7.5.4) to (7.5.6) result in

$$-\overline{uv} = \nu_t\frac{dU}{dy},\quad -\overline{cv} = \nu_s\frac{dC}{dy},\quad \varepsilon = \gamma^4 k^{3/2} l^{-1},\quad \nu_t = lk^{1/2} \text{ and } \nu_s = \mathrm{Sc}^{-1} lk^{1/2} \tag{7.5.7}$$

in which k is the turbulent kinetic energy, $l = \kappa\gamma y$ is the mixing length considering the sediment effect, κ is the von Karman constant and $\gamma \approx 0.5$ is the coefficient, $\mathrm{Sc} = \nu_t/\nu_s$ is the turbulent Schmidt number and ν_t and ν_s are turbulent viscosity and turbulence diffusion coefficient of sediment, respectively. From Eqs. (7.5.4) to (7.5.6), distributions of the turbulent energy, sediment concentration and mean velocity are obtained as follows:

$$k = (U_*/\gamma)^2 (1 - \mathrm{Ko})^{1/2} \tag{7.5.8}$$

$$C(y) = C(y_r)(y/y_r)^{-\omega},\quad \omega = v_0/(\kappa \mathrm{Sc}^{-1} U_*) \tag{7.5.9}$$

$$U/U_* = (1/\kappa)\ln(y/y_0) \tag{7.5.10}$$

in which y_0 is the roughness height. Eq. (7.5.8) leads the turbulent energy near the bottom, $k_{cw} = (U_*/\gamma)^2$, in the clear water condition, i.e., $\mathrm{Ko} = 0$. Therefore, a ratio of turbulent energy in the sediment flow to that in the clear water flow, $k/k_{cw} = (1 - \mathrm{Ko})^2$, depends on the Kolmogorov number, and the turbulence is modulated by the suspended sediment. Eq. (7.5.9) is approximated well to the Rouse equation, which is explained later, as long as y/H is not large.

According to the field measurement with the acoustic Doppler velocimetry (ADV) by Nikora and Going (2002), $\mathrm{Sc}^{-1} = \nu_s/\nu_t$ under a high concentration condition over 100 mg/l is almost 1 in the lower layer and increases toward the free-surface where $y/H > 0.5$, whereas under the low concentration condition, less than 10 mg/l, it is slightly less than 1 in the lower layer and slightly greater than 1 where $y/H > 0.5$.

Rouse equation

The Rouse equation expresses the concentration profile of suspended sediment, as shown in FIG. 7.26. Here, assuming that the concentration decreases toward the free surface, the amount of suspended sediment transported with

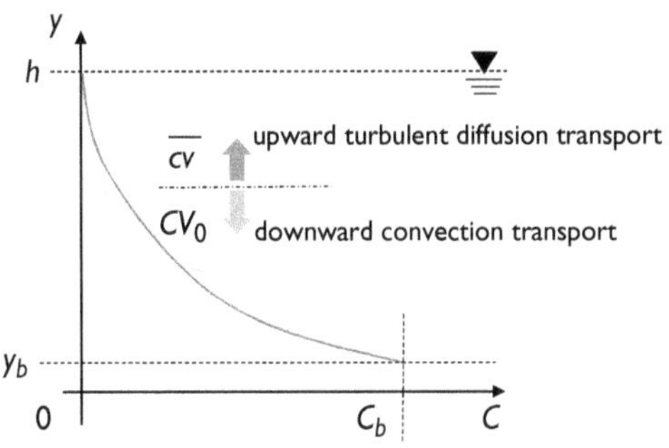

Figure 7.26 Vertical transport of suspended sediment by convection and turbulent diffusion.

upward direction is considered as the turbulent transport flux, $\overline{cv} = -\nu_s \frac{dC}{dy}$. In contrast, the downward sediment transport is attributed to the settling velocity, V_0, and expressed as CV_0.

$$CV_0 = -\nu_s \frac{dC}{dy} \tag{7.5.11}$$

Integrating both sides from $y = y_b$ to y, the following equation is obtained:

$$\ln \frac{C(y)}{C_b} = -\int_{y_b}^{y} \frac{V_0}{\nu_s} dy \tag{7.5.12}$$

in which C_b is the reference concentration at the near-bottom elevation, $y = y_b$.

The Reynolds stress in the two-dimensional uniform flow can be expressed as follows from Eq. (3.5.5):

$$-\overline{uv} = \nu_t \frac{\partial U}{\partial y} \tag{7.5.13}$$

On the other hand, in a region away from the bottom wall where the influence of the viscosity is small, the following relation holds from Eq. (3.7.1):

$$-\overline{uv}/U_*^2 = 1 - y/h \tag{7.5.14}$$

Therefore, using these two equations and Eq. (3.6.4) of the log law, the eddy viscosity can be expressed as follows:

$$\nu_t = U_*^2\left(1 - \frac{y}{h}\right)\left(\frac{\partial U}{\partial y}\right)^{-1} = \kappa U_* y\left(1 - \frac{y}{h}\right) \tag{7.5.15}$$

Substitution of Eq. (7.5.15) into Eq. (7.5.12) with $\nu_s = \mathrm{S_c}^{-1}\nu_t$ yields the following concentration distribution of suspended sediment called the Rouse equation (Rouse 1937):

$$\frac{C(y)}{C_b} = \left(\frac{h-y}{y}\frac{y_b}{h-y_b}\right)^Z, \quad Z = \mathrm{Sc}\,V_0/(\kappa U_*) \tag{7.5.16}$$

7.5.3 Bursting and Sediment Entrainment

The diagrams in FIG. 7.23 show that an increase of the bed shear stress and corresponding mean velocity promote sediment transport significantly. In contrast, the coherent structure of the turbulence flow plays a major role in how the particles at the bottom are incorporated into the flow, i.e., particle entrainment. Experiments and numerical analysis so far have revealed the entrainment mechanism by the bursting.

Sumer & Deigaard (1981) observed three-dimensional behavior of light-density particles by stereo photography, and terminal elevations of upward and downward particles were evaluated, including rising time and bursting duration. They have shown the particles roll up to a height of $y^+ \sim 100$ in the smooth bottom. The rise time of particle normalized by the maximum mean velocity, U_{max}, and the flow depth, h, is $TU_{max}/h = 1 \sim 3$, which is the same order as that of the fluid obtained by Jackson (1976). From these results, it may be concluded that the followability to the bursting of light particles is good.

Further, they examined the influence of bottom roughness, and obtained the following finding. The rough bottom has a shorter rise time than the smooth bottom, but the burst is stronger and the reattachment distance in the streamwise direction is shorter with higher particle rollup. However, basically, a similar trajectory of particles is observed in both the rough and smooth bottom surfaces.

The high-speed fluid descends from the upper layer and spreads in the transverse direction on the smooth bottom surface. The low-speed streak is formed between the adjacent neighboring high-speed zones. The particles deposited there wait for a chance to be lifted up again. On the other hand, on the rough bottom, such a lateral current is attenuated by the drag force of the roughness element, and downward high-speed flow is easily trapped in the gap, so it is difficult to form the streak compared to the smooth bottom.

The particles lifted up from the bottom surface cannot be maintained in the ejection fluid. Gravity pulls relatively heavy particles away from the ejection fluid before the accompanying bursting breaks up. This is a known as the crossing trajectories effect, described later, in which the particle falls out of the coherent vortex before it disappears.

Beginning with the visual study by Kline et al. (1967), the elucidation of bursting or streak structure near the wall has progressed. Robinson (1991) modeled a series of cycles of organized motions composed of low-speed streak, streamwise vortex, pumping effect, ejection and sweeps.

A common understanding has also been gained regarding the relationship between sediment transport and such three-dimensional organized motions. In wall turbulence, the particles are ejected by the low-speed streak and accumulate in the high-speed zone between the adjacent low-speed streaks. In addition, the streamwise vortex transports particles from the high-speed region to the low-speed streak in the transverse direction. But when the particle size is larger than viscous sublayer thickness, the relationship between these organized motions and the particle becomes unclear (Nino & Garcia 1996).

FIG. 7.27 shows a sketch of bottom particle motion governed by the organized turbulent structure (e.g., Nino & Garcia 1996). Particles accumulated in the low-speed streak are rolled up to a certain height by ejection. They fall to the bottom surface by a strewamwise vortex or sweep motion. Particles accumulated in the high-speed region are transported to a low-speed streak due to a cross flow. The streamwise vortex that contributes to the particle transport repeats generation and decay quasi-periodically. The scale of the streamwise vortex tube depends on the presence of the particles. Nino and Garcia reported that the length in streamwise direction is about 500 wall units ($L^+ \sim 500$) without particles. In contrast, the presence of fine particles smaller than a viscous sublayer increases their size by a factor of two or three ($L^+ \sim 1000$). It is known that the fine particles stabilize the coherent structure, while the large particles destabilize one (Rashidi et al. 1990).

Streak structure is known to appear only near the wall. For example, in Klein et al.'s experiment, the upper limit is $y^+ \sim 5$. The low-speed streak is

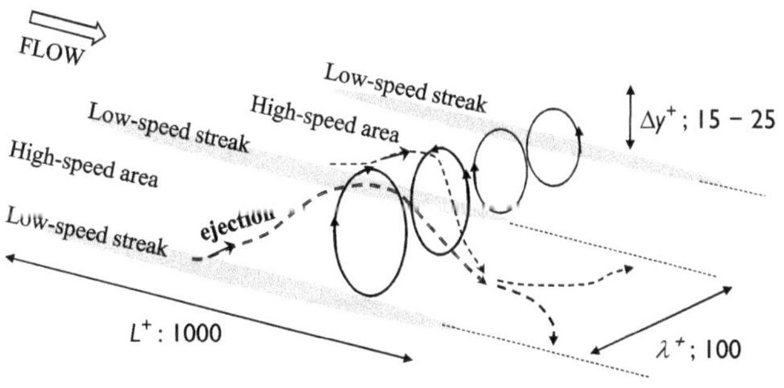

Figure 7.27 Three-dimensional particle motion near the bottom wall related to the low-speed streak and bursting, which is created with reference to Nino & Garcia (1996).

not observed above $y^+ = 10$ in DNS results (Moin & Kim 1982). Smith and Schwartz (1983) reported that the low-speed streak is still observed at a height of $y^+ \sim 5$ or above, but is irregular due to the mixture of intermittent outward motions. Furthermore, the behavior is energetic and has a regular ration in up to the height of $y^+ \sim 25$ or less. This suggests that there is a streamwise vortex involved in the formation of the low-speed streak. The cross-flow induced by the streamwise vortex is well organized in the viscous sublayer, which causes the small particles to be carried in the lateral direction to the low-speed streak. On the contrary, large particles are not so affected by the cross flow.

Kaftori et al. (1995a) considered the organized structure in the bottom as a funnel vortex. They explained that particles are swept by high-speed stream and, when they hit a bottom wall, the particles are pushed out of the way. If the particles are light in weight, they may roll up again due to the ejection event.

Pedinotti et al. (1992) performed DNS using very light particles with a density of 1.03. They investigated the conditions under which sediment particles make a group along the low-speed streaks. In particular, it was shown that particles are most likely to be sorted when the dimension-less particle time scale, $T_{p+} \equiv (\rho_s d^2 U_*^2)/(18\nu^2\rho)$, is about 3. When T_{p+} is greater than or less than 3, the particles are evenly distributed without bias on the bottom surface. In order for the particles to accumulate along the low-speed streak, the particle time scale must be small enough to follow the streak motion caused by the streamwise vortex. Moreover, if the time scale is too small, the motion of the streamwise vortex will have a relatively high frequency, and the particles will not follow it.

The quantitative value of T_{p+} with particle accumulation varies with the researchers. Nino and Garcia suggested $T_{p+} = 0.05$ to 4. There is no segregation when $T_{p+} = 10$, similar to Pedinotti et al.

Now, the bottom roughness also affects the organized turbulence structure. The formation of streaks is not clear on the rough bed, and particles do not group along a specific preferential line (Nino & Garcia 1996). The cross-flow that is expected to be formed between the high-speed regions is wakened by the form drag of roughness element. It is difficult for wall streaks to form in the rough beds (Sumer & Deigaard 1981). On the other hand, Grass et al. (1991) showed that wall streaks also occur in the rough beds. In particular, the larger the roughness size, the greater the span of the streaks. However, the correlation in the streamwise direction becomes smaller.

Nino and Garcia reported that wall streaks may remain on the rough bed where a viscous sublayer collapses; the organization such as coherence, persistence and spatial extent is reduced. The cross-flow is attenuated by the roughness element. It is verified that particles smaller than the roughness scale are trapped in the gap. Larger particles move on the roughness element and no particular accumulation is seen.

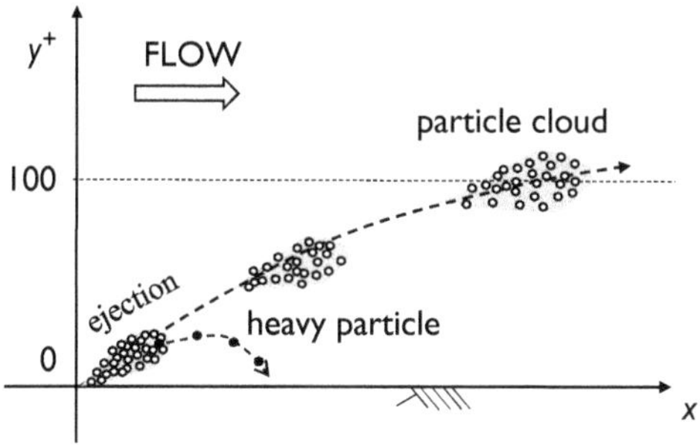

Figure 7.28 Crossing-trajectories' effect sketched with reference to Nino and Garcia (1996).

Typical particle orbital angles by ejections are 10 to 20 degrees (Yung 1989, Nino & Garcia 1996). In the experiment by Nino and Garcia, several particles are rolled up to $y^+ = 100$, having a launch angle of 12 degrees. However, some can only reach a height up to $y^+ = 30$. Hence, the interaction between ejection and particles is not always effective in terms of particle lifting. Some particles are actually captured by the ejection and lifted up, while others are not captured by the ejections and cannot rise (FIG. 7.28). This phenomenon is called the crossing-trajectories effect (Wells & Stock 1983, Zhuang 1989). When the bottom shear stress is large, the ejection is also large and, as a result, the number of particles lifted up by ejections increases. The dimensionless bottom shear stress is less than 0.12, and this number decreases sharply (Nino & Garcia 1996).

7.5.4 Particles and Fluid Interaction

Rashidi et al. (1990) conducted flow visualization of varying particles' diameter, density, concentration and Reynolds number. They reported the particles accumulate in the low-speed streak and low concentration in the high-speed region as well as Sumer & Deigaard (1981), and Nino & Garcia (1996), as shown in FIG. 7.27. Further, they studied stabilization or destabilization of the near-wall turbulence by the particles. Comparatively smaller polystyrene particles with 120 μm reduce the turbulence intensity and Reynolds stress. In contrast, the larger particles with 1,100 μm increase those. This tendency is more remarkable under higher concentration conditions. However, comparatively heavier glass particles do not induce such a turbulence modulation. They concluded that the light particles smaller than the viscous sublayer thickness stabilize turbulence and the larger particles destabilize one. It was also found by their experiments that the quasi-periodical bursting process controls the particles transported including lift up, and then the particles are affected significantly in the turbulence energy transport from the bottom toward the bulk layer.

Several other researches also focused on the modulation of mean velocity and turbulence modulations. Lyn (1991, 1992) measured mainly on the outer layer in the sediment-laden flow, and reported that the velocity profile was shifted downward from the log law for the clear-water open-channel flow. Furthermore, the von Karman constant decreases as the increase of the sediment concentration. In contrast, Cioffi & Gallerano (1991) measured particle velocity and concentration simultaneously and there is no variation in the von Karman constant, regardless of particle concentration. Hence, the influence of sediment particles on the velocity profile may depend on various hydraulic factors including turbulence level.

Best et al. (1997) showed that the turbulence could be enhanced near the flume bottom, depending on the Stokes number of particles. Bennett et al. (1998) reported that the von Karman constant in the sedimen-laden flow on the mobile bed is about $\kappa = 0.33$, which is lower than $\kappa = 0.41$ observed in a no-sediment flow with the fixed bed. In addition, they pointed to the bottom turbulence in the mobile bed is enhanced by the wake induced by low-relief bed waves. Muste and Patel (1997) suggested the occurrence of the relative velocity between the particle and the water, in which mean particle velocity is smaller than that of the fluid in the outer region as well as the results of Best et al. (1991) and Bennett et al. (1998).

In contrast, Kaftori et al. (1995b) and Nezu and Azuma (2005) reported that the particle velocity is larger than the fluid velocity near the bottom of $y^+ \leq 15$. Nezu and Azuma pointed out the turbulence intensity of particle velocity is larger than that of the fluid in the wall region lower than $y^+ = 15$. Furthermore, their quadrant analysis supported sweeps contribute to particle motion more than ejections and, thus, these findings may make the particle velocity exceed compared to the fluid velocity near the bottom wall.

7.6 VERTICAL 2-D DEBRIS FLOW

7.6.1 Introduction

A debris flow is a highly unsteady flow that occurs momentarily when loose ground collapses. A solid-liquid flow containing a large amount of sediment and driftwood has a very large kinetic energy, because it flows through a high head in a short time. Therefore, it causes enormous damage (FIG. 7.29).

Not much is known about the generation, development and deposition processes of the debris flow, which is a complicated high-speed sediment current on a steep slope. This phenomenon causes severe damages and so many researchers have been studying it, as reviewed by Takahashi (2019). Although unpredictable occurrence of debris flow makes it challenging to carry out a field survey, some researches reported the velocity data measured by ultrasonic sensors, e.g., Arattano & Marchi (2005). Also, many experiments have been conducted to reveal hydrodynamic properties and to

Figure 7.29 The state of the mountain surface where the debris flow occurred in Hyogo prefecture, Japan in 2018 (a torrential rain for a short period of time collapsed a 250 m long mountain surface, causing a large amount of sediment and driftwood to flow downstream).

classify such complicated natural events, e.g., due to Parsons et al. (2001), and Chen et al. (2017).

Fundamental dynamics of the debris flow have also been established as explained by Iverson (2012) and Takahashi (2019). However, many parameters are required for physical modeling and, thus, how we decide them is one of the important future issues. This section focuses on vertical two-dimensional debris flow and the velocity profile introducing previous studies.

7.6.2 Dependency on Bottom Slope

It is known that the formation of uniform flow including sediment particles is significantly influenced by the bottom slope, as shown in FIG. 7.30. Sediment particles are individually conveyed downstream near the bottom

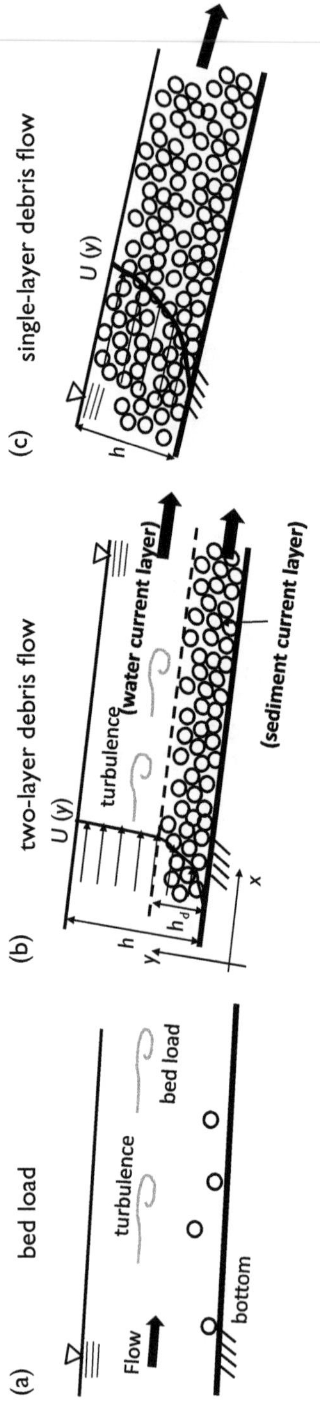

Figure 7.30 Formation of sediment flow depending on bottom slope: (a) bed load; (b) two-layer debris flow; (c) single-layer debris flow.

in the mild bottom slope under 1 to 2 degrees. This is called the bed load. In contrast, a debris flow occurs in the steep bottom slope over 15 degrees, in which the whole depth region is occupied with traveling sediment. In the former case, the turbulent current governs except for near the bottom, and the coherent structure like the bursting contributes significantly to the sediment transport. The latter case includes highly concentrated sediment particles, in which the energy dissipation and production of fluid production are of importance. In the transition of them, we can see two-layer flow consists of lower debris current and upper water current. Particularly, the upper layer includes significant turbulent motions, and it is also important for the accurate prediction in the debris flow to model the turbulence structure properly.

7.6.3 Velocity Profile in the Single-Layer Debris Flow

Egashira et al. (1989) derived a theoretical formula of time-mean velocity profile in the vertical 2-D single-layer debris flow based on the following momentum equation and kinetic energy equation. They dealt a mixed current that consists of fluid and sediment as a single-layer current. Here, the debris flow is assumed to be a laminar-like stream, and so that instantaneous velocity can be replaced by the time-mean value, U_i. The momentum equation could be given by

$$\frac{\partial U_i}{\partial t} + U_j \frac{\partial U_i}{\partial x_j} = -\frac{1}{\rho_m}\frac{\partial P}{\partial x_i} + \frac{1}{\rho_m}\frac{\partial \tau_{ij}}{\partial x_j} + F_i \tag{7.6.1}$$

in which ρ_m is a density of the mixed current, τ_{ij} is the stress tensor and F_i is the external forces exerted on the mixed current. The kinetic energy equation could be given by

$$\frac{\partial K_m}{\partial t} + U_j \frac{\partial K_m}{\partial x_j} = -\frac{\partial P U_j}{\partial x_j} + \frac{\partial \tau_{ij} U_j}{\partial x_j} + \rho_m F_i U_i - \Phi \tag{7.6.2}$$

in which $K_m \equiv \frac{1}{2}\rho_m U_i U_i$ is the kinetic energy of the mixed current and Φ is an energy dissipation rate per second and volume. Eqs. (7.6.1) and (7.6.2) lead to

$$\Phi = \tau_{ij} \frac{\partial U_i}{\partial x_j} \tag{7.6.3}$$

Egashira et al. considered that the energy dissipation consists of three components like

$$\Phi = \Phi_s + \Phi_g + \Phi_f \tag{7.6.4}$$

in which Φ_s is the energy dissipation due to the collision of neighboring sediment particles, Φ_g is the energy consumption for the inelastic collision and Φ_f is the energy consumption used for the turbulence production in the pore water. Although these components are not explained in detail in this book, Egashira et al. suggested the following forms:

$$\Phi_s = P_s \tan\phi_s \frac{\partial U}{\partial z} \tag{7.6.5}$$

in which P_s is a particle skeletal stress as mentioned later and ϕ_s is a frictional angle of the sediment particle:

$$\Phi_g = k_g(1 - e^2)\sigma d^2 C^{1/3}\left(\frac{\partial U}{\partial y}\right)^3 \tag{7.6.6}$$

in which $k_g = \frac{\pi}{12}\left(\frac{\pi}{6}\right)^{-1/3}(\sin^2 a_i)$, d is a sediment diameter, e is an elastic modulus, σ is a density of sediment particle and a_i is an impact angle.

$$\Phi_f = \rho k_f d^2 \frac{(1 - C)^{5/3}}{C^{2/3}}\left(\frac{\partial U}{\partial y}\right)^3 \tag{7.6.7}$$

in which $U = U_1$, $y = x_2$ and C is a sediment concentration.

They result in

$$\Phi = \Phi_s + \Phi_g + \Phi_f = P_s \tan\phi_s \frac{\partial U}{\partial z} + k_g(1 - e^2)\sigma d^2 C^{1/3}\left(\frac{\partial U}{\partial y}\right)^3 + \rho k_f d^2 \frac{(1 - C)^{5/3}}{C^{2/3}}\left(\frac{\partial U}{\partial y}\right)^3 \tag{7.6.8}$$

Focusing on the shear stress in the vertical $x - y$ plane, $\tau = \tau_{12}$, a following relation is obtained:

$$\Phi \sim \tau_{12} \frac{\partial U}{\partial y} \tag{7.6.9}$$

Therefore, the shear stress is expressed by

$$\tau = P_s \tan \phi_s + k_g (1 - e^2)\sigma d^2 C^{1/3} \left(\frac{\partial U}{\partial y} \right)^2 + \rho k_f d^2 \frac{(1 - C)^{5/3}}{C^{2/3}} \left(\frac{\partial U}{\partial y} \right)^2 \tag{7.6.10}$$

Applying a uniform flow condition to Eq. (7.6.1), the force balance relations for x and y directions are obtained as follows:

$$g \sin \theta - \frac{1}{\rho_m} \frac{\partial \tau}{\partial y} = 0 \tag{7.6.11}$$

$$g \cos \theta + \frac{1}{\rho_m} \frac{\partial P}{\partial y} = 0 \tag{7.6.12}$$

in which $\rho_m = \rho\{(\sigma/\rho - 1)C + 1\}$.

Egashira et al. explained the pressure in the debris flow consists of three parts, as follows:

$$P = P_w + P_s + P_d \tag{7.6.13}$$

in which P_w is a pressure of the pore water, P_s is a skeletal stress of sediment particles and P_d is a dynamic pressure due to particle collision. Assuming the static water pressure distribution, a following form is obtained:

$$\frac{\partial P_w}{\partial y} = -\rho g \cos \theta \tag{7.6.14}$$

Further, Egashira et al. proposed the following formulae about P_d and P_s:

$$P_d = k_g e^2 \sigma d^2 C^{1/3} \left(\frac{\partial U}{\partial y} \right)^2 \tag{7.6.15}$$

$$P_d = \alpha P_s \tag{7.6.16}$$

in which α is a constant coefficient.

Substitution of Eqs. (7.6.13), (7.6.14), (7.6.15) and (7.6.16) into Eq. (7.6.12) gives the following equations:

$$\frac{\partial P_s}{\partial y} = -\rho \frac{1}{1+\alpha}\left(\frac{\sigma}{\rho} - 1\right) Cg \cos\theta \tag{7.6.17}$$

$$\frac{\partial P_d}{\partial y} = -\rho \frac{\alpha}{1+\alpha}\left(\frac{\sigma}{\rho} - 1\right) Cg \cos\theta \tag{7.6.18}$$

Substitution of Eq. (7.6.10) into an integral of Eq. (7.6.11) from $y = y$ to h gives

$$\left\{\sigma k_g (1 - e^2) d^2 C^{1/3} + \rho k_f d^2 \frac{(1-C)^{5/3}}{C^{2/3}}\right\}\left(\frac{\partial U}{\partial y}\right)^2$$
$$= \int_y^h \rho \left\{\left(\frac{\sigma}{\rho} - 1\right) C + 1\right\} g \sin\theta dy - P_s \tan\phi_s \tag{7.6.19}$$

Removing P_s with Eq. (7.6.17), the following equation is obtained:

$$\left\{k_g \frac{\sigma}{\rho}(1 - e^2) d^2 C^{1/3} + k_f d^2 \frac{(1-C)^{5/3}}{C^{2/3}}\right\}\left(\frac{\partial U}{\partial y}\right)^2$$
$$= \int_y^h \left[\left\{\left(\frac{\sigma}{\rho} - 1\right) C + 1\right\} g \sin\theta - \left\{\frac{1}{1+\alpha}\left(\frac{\sigma}{\rho} - 1\right) Cg \cos\theta \tan\phi_s\right\}\right] dy \tag{7.6.20}$$

We can calculate the mean velocity profile using Eq. (7.6.20). First, velocity and coordinates are normalized as follows: $U_o = U/\sqrt{gh}$ and $y' = y/h$. Eq. (7.6.20) is rewritten by

$$\frac{\partial U_o}{\partial y'} = \frac{h}{d}\left\{\frac{1}{f_f + f_g}\int_{y'}^{1} (G - Y) dyz\right\}^{1/2} \tag{7.6.21}$$

in which $f_f = k_f \frac{(1-C)^{5/3}}{C^{2/3}}$, $f_g = k_g \frac{\sigma}{\rho}(1 - e^2) C^{1/3}$, $G = \{(\sigma/\rho - 1) C + 1\} \sin\theta$ and $Y = \frac{1}{1+\alpha}(\sigma/\rho - 1) C \cos\theta \tan\phi_s$. When C is constant, a following solution is obtained:

$$U_o = \frac{2}{3}\frac{h}{d}\left\{\frac{G - Y}{f_f + f_g}\right\}^{1/2} \{1 - (1 - y')^{3/2}\} \tag{7.6.22}$$

Chen et al. (2017) conducted flume experiments using machine oil and white oil, in which diluted debris flow was generated with a high viscosity. They measured the mean velocity profile by using a PIV technique in several hydraulic conditions, varying sediment diameter and bottom slope. Here, they focused on single-layer debris flow, where a sediment concentration is almost constant in the whole depth. Also, they proposed an experimental formula of velocity in the debris flow based on that of the sediment-laden flow, and it was compared with the experimental data.

Wang el al. (2001) proposed the following velocity profile in the sediment-laden flow:

$$U^+\left(=\frac{U}{U_*}\right)=\frac{1}{\kappa}\ln\frac{y}{k_s}+B,\ \left(\mathrm{Re}_*=\frac{U_*d}{\nu}>70\right) \tag{7.6.23}$$

in which κ is the von Karmant constant, k_s is a roughness height, B is an integral constant and d is a sediment diameter. Assuming the appearance of the maximum velocity, $U_{\max}$, at the elevation of $y = h$:

$$\frac{U_{\max}}{U_*}=\frac{1}{\kappa}\ln\frac{h}{k_s}+B \tag{7.6.24}$$

These two equations result in

$$\frac{U_{\max}-U}{U_*}=\frac{1}{\kappa}\ln\frac{h}{y} \tag{7.6.25}$$

It is expected in the debris flow that collisions of neighboring particles and interaction of fluid and particles make the flow resistance and energy consumption larger compared to the sediment-laden flow. Hence, Chen et al. suggested the following form of the mean velocity profile in the debris flow using correction factors α and β:

$$\frac{U_{\max}-U}{U_*}=\alpha\left(\frac{h}{y}\right)^{\beta}\frac{1}{\kappa}\ln\frac{h}{y} \tag{7.6.26}$$

The experiment by Chen et al. obtained the best fitting combination, i.e., $\alpha = 0.92$ and $\beta = -1/6$. FIG. 7.31 shows the measured velocity profile in which theoretical equations are also plotted for the composition. It is found that the experimental results are closer to Eq. (7.6.26), corrected for the debris flow rather than Eq. (7.6.25) for the sediment-laden flow.

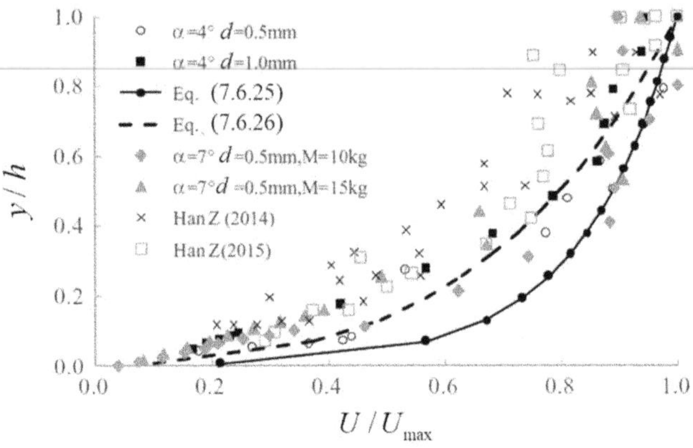

Figure 7.31 Mean velocity profile in the vertical two-dimensional debris flow: This figure is drawn based on FIG. 9 in Chen et al. (2017).

7.6.4 Velocity Profile in the Two-Layer Debris Flow

In transition, the bottom slope from the bed load to the single-layer debris flow, a two-layer current is formed that consists of a lower sediment-current layer and an upper turbulent water-current layer. Some previous works derived a theoretical formula by an integral of force balance equation with a steady flow assumption (Takahashi 1982, Hashimoto & Hirano 1996).

7.6.4.1 Velocity profile in sediment current layer

Assuming a steady flow, the streamwise components of the gravity force and the friction force are balanced. Focusing on the debris current layer of $y < h_d$, the force balance equations with the x and y directions are given by

$$x:\ s_{yx}(y) + \tau_w(y) = \int_{h_d}^{y} \rho g \sin\theta dy + \int_{y}^{h_d} \{\sigma C + \rho(1-C)\} g \sin\theta dy \quad (7.6.27a)$$

$$y:\ s_{yy}(y) - p = -\int_{y}^{h_d} (\sigma - \rho) C g \cos\theta dy \quad (7.6.27b)$$

in which $\frac{s_{xy}}{s_{yy}} = -\alpha$, $\alpha = \frac{\gamma}{1 + \rho/(2\sigma)}$, $\gamma = \frac{0.0762 + 0.102\mu}{0.0898 - 0.067\mu}$ and $\mu = 0.1$ (Tsubaki et al. 1982). τ_w is a turbulence shear stress of the pore water and s_{yx} and s_{yy} are collision stresses of neighboring sediment particles in the x and y directions, respectively.

Tsubaki et al. (1982) expressed τ_w in the following way, correcting the mixing length model:

$$\tau_w(y) = -\rho(1 - C_d)\overline{u_w w_w} = \rho(1 - C_d)l^2\left(\frac{dU}{dy}\right)^2 \tag{7.6.28}$$

in which l is the mixing length and $\overline{u_w w_w} = l^2\left(\frac{dU}{dy}\right)^2$ is the Reynolds stress of the pore water. C_d is a depth-averaged sediment concentration.

The collision stress is expressed by Egashira et al. (1989) using a function of C_d:

$$s_{yx} = \kappa_{yx}\sigma d^2 F(C_d)\left(\frac{dU}{dy}\right)^2 \tag{7.6.29}$$

in which $\kappa_{yx} = \frac{\pi}{6}(0.076 + 0.102\mu)\beta^2 k_M$ and $F(C_d) = \frac{(C_d/C_*)^2}{1 - C_d/C_*}$, $\beta = 1.15$, $k_M = 5$, $\mu = 0.1$. C_* is a closest packing concentration. Substitution of Eqs. (7.6.28) and (7.6.29) into (7.6.27a) gives

$$\begin{aligned}&(\kappa_{yx}\sigma d^2 F(C_d) + \rho(1 - C_d)l^2)\left(\frac{dU}{dy}\right)^2 \\ &\quad = \{\rho h + \sigma C_d h_d - \rho C_d h_d + (\rho C_d - \sigma C_d - \rho)y\}g \sin\theta\end{aligned} \tag{7.6.30}$$

The mixing length is assumed to be proportional to the height of the sediment current layer, i.e., $l = \alpha_l h_d$, in which $\alpha_l = 0.1$ is an experimental constant. Hence, it results in

$$\left(\frac{dU}{dy}\right)^2 = \frac{(\rho h + \sigma C_d h_d - \rho C_d h_d)g \sin\theta}{A} + \frac{(\rho C_d - \sigma C_d - \rho)g \sin\theta}{A}y \tag{7.6.31}$$

in which $A = \kappa_{yx}\sigma d^2 F(C_d) + \rho(1 - C_d)l^2$. The mathematical solution is as follows:

$$U = \frac{2}{3}\frac{A}{(\rho C_d - \sigma C_d - \rho)g\sin\theta}\left\{\frac{(\rho h + \sigma C_d h_d - \sigma C_d h_d)g\sin\theta}{A} + \frac{(\rho C_d - \sigma C_d - \rho)g\sin\theta}{A}y\right\}^{3/2} + D$$

$$= \frac{2}{3}\frac{A}{(\rho C_d - \sigma C_d - \rho)g\sin\theta}\left(\frac{g\sin\theta}{A}\right)^{3/2}$$

$$\{(\rho h + \sigma C_d h_d - \sigma C_d h_d) + (\rho C_d - \sigma C_d - \rho)y\}^{3/2} + D$$

$$= \frac{2}{3}\frac{A^{-1/2}(g\sin\theta)^{1/2}}{\rho C_d - \sigma C_d - \rho}\{(\rho h + (\sigma - \rho)C_d h_d) + (-(\sigma - \rho)C_d - \rho)y\}^{3/2} + D$$

$$\leftrightarrow \frac{U}{U_*} = \frac{2}{3}\frac{A^{-1/2}}{\rho h^{1/2}\left\{\left(\frac{\rho-\sigma}{\rho}\right)C_d - 1\right\}}\left\{\rho h\left(1 + \frac{(\sigma - \rho)C_d h_d}{\rho h}\right) - \rho h\frac{(\sigma - \rho)C_d + \rho}{\rho}\frac{y}{h}\right\}^{3/2} + D \tag{7.6.32}$$

in which D is an integral constant. Introducing normalized factors, $\eta = \frac{y}{h}$, $\eta_d = \frac{h_d}{h}$, $s = \frac{\sigma - \rho}{\rho}$, Eq. (7.6.33) is obtained:

$$U^+ = \frac{U}{U_*} = \frac{2}{3}\frac{A^{-1/2}}{\rho h^{1/2}(-sC_d - 1)}\{\rho h(1 + sC_d\eta_d) - \rho h(sC_d + 1)\eta\}^{3/2} + D' \tag{7.6.33}$$

Calculating a constant D' using a boundary condition, $U = 0$ at $\eta = 0$, the following normalized velocity profile is derived:

$$U^+ = \frac{2}{3}\frac{\rho^{1/2}A^{-1/2}}{sC_d + 1}\{(1 - sC_d\eta_d)^{3/2} + (-1 - sC_d\eta_d + (sC_d + 1)\eta)^{3/2}\} \tag{7.6.34}$$

7.6.4.2 Velocity profile in water-current layer

In the water-current layer of $y > h_d$, there are no sediment particles and, thus, $s_{yx} = 0$ unlike the sediment-current layer. Hence, the friction force is only the turbulence shear stress that is expressed by the mixing length model:

$$\tau_w(y) = \rho l^2\left(\frac{dU}{dy}\right)^2 \tag{7.6.35}$$

The gravity force exerted on a control volume with height from $y = y$ to h is given by

$$\int_y^h \frac{\rho g}{\sin\theta}dy = (h - y)\rho g\sin\theta \tag{7.6.36}$$

Assuming zero shear stress in the free surface, the force balance equation is as follows:

$$\rho l^2 \left(\frac{dU}{dy} \right)^2 = \rho g \sin \theta (h - y) \tag{7.6.37}$$

The mixing length in this layer is assumed to be proportional to the distance from the wall, like the typical boundary layer, i.e., $l = \kappa (y - h_d) + \alpha_l h_d = (\alpha_l - \kappa) h_d + \kappa y$. Here, we define $\alpha = (\alpha_l - \kappa) h_d$. Further, the bottom shear stress τ_b is given by

$$\tau_b \equiv \rho U_*^2 = \rho g h \sin \theta \tag{7.6.38}$$

Therefore, the following relation is obtained:

$$l^2 \left(\frac{dU}{dy} \right)^2 = U_*^2 \left(1 - \frac{y}{h} \right) \leftrightarrow \frac{dU^+}{dy} = \frac{1}{\alpha + \kappa y} \left(1 - \frac{y}{h} \right)^{1/2} \tag{7.6.39}$$

Introducing a non-dimensional factor, $\psi = \left(1 - \frac{y}{h} \right)^{1/2}$, $\alpha + \kappa y = \alpha + \kappa h (1 - \psi^2)$. Hence, Eq. (7.6.39) is rewritten by

$$\frac{dU^+}{dy} = \frac{1}{\alpha + \kappa h (1 - \psi^2)} \psi = \frac{1}{(\alpha + \kappa h) - \kappa h \psi^2} \psi \tag{7.6.40}$$

Also, $dy = -2h\psi d\psi$ is obtained. Defining $A = \alpha + \kappa y$, $B = \kappa h$ results in

$$\frac{dU^+}{dy} = \frac{1}{A - B\psi^2} \psi = \frac{\psi / B}{A/B - \psi^2} \leftrightarrow dU^+ = \frac{D\psi^2}{\psi^2 - C} d\psi \tag{7.6.41}$$

in which $C = \frac{A}{B}$, $D = \frac{2h}{B}$. An integral of Eq. (7.6.41) results in

$$\begin{aligned} \int dU^+ &= D \int \frac{\psi^2}{\psi^2 - C} d\psi \\ \leftrightarrow U^+ &= -\{(1 - \psi - \sqrt{C}) \ln (\psi + \sqrt{C})^{1/2} + (1 - \psi + \sqrt{C}) \ln (\psi - \sqrt{C})^{1/2} \\ &\quad + \psi\} \frac{2h}{B} + \Pi \end{aligned} \tag{7.6.42}$$

in which Π is an integral constant. When the value of Π is given that makes a smooth connection of Eqs.(7.6.34) and (7.6.42) at $y = h_d$, the velocity profile in the whole depth could be calculated in the two-layer debris flow.

7.6.4.3 Depth-averaged sediment concentration

The sediment concentration is an important parameter that controls current frictions. Hashimoto and Hirano (1996) derived a theoretical formula of the depth-averaged concentration C_d in the following. They suggested force balance equations with x and y directions, respectively.

$$\begin{aligned} x:\ s_{yx0} + \tau_{w0} &= (h - h_d)\rho g \sin\theta + \int_0^{h_d} \{(\sigma - \rho)Cg + \rho g \sin\theta\} dy \\ &= \rho g h \sin\theta + (\sigma - \rho) g h_d C_d \\ \leftrightarrow s_{yx0} + \tau_{w0} &= \rho U_*^2 + (\sigma - \rho) g h_d C_d \end{aligned} \tag{7.6.43}$$

$$y:\ s_{yy0} = p_0 - (\sigma - \rho) g \cos\theta h_d C_d \tag{7.6.44}$$

in which $C_d = \int_0^{h_d} C dy / h_d$ and the subscript 0 means the bottom, i.e., $y = 0$.

Hashimoto and Hirano suggested the relation of s_{yx0} and s_{yy0} as follows:

$$s_{yx0} - \alpha s_{yy0} = -\alpha p_0 + \alpha(\sigma - \rho) g \cos\theta h_d C_d \tag{7.6.45}$$

Substitution of Eqs. (7.6.44) and (7.6.45) into Eq. (7.6.43) gives

$$\begin{aligned} &-\alpha p_0 + \alpha(\sigma - \rho) g \cos\theta h_d C_d + \tau_{w0} = \rho U_*^2 + (\sigma - \rho) g h_d C_d \\ &\leftrightarrow (\alpha \cos\theta - 1)\frac{(\sigma - \rho) g h_d C_d}{\rho U_*^2} = 1 - \frac{\tau_{w0} + \alpha p_0}{\rho U_*^2} \\ &\leftrightarrow (\alpha \cos\theta - 1)\frac{s\eta_d C_d}{\sin\theta} = 1 - \frac{\tau_{w0} + \alpha p_0}{\rho U_*^2} \leftrightarrow C_d = \frac{\tan\theta}{s\eta_d(\alpha - \tan\theta)}\left(1 - \frac{\tau_{w0} + \alpha p_0}{\rho U_*^2}\right) \end{aligned} \tag{7.6.46}$$

7.6.5 End remarks

The present section focuses mainly on the prediction of velocity profile in vertical two-dimensional debris flow. As mentioned previously, the debris flow is very complicated and many parameters are needed. They must be given by laboratory experiments and there are many issues that should be solved, especially a reliable evaluation or modeling of turbulence, e.g., Reynolds stress, is one of the urgent issues. Furthermore, it is hard to measure the debris flow. The velocity measurement is limited to the free surface in field surveys and near the side glass walls in laboratory experiments. It is expected in future advanced experiments to conduct flow visualization within the debris flow using transparent sediment particles.

7.7 UNSTEADY TURBULENT OPEN CHANNEL

7.7.1 Introduction

It is one of the most important topics in hydraulic engineering to investigate velocity profiles and turbulence characteristics in depth-varying and time-dependent unsteady open-channel flows, i.e., rivers in flood. They consist of the rising stage and falling stage. Generally speaking, the discharge $Q(t)$ in a flood does not show a single curve against the flow depth $h(t)$. A peak discharge is known to appear before a time of peak depth. This is called loop property, and time-dependent depth profiles could be predicted reasonably by one-dimensional unsteady equation of motion. Nezu et al. (1997) revealed turbulence intensities in depth-varying unsteady open-channel flows became larger in rising (accelerating) stages than in falling (decelerating) ones.

In contrast, unsteady turbulence properties have been more intensively studied in the closed channel like the oscillatory boundary layer than the unsteady open channel. Jensen et al. (1989) and Sana & Tanaka (1995) conducted experiments and numerical calculations of zero-mean oscillatory flow.

Also, there is a lot of knowledge about non-zero mean unsteady wall-bounded flows e.g., Michel et al., (1981). Tu & Ramaprian (1983) measured pipe flows with non-zero means by using a single component LDA, and suggested that the mean velocity might not obey the log law of the von Karman constant. Brereton et al. (1995) conducted velocity measurements, and reported that the time-averaged velocity profiles obeyed the log law of $\kappa = 0.41$. Tardu et al. (1994) also verified that the time-averaged velocity profiles obeyed the standard log law. Brereton & Mankbadi (1995) concluded that the turbulence structure normalized by the inner variables is not affected significantly by the moderate unsteadiness, and it results in the ensemble-averaged velocity profiles obey still the log law of $\kappa = 0.41$ in all stages of oscillations.

7.7.2 Two-Dimensional Unsteady Open-Channel Flow

Two-dimensional unsteady flow has been studied to reveal the unsteadiness effects of velocity profile, turbulence structure and bottom shear stress. FIG. 7.32 shows parameters controlling an unsteady open channel. h_b and h_p are base and peak water depths, respectively. T_d is a flood

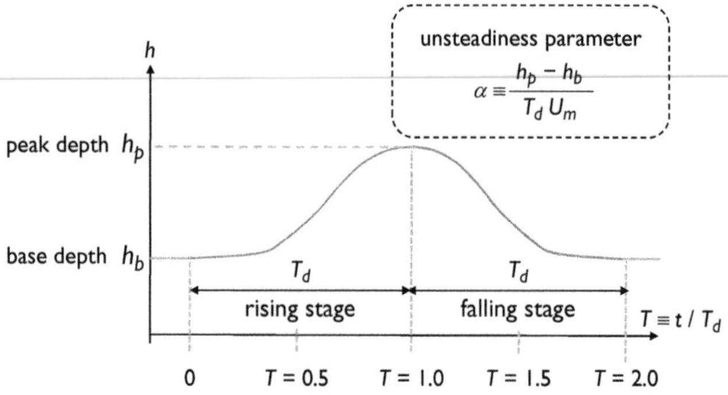

Figure 7.32 Idealized depth curve in unsteady open-channel flow.

duration time from the base flow to the peak flow. α is an unsteadiness parameter that is defined by

$$\alpha \equiv \frac{V_s}{U_m} = \frac{h_p - h_b}{T_d U_m} \qquad (7.7.1)$$

V_s is the rising velocity of water surface in a flood, and is approximated by $V_s = (h_p - h_b)/T_d$. U_m is the bulk mean velocity during the flood, and is approximated by $(U_{mb} + U_{mp})/2$. The subscripts b and p mean the base and peak flows, respectively. $T \equiv t/T_d$ is the normalized time, as explained later.

7.7.2.1 Velocity profiles in viscous sublayer

The friction velocity, U_*, is usually evaluated from the following log law in the inner layer, i.e., $y/h \le 0.2$, as explained in section 3.6.3.

$$U/U_* = \frac{1}{\kappa}\ln(yU_*/v) + A \quad \text{for} \quad y^+(\equiv yU_*/v) > 30 \qquad (7.7.2)$$

in which ensemble-averaged mean velocity, U. In steady open-channel flows, this log-law method is very useful because the von Karman constant is the universal value of $\kappa = 0.412$, as pointed out by Nezu & Rodi (1986). However, it is not guaranteed whether the value of κ remains 0.41 even at high unsteadiness without detailed examination. Hence, U_* must be evaluated by the following viscous sublayer profile without any model constants:

$$U/U_* = yU_*/v \quad \leftrightarrow \quad U_* = \sqrt{Uv/y} \qquad (7.7.3)$$

FIG. 7.33 shows distributions of $U^+ \equiv U/U_*$ in the viscous sublayer, in which both calculated data (Nezu & Sanjou 2006, numerical simulation

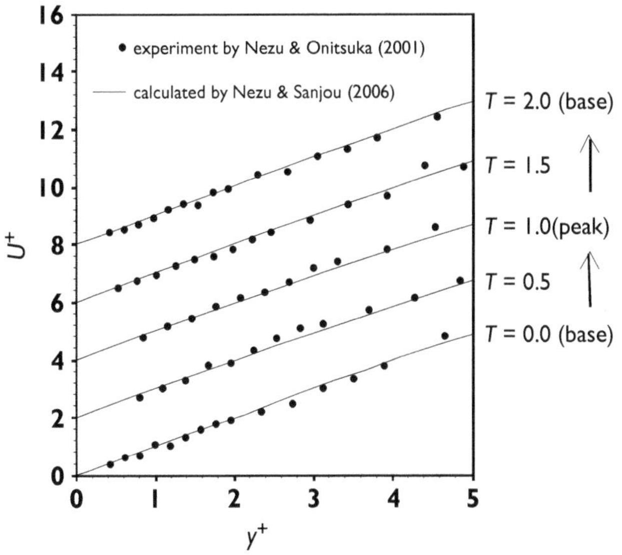

Figure 7.33 Normalized velocity profiles in viscous sublayer at each phase: This is drawn referring to Nezu & Sanjou (2006).

with a low Reynolds number $k - \varepsilon$ model) and measured data (Nezu & Onitsuka 2001, experiment with a laser Doppler anemometer) are included. U_* was evaluated from Eq. (7.7.3).

The hydraulic case has the unsteadiness parameter of $\alpha = 6.3 \times 10^{-3}$ with symmetrical flood wave. The normalized time $T \equiv t/T_d$ is indicated as a phase time in Fig. 7.32, and the figure at each phase is shifted upward in order to avoid confusion. $T = 0.0$ and 2.0 are the base-flow phases, whereas $T = 1.0$ is the peak-flow phase. $T = 0.5$ and 1.5 are the typical phases of the rising and falling stages, respectively. Of particular significance is that both of calculation and experiment support Eq. (7.7.3) holds even in the unsteady open-channel flow. Consequently, the friction velocity, U_*, can be evaluated accurately from the corresponding velocity profiles at each phase time T.

7.7.2.2 Time variations of velocity and flow depth

FIG. 7.34 shows some examples of the time variations of ensemble-averaged U in the unsteady open-channel flow with $\alpha = 3.4 \times 10^{-3}$. These are normalized by the time-averaged velocity, $\overline{U}$, which is defined as

$$\overline{U} = \frac{1}{T_{all}} \int_0^{T_{all}} U(t)\,dt = \frac{1}{2} \int_0^2 U(T)\,dT \qquad (7.7.4)$$

where $T_{all} = 1/f = 2T_d$ is the period of the oscillations in unsteady flows.

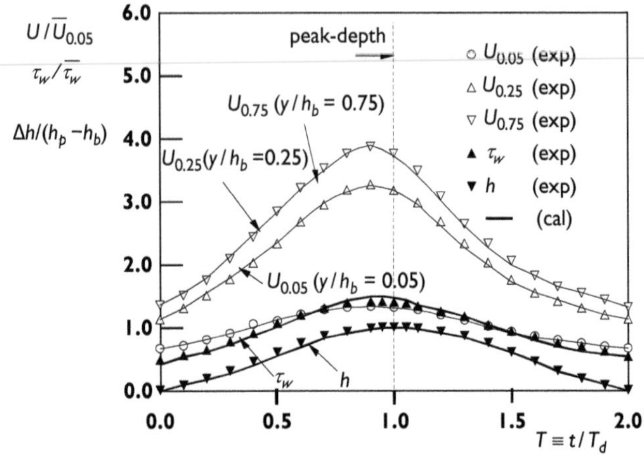

Figure 7.34 Time variations of velocity, bottom shear stress and water depth: This is drawn referring to Nezu & Sanjou (2006)

Three kinds of U are chosen at different elevations, i.e., near the wall, $(y/y_b = 0.05)$, in the middle $(y/y_b = 0.25)$, and near the free surface $(y/y_b = 0.75)$ for comparison. The time variation of the water surface $\Delta h/(h_p - h_b)$, in which $\Delta h(T) \equiv h(T) - h_b$ is the increment from the base depth, is also included in FIG. 7.34. The hydrograph of the flow depth h vs. time T is almost symmetrical in respect to the peak-depth phase, i.e., $T =$ 1.0. In contrast, the peak of velocity appears before the peak-depth phase, irrespective of the elevation y/h_b. This time lag between the velocity and flow-depth peaks is very important to consider a loop property of the discharge, Q, vs. flow depth, h. Although the velocity variation amplitude near the free surface $(y/y_b = 0.75)$ is much larger than that near the wall $(y/y_b = 0.05)$, the time lag is not observed significantly. This is quite different from unsteady wall-bounded flows with zero means, such as oscillatory boundary layers.

7.7.2.3 Bottom shear stress against time

FIG. 7.34 includes the time variation of bottom shear stress, $\tau_w \equiv \rho U_*^2$. They are normalized by the time-averaged value, $\bar{\tau}_w$. This implies that the bottom shear stress increases with time and attains the maximum at about $T = 0.9$ in the rising stage. This maximum value, $\tau_{w,\max}$, attains about three times as large as the base-flow value, $\tau_{w,b}$, at $T = 0.0$. It is expected from this fact that turbulence and suspended sediment transport increase significantly in rising flood rivers.

7.7.2.4 Log law in the unsteady flow

FIG. 7.35 show sthe normalized velocity profile in the whole-depth region in the case of $\alpha = 3.4 \times 10^{-3}$. By assuming that the wake strength parameter Π might be reduced to zero at low Reynolds numbers, Nezu & Onitsuka (2001) have evaluated the value of κ from LDA data in the inner and outer layers ($30 \le y^+ \le R_*$), and pointed out that κ changes slightly with time at a much higher unsteadiness. However, this assumption of $\Pi = 0$ is not guaranteed, even at low Reynolds numbers when the unsteadiness becomes high. The unsteadiness effect is combined with the pressure-gradient effect in unsteady open-channel flow. Then, it is well known that Π is a function of the pressure-gradient parameter, as pointed out by Song & Graf (1994) and Nezu et al. (2001). This suggests strongly that Π changes with time in unsteady flows even at low Reynolds numbers when the unsteadiness is high. Therefore, Nezu & Sanjou (2006) concluded that the von Karman constant remains unchanged, even at high unsteadiness, is much more

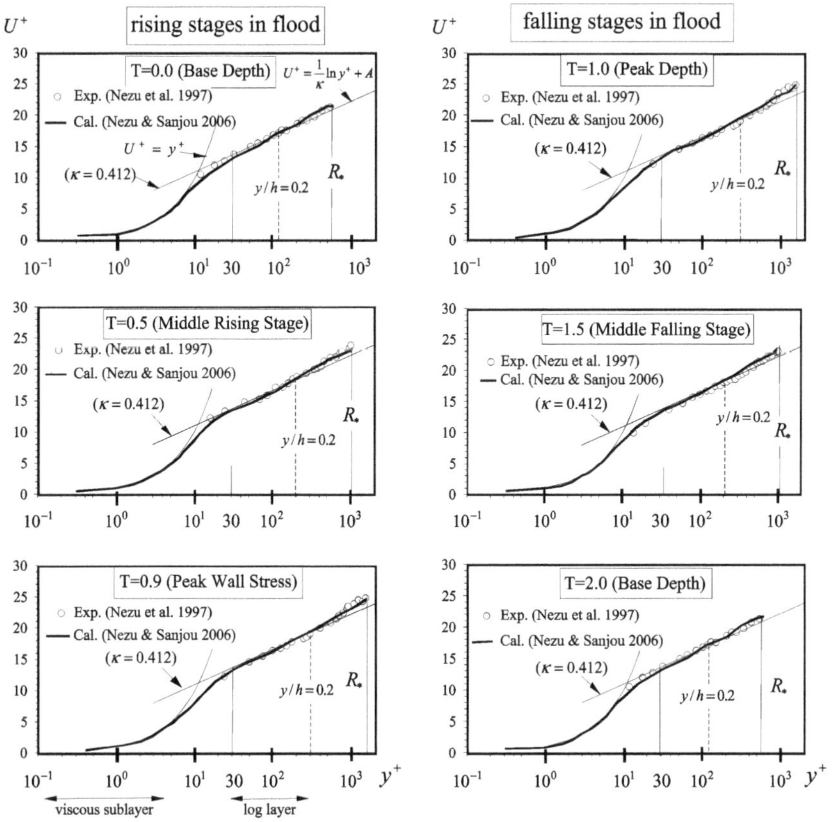

Figure 7.35 Normalized velocity profiles in whole-depth layer at each phase: This is drawn referring to Nezu & Sanjou (2006).

reasonable than the assumption that the value of Π remains zero at a low Reynolds number, because $\kappa = 0.41$ is satisfied in time-averaged velocity profiles at much higher unsteadiness, as mentioned previously. Therefore, the log law of $\kappa = 0.412$ is shown by a straight line at each phase T. The Π-value increases from about 0.2 in the rising stage and attains the maximum at $T = 0.9$. At this time, $T = T_{max} = 0.9$, the bottom shear stress also attained the maximum as discussed before. Then, Π decreases with time in the falling stage (Nezu & Sanjou 2006).

7.7.2.5 Unsteadiness effect on turbulent energy

Nezu & Sanjou (2006) suggested that the value of k/U_*^2 is not significantly influenced by the phase T at the low unsteadiness. Hence, the profile of the normalized turbulent kinetic energy is described well by the same universal function as used in steady open-channel flow by Nezu & Nakagawa (1993). In contrast, a phase effect appears significantly in the near-wall region except for the viscous sublayer at a high unsteadiness. The value of k/U_*^2 increases in the rising stage and decreases in the falling stage. Because U_* increases in the rising stage, the absolute value of k itself increases much more significantly in the rising stage. FIG. 7.36 shows the time variations of k/U_*^2 at some positions y^+ near the wall. As mentioned before, k/U_*^2 for the low unsteadiness remains almost constant in all phases at each position. In contrast, k/U_*^2 for the high unsteadiness changes with the phase T. Of particular significance is that the turbulent energy increases most significantly in the rising stage of $T = 0.5$.

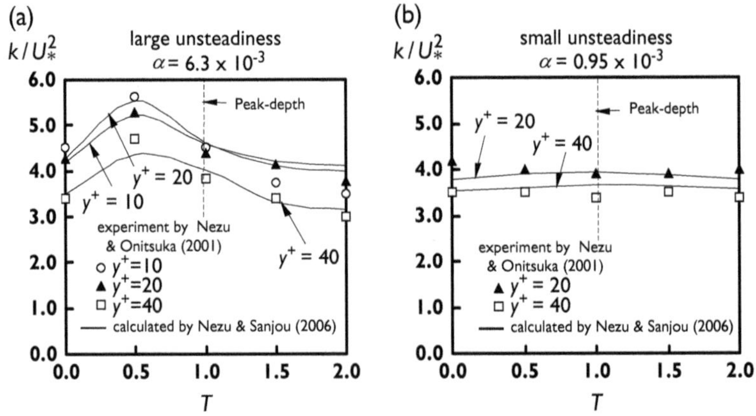

Figure 7.36 Time variation of normalized turbulent kinetic energy: This is drawn referring to Nezu & Sanjou (2006).

7.7.3 Unsteady Compound Open-Channel Flow

Unsteady turbulence flow is also observed in a compound open-channel during flood stage, as shown in FIG. 7.37(a). Sanjou & Nezu (2003) considered unsteady properties of cross-sectional distribution of secondary currents and primary velocity. FIG. 7.37(b) shows a time variation of spanwise profile of bottom shear stress. They are normalized by the spanwise-averaged value at the base flow, $\overline{\tau}_{w,b}$. Sanjou & Nezu (2003) evaluated approximately the friction velocity related to the bottom shear stress from the log law. In all time stages, the bottom shear stress decreases locally near the junction edge, i.e., $z/B = 0.5$. This feature is the most remarkable at the peak depth, $T = 1.0$. With an increase of the water depth, the bottom shear stress on the floodplain becomes larger rapidly, and is comparable with one on the main channel at $T = 1.0$. This implies that deeper relative water depth weakens the hydrodynamic characteristics of the compound open-channel flow.

FIG. 7.38(a) shows the time variation of the secondary currents, $(V, \ W)$, normalized by the maximum mean velocity. In all stages, there are typical secondary currents observed in steady compound open-channel flows, as mentioned in section 7.3. The direction of the velocity vectors that run from the junction edge toward the free surface of the main channel becomes almost vertical in the rising stage ($T = 0.5$). FIG. 7.38(b) is the time variations of the vector angle, θ, in the secondary flows from the junction edge toward the free surface for the cases HH60 ($\alpha = 1.80 \times 10^{-3}$) and HH120

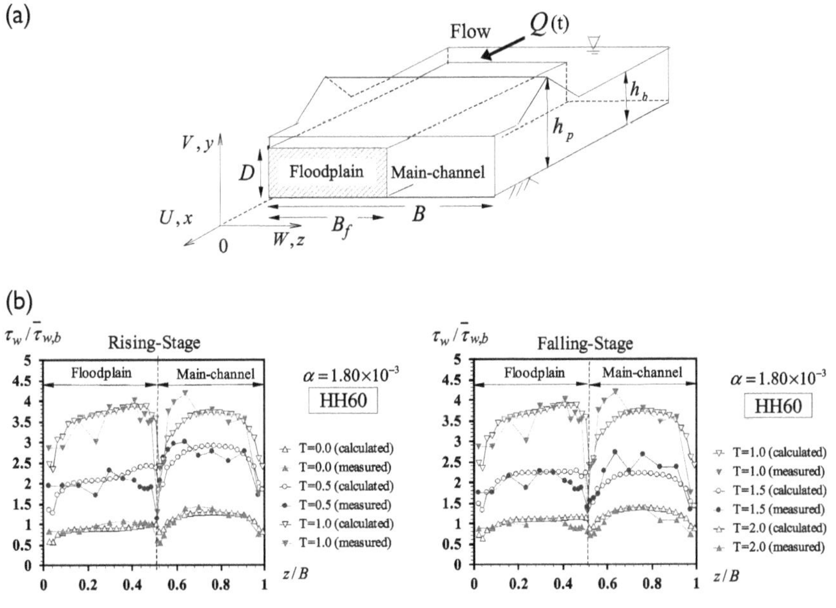

Figure 7.37 Unsteady bottom shear stress: (a) coordinate system and (b) spanwise profiles: This is drawn referring to Sanjou & Nezu (2003).

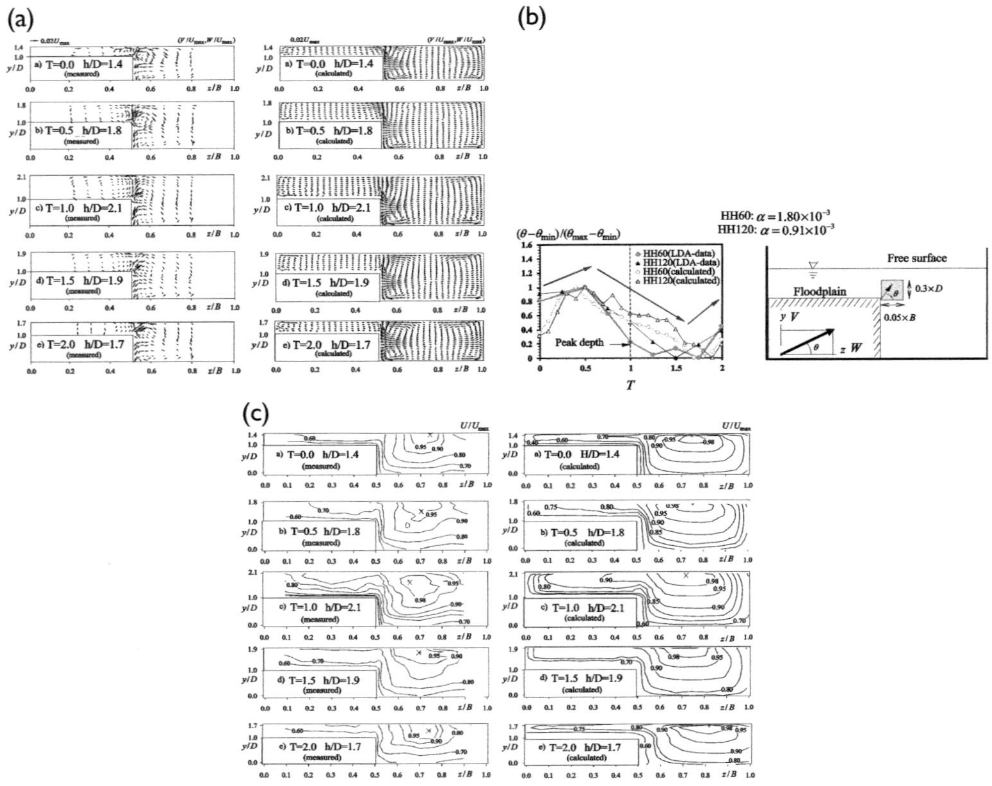

Figure 7.38 Crosse-sectional distributions of velocities in a depth-varying compound open channel: (a) distribution of velocity vectors; (b) vector angle near the junction edge; (c) corresponding cross-sectional contour of primary velocity.

($\alpha = 0.91 \times 10^{-3}$). θ is a space-averaged value in the finite region indicated on right side of FIG. 7.38(b). It is found that the value increases in the rising stage and decreases in the falling stage. This tendency is more remarkable as the unsteadiness is larger in both the measured and computed results.

Fig. 7.38(c) shows a distribution of primary velocity normalized by the maximum velocity. The isovel lines of U, the bulge patterns from near the junction toward the free surface, are the same manner as those in steady compound channels. Furthermore, it is considered that these bulge features are associated with the secondary currents in the same mechanism, as discussed by Naot et al. (1993). At the base depth time ($T = 0$), the velocity of the main channel is much larger than that over the floodplain. As the floodplain depth increases with time, the differences of velocities between the main channel and floodplain become smaller.

7.4.4 Inundation on a Floodplain

It is also important to consider an unsteady channel transition from the low-depth single channel to the high-depth compound channel. An actual inundation flood accompanies such a channel transition. Nezu & Sanjou (2004) showed the time variation of cross-sectional distribution of secondary currents and primary velocity.

FIG. 7.39 shows the time variations of friction velocity, U_*, at the floodplain center ($z/B = 0.25$) and main-channel one ($z/B = 0.75$). The result is

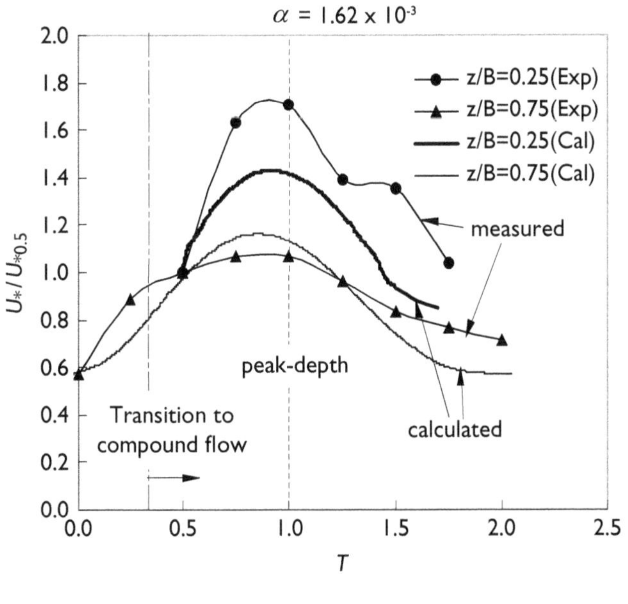

Figure 7.39 Time variation of friction velocity with channel transition: This is drawn referring to Nezu & Sanjou (2004).

normalized by the rising-stage value at $U_{*,0.75}$. Nezu & Sanjou (2004) evaluated the friction velocity by weight average of log law and linear formula in the viscous sublayer. It is found that U_* increases and decreases with the time variation of water depth, although there are some discrepancies with the experimental data. Of particular significance is that U_* on a floodplain increases rapidly in the rising stage after the transition time.

FIG. 7.40(a) shows the time-dependent velocity vector descriptions of the secondary currents. At the rising stage, (T = 0.5), there are one-way flows from the junction toward the floodplain. At the peak depth stage, (T = 1.0), there are typical secondary currents that are also observed in steady compound channel flows. In the falling stage, (T = 1.5), contrary to the rising

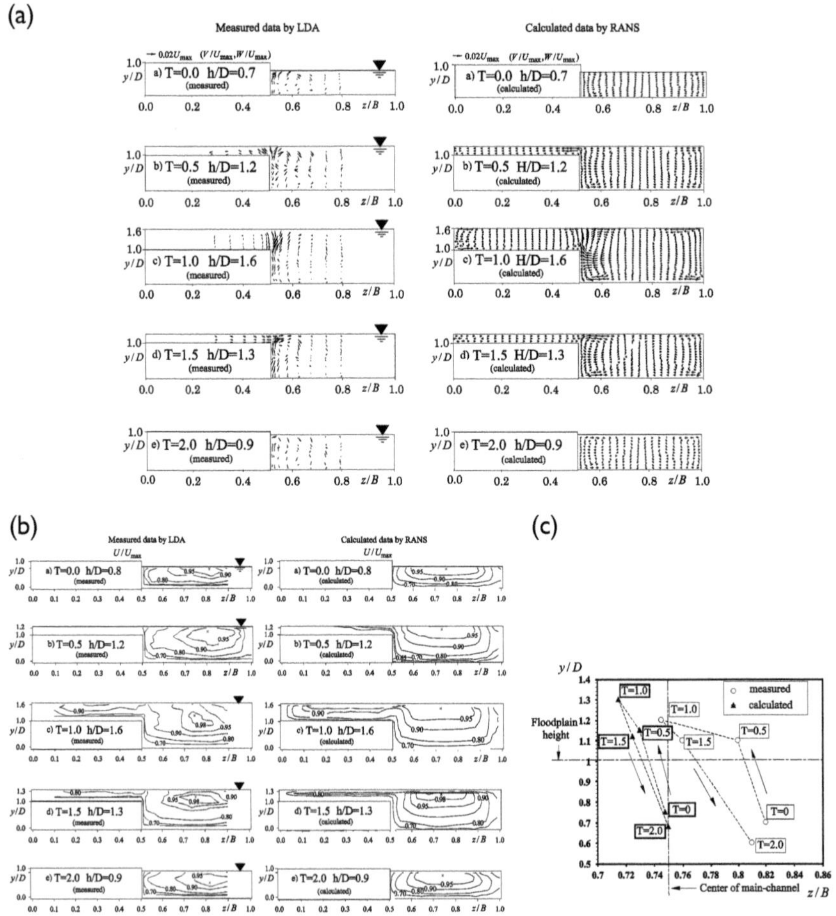

Figure 7.40 Time-dependent distributions of velocities in inundation flow with channel transition: (a) secondary currents; (b) primary velocity; (c) space–time variations of position of maximum primary velocity: This is drawn referring to Nezu & Sanjou (2004).

stage, one-way flows from the floodplain toward the junction are observed in both the calculated and experimental results. FIG. 7.40(b) shows the time-dependent isovel lines of primary velocity normalized by U_{max}. In the floodplain stages, (T = 0.5, 1.0, 1.5), secondary currents from the junction toward the main-channel surface have a strong influence upon the primary velocity; that is to say, the distributions of bulge from the junction to the main channel. This hydrodynamic feature is the most remarkable at the peak depth stage (T = 1.0). In all time stages, the bulk maximum velocity, U_{max}, appears under the free surface in both of the present calculation and measurement results. FIG. 7.40(c) shows the time-dependent spacing positions, (z, y), of U_{max}. Of particular significance in this figure is that the U_{max} position in the cross section varies in not only the vertical direction but also spanwise one. In the rising stages, U_{max} position rises up toward free surface and becomes close to the floodplain side. On the other hand, in the falling stage, U_{max} position falls down and moves to the main-channel side.

7.8 ACCELERATED FLOW

7.8.1 Introduction

Bottom deformation due to sedimentation and scouring induces local flow acceleration and deceleration, and they influence turbulence structure. Especially, a contraction flow is observed over the dune, in which water depth reduces with streamwise direction. Here, a primary current is accelerated in space with streamwise gradient of the flow velocity.

Generally speaking, relarminarization, which suppresses turbulence generation, occurs in the accelerated flow. Many studies has revealed the relarinarization by wind tunnel experiment and numerical simulation in aerodynamics, e.g., Narayanan & Ramjee (1969), Escudier et al. (1998), Warnack & Fernholz (1998), Ranjan & Narasimha (2017).

Note that laminarization highlighted here does not mean retransition from turbulence to laminar but reduction of turbulence normalized by a characteristic velocity scale. That is to say, this book defines the relaminarization as the phenomenon in which normalized turbulence intensity or turbulence kinetic energy decrease compared to values expected.

In contrast, velocity profile and bottom friction have been studied experimentally in the open-channel hydraulics. Particularly, pressure gradient parameter and its formulation were focused, e.g., Song & Grapf (1994), Onitsuka & Nezu (2000). Such a special velocity change and related turbulence modulation have a great impact on bottom shear stress and sediment transport in rivers. The author's group has conducted velocity measurement in two kinds of open-channel accelerated flows, as shown in FIG. 7.41. One is a linear slope case, which is comparable with previous wind tunnel experiments. Another is a continuous dunes case that is observed in rivers. In the following sections, velocity profile, mechanism of relaminarization and phenomenological models are introduced.

7.8.2 Distributions of Velocity and Reynolds Stress

The authors carried out flume experiments by using the PIV to reveal a turbulence structure in monotonically accelerated flow and in flow over continuous dunes.

FIGS. 7.41(a) and (b) show the distributions of time-averaged streamwise velocity and Reynolds stress, in which they are normalized by a bulk-mean velocity at the starting section of flow acceleration, $U_m(x/L = 0)$. $U_m(0)$ is 10 cm/s for both cases. In the linear slope case, the bulk-mean velocity increases with the streamwise direction due to the reduction of water depth. The logarithmic velocity profile is formed at $x/L = -0.16$ without flow acceleration, and the Reynolds stress decreases linearly toward the free surface, corresponding to the typical open-channel turbulence. The flow is accelerated in $x/L = 0$ to 1.0 over the slope due to the contraction effect. It is noted that a negative Reynolds stress is observed in $x/L = 0.5$ to 1.0. This is related to suppression of bottom-oriented turbulence production, discussed later. In the flat section after the flow acceleration, $x/L = 1.0$ to 2.0, the streamwise velocity is decelerated near the bottom and accelerated near the free surface. Hence, the velocity profile returns to that observed in the boundary layer, i.e., logarithmic profile. Also, by increasing the absolute value of Reynolds stress, turbulence redevelops with the streamwise traveling.

In the dune case, although a recirculation zone accompanied with flow separation is locally formed behind a crest, the bulk-mean flow is accelerated with streamwise direction as well as the slope case. In contrast, the separation shear layer that appeared behind the crest promotes turbulence production, and it results in a larger Reynolds stress in $y/H = 0.5$ at $x/L = 0.5$.

7.8.3 Pressure Gradient Parameter

Many researchers have been studying the accelerated boundary layer. Particularly, Narayanan & Ramjee (1969) are one of the pioneer works that measured turbulence accelerated on a wedge in the wind tunnel. They revealed that a mean velocity profile obeys the log law in the flat boundary layer and streamwise variation of turbulence intensity normalized by a characteristic velocity scale.

A proper scaling analysis gives some universal properties without dependency on Reynolds number in developed flat-bottom boundary layer or open channel. A strengthen of flow acceleration is generally evaluated by following a pressure gradient parameter K. U_∞ is the time-mean velocity of the free stream over the boundary layer in the wind tunnel. In contrast, U_s is the time-mean velocity in the open-channel $\nu$is the kinetic viscosity.

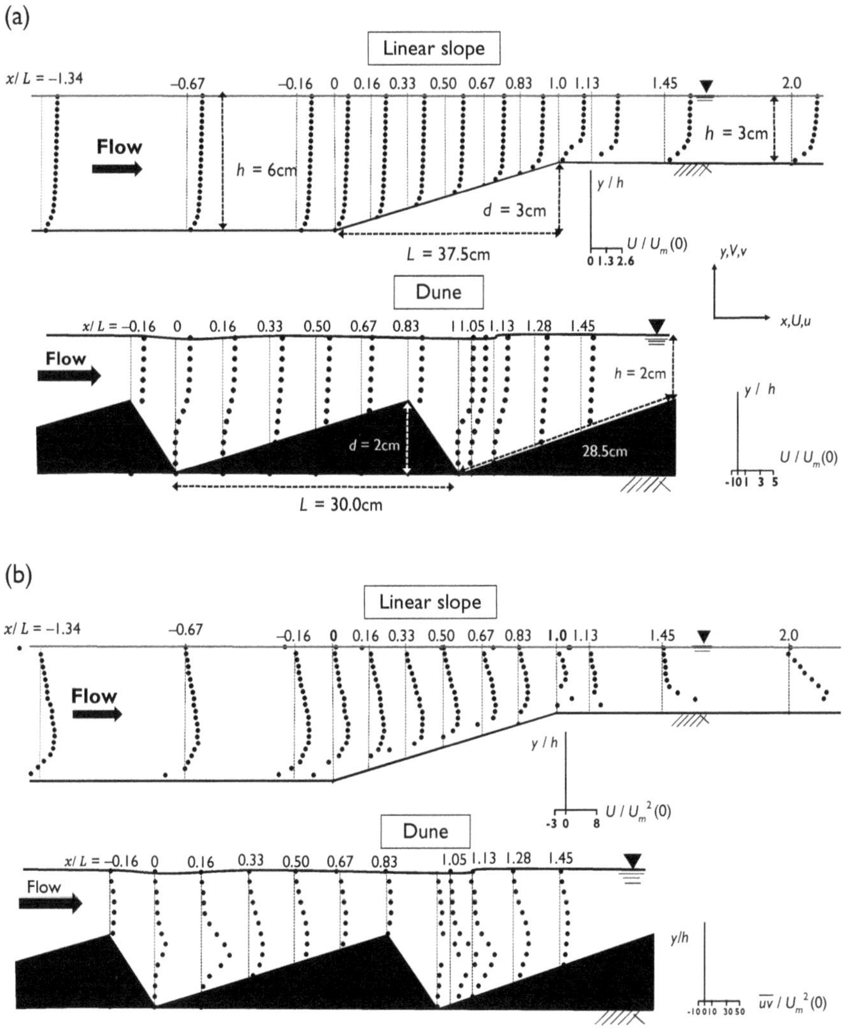

Figure 7.41 Measured flow acceleration over the linear slope and the continuous dunes: (a) spatial distributions of time-averaged velocity; (b) spatial distribution of Reynolds stress.

$$K \equiv \nu / U_\infty^2 \frac{dU_\infty}{dx} \sim \nu / U_s^2 \frac{dU_s}{dx} \tag{7.8.1}$$

FIG. 7.42 shows the streamwise variation of the pressure gradient parameter in the linear slope cases with different slope lengths, i.e., L = 24, 37.5 and 75 cm. Previous results of the wind tunnels by Warnack and Fernholz (WF), Escudier et al. (Es) are plotted for comparison. The accelerated open channel is found to have a peak value in the acceleration

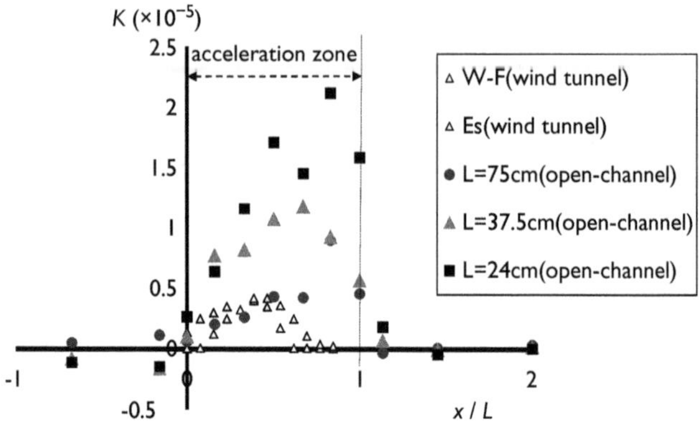

Figure 7.42 Streamwise variation of pressure gradient parameter.

zone as well as the accelerated wind tunnel. Also, the acceleration strength evaluated by this parameter is the same order as the wind tunnel.

7.8.4 Mechanism of Relaminarization

FIG. 7.43 shows a streamwise variation of surface turbulent intensity normalized by mean surface velocity, u'_s/U_s. In the slope cases, normalized turbulent intensity reduces in the acceleration zone of $0 < x/L < 1$. It should be noted that as acceleration becomes larger, this means a shorter slope length, L; this trend is more remarkable. Such a local reduction of normalized turbulence is defined as relarminarization here. In the dune case, the normalized turbulence increases in the upstream part of the acceleration zone, i.e., $0 < x/L < 0.5$. This may be because that turbulence generated by

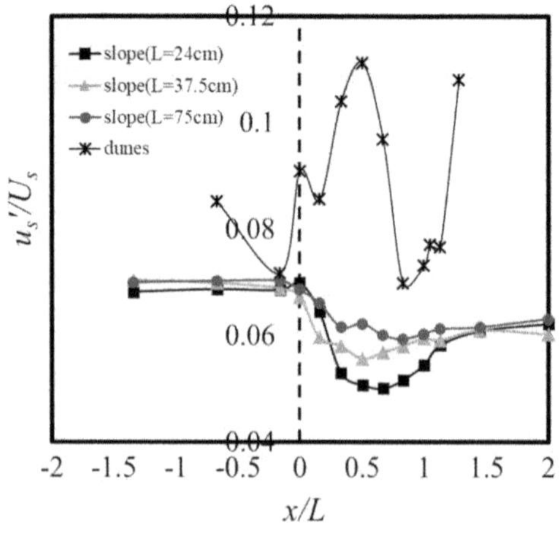

Figure 7.43 Streamwise variation of surface turbulent intensity normalized by mean surface velocity, u'_s/U_s.

the flow separation layer formed behind the crest is transported toward the free surface. In the downstream part, $0.5 < x/L < 1$, the normalized turbulence decreases significantly. The upward transport of bottom-oriented turbulence is suppressed by the flow acceleration accompanied with relaminarization as follows.

Let's consider the mechanism of relaminarization based on production of the turbulent kinetic energy in the acceleration zone. A generation term in the two-dimensional transport equation of the turbulent kinetic energy is expressed by

$$G \equiv -\overline{uu}\frac{\partial U}{\partial x} - \overline{uv}\frac{\partial U}{\partial y} \tag{7.8.2}$$

Although the first term can be removed in the uniform flow, we need to consider both the first and second terms in Eq. 7.8.2 in the accelerated flows. This is because the streamwise gradient of mean velocity occurs. Here, the first and second terms are expressed by G_{uu} and G_{uv}, respectively. FIG. 7.44 shows the measured vertical profiles of G_{uu} and G_{uv} at $x/L = 0.5$ for the slope and dune cases. G_{uv} is positive in the whole-depth region as well as the flat-bottom open-channel flow. It is noted that negative G_{uv} is distributed for both cases. Especially, this tendency is more remarkable in the slope case. It reduces the production of turbulent kinetic energy in the accelerated flows, and so the relaminarization occurs as shown in FIG. 7.43.

7.8.5 Phenomenological Flow Model

FIG. 7.45 shows phenomenological flow models proposed by experimental results. There is not a significant velocity shear $\frac{\partial U}{\partial y}$ in the free surface so that a large turbulence generation is not observed compared to the bottom

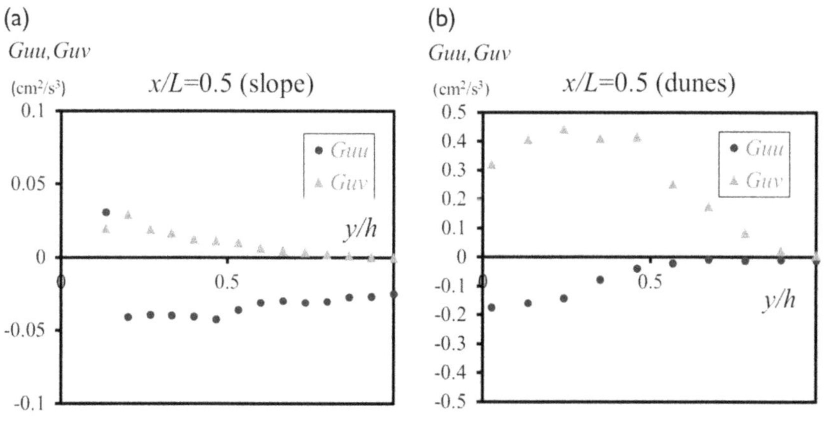

Figure 7.44 Profiles of turbulence production terms for the slope and dune cases.

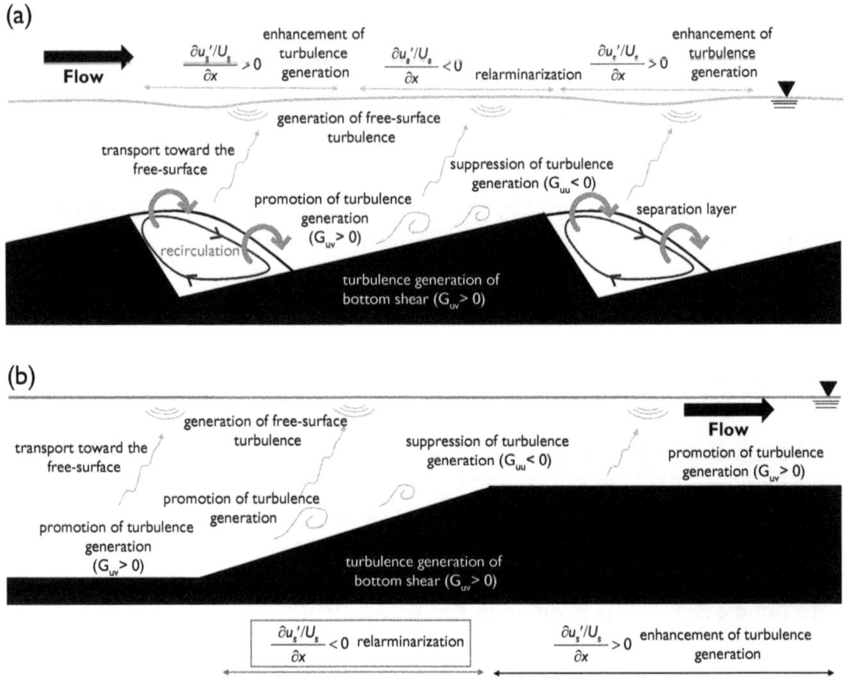

Figure 7.45 Phenomenological flow models: (a) slope case; (b) dune case.

region. However, the bottom-oriented turbulence is conveyed toward the free surface, and it can be detected by the velocimetry in the free surface. In the slope case, the generation of bottom-oriented turbulence is suppressed by the flow acceleration, as explained by the previous section. The results in the free-surface turbulence reduce in the acceleration zone. In the dune case, the turbulence source is not only the flume bottom but also the separation layer behind the crest. Hence, the upstream part of the acceleration zone enhances the free-surface turbulence. In contrast, further downstream of the separation layer, the flow acceleration reduces the free-surface turbulence. Particularly, the increase and decrease of the turbulence production are repeated over the continuous dunes.

7.9 ENERGY DISSIPATION

7.9.1 Introductory Remarks

High-speed water currents with high energy can sometimes cause great damage to rivers and river structures. How to calm such an enormous energy is important in river management. This section focuses on the energy dissipations in the spillway of the dam and in the flood protection forest. As we can understand from the energy transport equation of the mean current expressed by Eq. (3.7.9), the turbulent energy is supplied from the mean current; that is, the turbulence is generated and the energy of mean current is consumed. This characteristic is used in the practical energy dissipation.

7.9.2 Energy Dissipation in a Spillway

7.9.2.1 *Still basin*

Since the large amount or water stored in the dam has great potential energy, the energy discharged from the dam is also very large. If it is flowed downstream; as it is, there is a huge risk that the river channel, power plant, houses, roads and bridges downstream of the dam will be seriously damaged. There is also the danger of scouring the riverbed and the accompanying tilt of the dam site. The energy dissipater is a structure built most downstream of the spillway in order to prevent these risks. There are several types of energy dissipater, and it is selected according to the dam type, geology and topographical conditions of the site.

The hydraulic jump type is the most common, which reduces the energy forcing the hydraulic jump in the horizontal still basin connecting to the spillway. The ski-jump type releases water flow into the air from the guide channel, and the energy is reduced by the diffusion of the discharged jet flow. The free-fall type reduces energy by the diffusion of falling water. However, both the ski-jump type and free-fall type also utilize the energy reduction due to the hydraulic jump.

In the commonly used hydraulic type, a still basin is built downstream of the dam, which reduces an incoming high-speed flow. To increase the water level in the still basin, an endsill or a secondary dam are often installed at the downstream end, as shown in FIG. 7.46. The high-speed supercritical flow

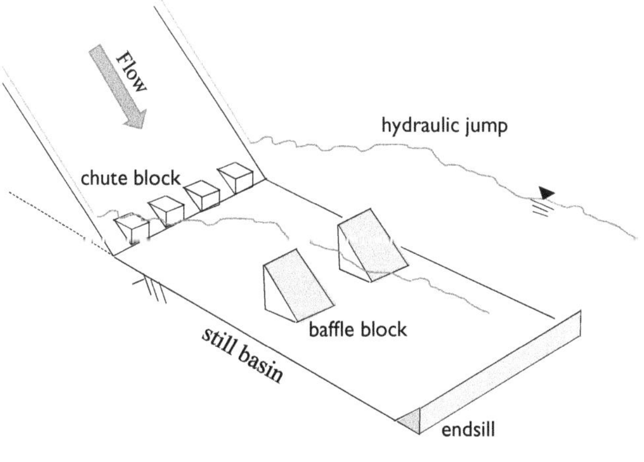

Figure 7.46 Still basin in spillway of dam inducing hydraulic jump for energy dissipation.

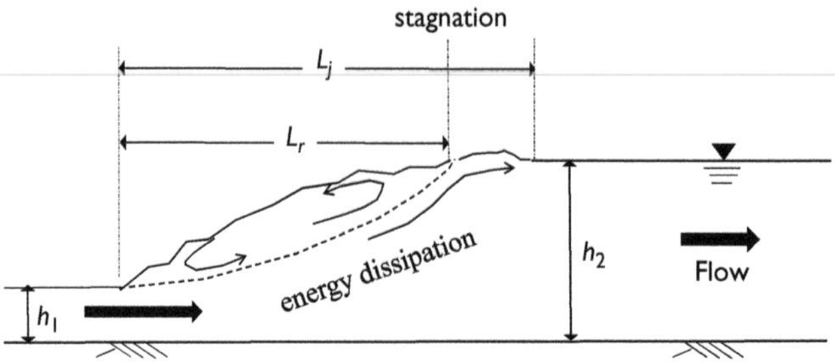

Figure 7.47 Side view of steady hydraulic jump.

comes from the upstream spillway in the still basin. On the other hand, since the water level on the downstream side is controlled relatively deeply by the endsill, the energy is attenuated by the hydraulic jump, as described later. Attempts have been made to shorten the length of the hydraulic jump, which is important in the design of the still basin using baffle blocks. FIG. 7.47 shows a steady hydraulic jump, in which L_j and L_r are jump length and roller length, respectively. A prediction of the jump length, roller length or effects of bottom roughness have been studied, e.g., Ali (1991), Ead & Rajaratnam (2002) and Carollo et al. (2007).

7.9.2.2 Hydraulic jump

The hydraulic jump is a very dynamic phenomenon that occurs during an abrupt transition from the supercritical flow to the subcritical flow. As shown in FIG. 7.47, a large roller-type reverse circulation and many vortices associated with it are generated. As a result, the energy decays.

A one-dimensional analysis of hydraulics derives a following form about the energy head loss, h_L:

$$h_L = \frac{(h_2 - h_1)^3}{4h_1h_2} \tag{7.9.1}$$

in which h_1 is a pre-jump depth and h_2 is an after-jump depth. Thus, the energy loss can be calculated only by the water depth before and after the jump. Gupta et al. (2013) developed an empirical formula of relative energy loss, i.e., ration of the energy loss due to the hydraulic jump to the energy of incoming fluid, based on the incoming Froude number Fr_1 and Reynolds number. Their result shows the relative energy loss is about 50%, with $\mathrm{Fr}_1 = 5$ corresponding to the steady hydraulic jump.

Cartigny et al. (2014) summarized the classification of the hydraulic jump depending on the incoming Froude number Fr_1, as follows. When incoming Froude numbers are between 1 and 1.7, undular jumps are observed with minor energy dissipation. As the incoming Froude numbers increase, the leading waves break and weak jump appears. The jump become unstable, accompanied with the free-surface oscillation when incoming Froude numbers are between 2 and 4. When the Froude number exceeds 4, i.e, $Fr_1 > 4$, the jump is stabilized and produces turbulence larg enough to dissipate fluid energy.

7.9.3 Energy Dissipation by Trees

7.9.3.1 Flood protection forest

A flood protection forest is one of a number of conventional river-management measures that have the advantage that its implementation can blend into the natural environment. Solid ground, strong foundations and an effective arrangement are very important in the construction of a flood protection forest, such that the trees can withstand the high hydraulic pressure and be effective in reducing fluid energy.

Nepf (1999) reviewed previous fundamental theories and experimental results considering the drag force acting on a single cylinder in uniform flow. The wake formed behind an isolated cylinder depends on the Reynolds number. Turbulent structures such as the coherent motions of the shed vortices contribute to the generation of the wake zone defining the drag force. The transition from laminar to turbulent flow observed in the wake zone of an isolated cylinder is also significant, as pointed by Williamson (1992).

Stone & Shen (2002) studied the total resistance in vegetated flows using channel-bed slope and plant spacing. The total resistance consists of the plant resistance and the bed shear resistance. The former means the drag force is due to plant stems, which can be evaluated by the drag coefficient. The latter, i.e., bed shear stress, controls sediment transport and so that is very important for prediction of bottom scouring. When the drag coefficient is known precisely, the bed shear stress can be obtained in vegetated flows using the total resistance and the stem drag force. Kothyari et al. (2009) proposed an experimental formula for the drag coefficient within a cylinder array zone, which is a function of the Froude number, stem Reynolds number and cylinder population. Sanjou et al. (2018a) conducted PIV and drag force measurements in a laboratory flume to examine the reduction efficiency of mean kinetic energy due to the tree members. In particular, two neighbouring high-speed streaks can merge and become uniform flow. The characteristic length of this process and incident velocity define the drag force.

7.9.3.2 Example of traditional flood protection forest

FIG. 7.48 shows an example of an extant flood protection forest that was built in the Yura River in Japan. The bamboo grove has been planted along the outer bank of the river in curved sections to slow the flood flow toward the nearby residential area. In particular, the bamboo trees reduce the fluid

Figure 7.48 Extant flood protection forest in Japan.

energy of inundation flows at the dike-break point and reduce runoff toward the houses, although flood damage is not necessarily prevented. Bamboo trees form the floodplain forest. They are about 15 m high, 10 cm thick and stand with 50–100 cm intervals.

Solid ground, strong foundations and an effective arrangement are very important in the implementation of a natural floodplain forest. The trees can withstand the high hydraulic pressure and be effective in reducing fluid energy.

7.9.3.3 Laterally placed tree models

Sanjou et al. (2018a) considered a simple model of a flood protection forest, in which cylinders were placed laterally across an open-channel flume, as shown in FIG. 7.49. In such a flow, the variation of depth was more substantial as the bulk mean velocity increased. The free surface dropped behind the cylinder, contrasting the increase in surface elevation on the upstream side of the cylinder. In particular, higher upstream velocities occurred behind the cylinders.

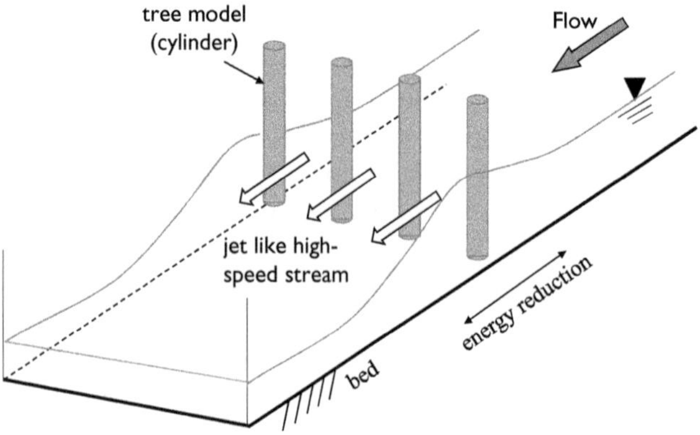

Figure 7.49 Simplified model of flood protection forest against flood.

A single cylinder placed in the flow is known to induce a typical symmetrical wake pattern in which boundary layer separation can be observed in the mean flow field. In contrast, a grove of laterally placed cylinders induces a flow similar to two-dimensional single jets. For both allocation patterns, high-speed and low-speed streaks are formed after passing the cylinders, as explained later. Separation zones accompanied by recirculating flows can be observed behind the cylinders. The drag force yields because of the pressure difference around the cylinder induced by flow separation. Sanjou et al. (2018a) reported the free-surface variation is very small; an increase of allocation density reduces the cross-sectional area of the passing current, which results in an increase of the velocity in the high-speed streak. Furthermore, the longitudinal length of the recirculating zone corresponding to flow separation becomes larger, and this property has a striking impact on the generation of the drag force, as discussed further.

Experiments of Sanjou et al. suggest that bulk mean velocity has a minor influence on the velocity vectors. A relatively fast current can be observed near the cylinder in the high-speed stream zone because the momentum is accumulated there. The velocity difference between the high-speed and low-speed streaks is reasonably large behind the cylinder and lateral mixing has not started. Lateral momentum exchange is promoted by the vortex streets traveling downstream, and a continuous velocity profile is formed in the lateral direction. These velocity distributions and turbulent structures are closely related to the drag force acting on the cylinders; thus, it is important to consider them in detail.

7.9.3.4 Potential core formed behind trees

Jet width increases linearly in the developing area for a single-hole jet phenomenon. In contrast, neighboring high-speed streaks meet some distance downstream in the parallel jetlike flows highlighted in the present model of a flood protection forest. This interference of several jet streaks makes the current structure more complex. FIG. 7.50 shows the wake flow behind the cylinders classified into four characteristic areas in the streamwise direction. The area in which the existence of the cylinders influences the upstream flow is defined as the upstream area. The downstream boundary of this zone corresponds to the downstream edge of the cylinders. The upstream boundary corresponds to an imaginary origin, which is discussed later. Immediately downstream of the cylinders, countercurrents ($U < 0$) in the area of flow separation and high-speed streaks are formed alternately, and this is called the converging zone. A potential core is generated in the high-speed streak in which the streamwise velocity profile is almost constant in the lateral direction. The high-speed streak provides momentum to the separation zone, and so the countercurrent region changes to a cocurrent flow with distance downstream. This is defined as

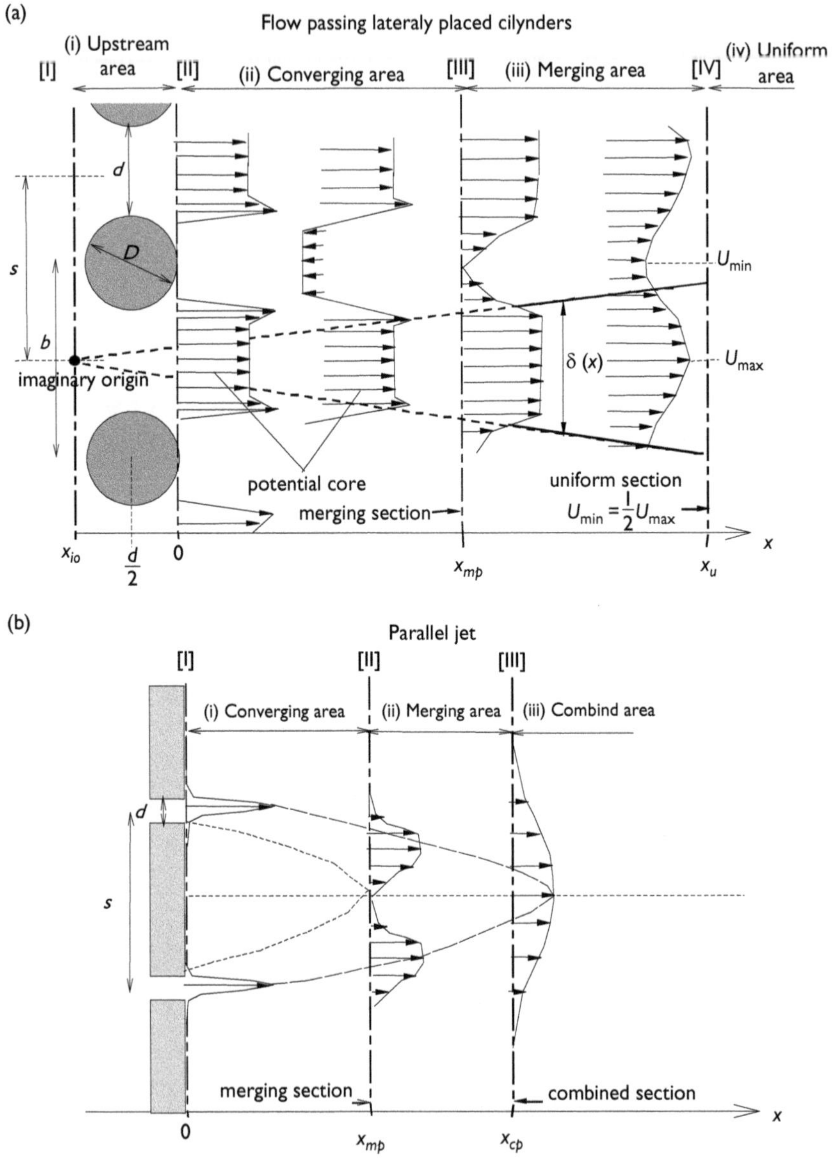

Figure 7.50 Phenomenological model of jet: (a) flow passing laterally placed cylinder; (b) two-dimensional parallel jet.

the merging area, where this mixing process of the potential core is promoted significantly.

The width of the high-speed streak is known to increase linearly for a typical single jet. A cross point can be obtained by extending the lines defining the edges of the high-speed streak toward the upstream side. This point can be defined as the imaginary origin ($x = x_{io}$) at which the jet, i.e., the high-speed streak, can be assumed to yield. This point is also considered the upstream boundary of the region influenced by the presence of the cylinders. The lateral profiles of velocity and turbulence in this zone are expected to be similar to the typical two-dimensional jet. The longitudinal lengths corresponding to the three zones defined previously are given by x_{io}, x_{mp} and x_{cp}, respectively. The low-speed streaks are accelerated and the high-speed streaks are decelerated while further downstream travel. This results in the minimum velocity becoming half the maximum velocity and the lateral profile of the streamwise velocity becomes flatter. This area is defined as the near uniform zone, in which the neighboring high-speed streaks on the upstream side interact significantly.

The present flow passing laterally placed cylinders is compared with the typical situation of two parallel jets, for which many researchers have defined three areas, as shown in FIG. 7.50(b). The widths of the single-jet structures diffuse in the spanwise direction in the converging zone just behind nozzles, in the same manner as the corresponding flow past the single-line trees. The lateral positions of the peak velocities begin to shift toward each other, which is markedly different from the flow past the single-line trees shown in FIG. 7.50(a). It results in the intersection of the two jets in the merging zone. Also, the two peaks meet in the combined section and a single jet structure newly appears in the combined area. The two parallel jets become unified into a single jet structure along an axis aligned between the two nozzles, unlike the flow past the single-line trees. Many previous experimental studies have suggested positions for the locations of the merging section, x_{mp}, and the combined section, x_{cp}. In particular, similarities are expected between the present flow past the single-line trees and the two parallel jets x_{mp}.

7.9.3.5 Flood energy reduction

Part of the kinetic energy the water current possesses is lost to the drag force acting on the cylinder. The total energy of the fluid passing the cross section far from the cylinders in unit time, E_o, can be defined as

$$E_O \equiv \rho g Q\left(\frac{U_m^2}{2g} + H\right) \qquad (7.9.2)$$

Considering the ratio of the work per unit time by the water stream to the total friction induced by cylinders ΔE, the energy reduction efficiency, η, can be given by

$$\eta \equiv \Delta E / E_O, \qquad (7.9.3)$$

in which $\Delta E = nFU_m$.

Sanjou et al. (2018a) reported that the energy reduction efficiency decreased as the slit width, b, indicated in FIG. 7.50(a) increased, implying that a denser arrangement of cylinders would be more efficient in decreasing the kinetic energy of the main stream. Their results allowed us to assume that energy reduction is dominated by two non-dimensional numbers, i.e., the Froude number and the relative slit length. Based on a fitting operation using the measured data, the following experimental formula was given:

$$\eta = 1.15F_r^2 (b/D)^{-2}, \quad (b/D > 0.5) \qquad (7.9.4)$$

Chapter 8

Measurements in Open-Channel Turbulence

8.1 INTRODUCTORY REMARKS

Previous chapters described fundamental knowledge in open-channel turbulence and applications to river flows. Many experimental works including field study by past researchers have explored complicated turbulence mechanisms and, also, they provided empirical constants. These days, the development of numerical simulation has been tremendous, but experiments still play a major role in elucidating phenomena. Many papers have been published on the experimental method. Tavoularis (2005) explains in detail the basic knowledge of the fluid measurement and the principle and usage of the related instruments. Final chapter explains topics related to the measurement techniques. Particularly, section 8.2 describes the data sampling including the sampling theorem and aliasing. Section 8.3 describes the principle and properties of existing velocimetry used for the both laboratory experiment and fieldwork. Section 8.4 describes methods about detection of organized motion such as bursting events. A quadrant analysis and variable-interval time-averaging (VITA) and proper orthogonal decomposition technique (POD) are introduced. Section 8.5 describes a setup of experimental flume and measurement procedure.

8.2 SAMPLING OF TURBULENCE

8.2.1 Sampling Theorem

Recording an intensity of measured analog signal with a constant time interval is called sampling. Further, quantization, which is discretizing the measured values, is carried out to record them as digital data strings. The measurement time interval is called a sampling cycle and its inverse is the sampling frequency. The quantity of the frequency is in Hz.

Can the analog signal be converted correctly to the digital signal? It is possible when we sample the analog signal with a frequency that is more than twice as large as the maximum frequency included in the signal. The

DOI: 10.1201/9781003112198-8

analog signal cannot be reconstructed by the digital signal sampled with an insufficient frequency.

8.2.2 Aliasing

A half value of the sampling frequency, f_s, is called the Nyquist frequency, f_N, i.e., $f_N = f_s/2$. The sampling theorem implies that the original analog signal can be reconstructed from discrete samples when the sampling frequency is larger than the Nyquist frequency.

FIG. 8.1 shows a time series of continuous analog signal, which is a sine curve with a frequency, f_0, drawn by the solid line. Here, a circle marker means the sampling timing with a constant sampling interval, $1/f_s$. When the sampling frequency does not meet the previous requirement, the original sine curve cannot be reproduced. In this case, another curve is drawn with a dotted line that has a smaller frequency, f_a, than that of the original curve, f_0. The original curve, including the larger frequency component than the Nyquist frequency, yields the aliasing shown in FIG. 8.1. The frequency component larger than the Nyquist frequency appears in the lower frequency region with an inversion symmetry about the Nyquist frequency.

Hence, a prediction of frequency scale in the vortices focused on is needed before the turbulence measurement. Also, it is important to conduct the velocity measurement with a high sampling frequency not to let the organized motion go out due to the data discretization.

8.3 VELOCIMETRY

8.3.1 Laboratory Study

Several types of velocimetry have been developed, as shown in Table 8.1. They are classified into intrusive and non-intrusive, or single-point or multi-point measurements. The tiny sensor or non-intrusive probe are preferred

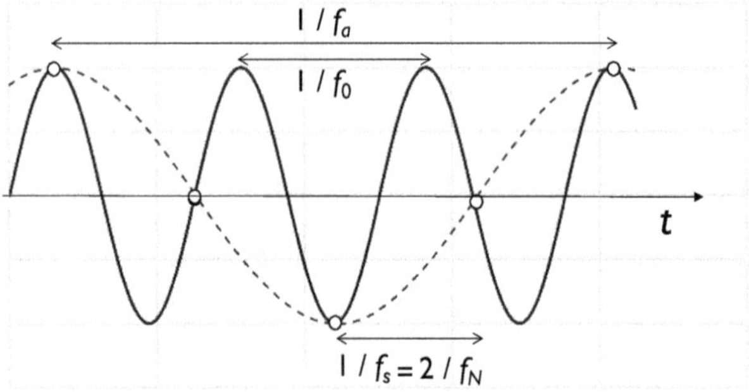

Figure 8.1 Continuous signal and discrete samples.

Table 8.1 Comparison of velocimetry for laboratory experiment

	Propeller current meter (KENEK VR401)	*EMV (KENEK VMT3–200-13P)*	*ADV (SonTek microADV)*	*UVP (Ubertone UB-Lab 2C)*	*LDA (MSE mini-LDV)*	*PIV*
type	intrusive	intrusive	semi-intrusive	semi-intrusive	non-intrusive	non-intrusive
sampling rate (Hz)	analog output	analog output	0.1–50	up to 100	depends on tracer concentration	>30 (depends on camera speed)
sampling volume	* propeller diameter: 5 mm to 20 mm	*probe dimameter: 13 mm	0.9 cm^3	*minimum cell size: 1.4 mm	0.024 mm^3	~ 0.5 mm^3 (depends on device setting)
velocty range (m/s)	0.03 to 4	0 to ±2	0 to ±2.5	–0.5 to 3	–50 to 600	—
accuracy	3 cm/s	2%	1% or 0.25 cm/s	0.2 to 1%	0.3%	—
component	1C point	3C point	2C or 3C point	2C profier	1C to 3C point	1C to 3C multi-point
remarks	not available for counter current detection	—	point measurement 5 cm under the probe	—	not suitable for turbid water	planar multi-point measurement not suitable for turbid water

for the turbulence measurements in a laboratory flume. Of particular significance is that researchers pay attention to the specification and limitation of their own velocimetry. The following subsection explains a basic principle of each velocimetry.

8.3.1.1 Pitot tube

A Pitot tube is a traditional velocimetry using Bernoulli's principle. It is composed of two tubes. One has a small hole in the water current, and another has an open end against a stream, as shown in FIG. 8.2. Applying the Bernoulli's principle at points S and D corresponding to the static hole and the open end, respectively, the current velocity, V, can be measured by the gap of water levels within two tubes. Note that we assume that the velocity is zero at point D. The pressure at points S and D are given by $P_S = \rho g h_s$ and $P_D = \rho g h_d$ using the water surface levels in two tubes. The Bernoulli equation is expressed by $\frac{V^2}{2g} + \frac{P_S}{\rho g} = \frac{P_D}{\rho g}$ between points S and D. Hence, $V = \sqrt{2g\Delta h}$ is obtained. Becker and Brown (1974) examined accuracy and limitation in the turbulent flows in detail.

8.3.1.2 Propeller current meter

The current velocity is measured by rotation speed of the propeller attached to a rod (FIG. 8.3). The velocity is theoretically proportional to the

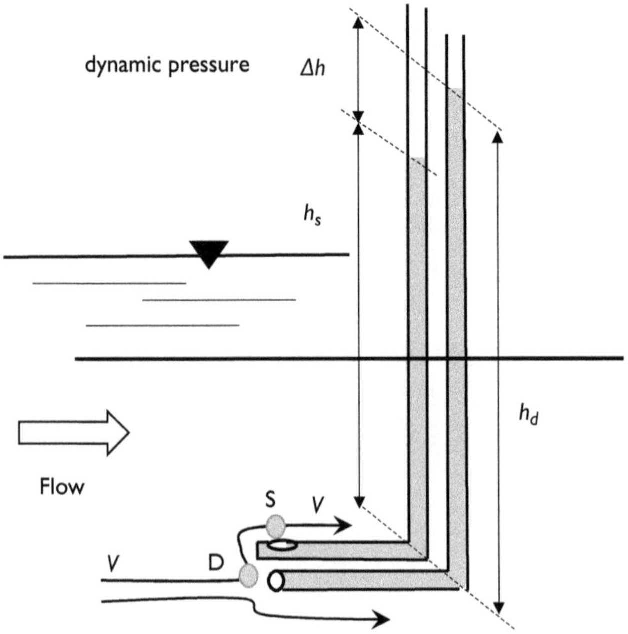

Figure 8.2 Pitot tube used in an open-channel flow.

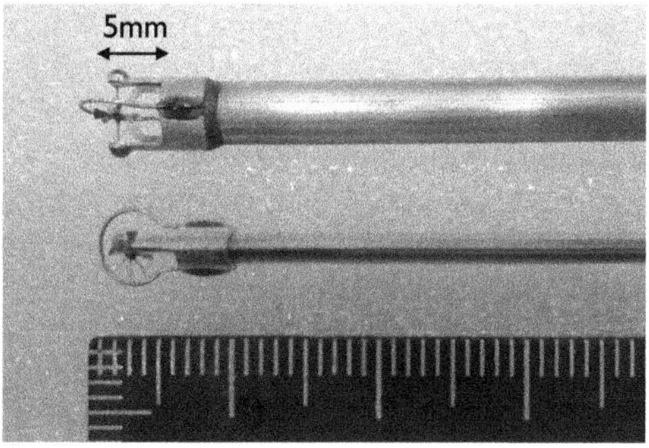

Figure 8.3 Probe of propeller current meter.

propeller speed and, thus, a calibration curve is required in a preliminary experiment. The rotation speed is detected by an encoder, which has several types, as follows.

a. Optical encoder

A slitting rotating disk is attached to the propeller axis; the only light through the slit is detected by a photo sensor. A photo pulse occurs by detection of the light, and the rotation speed can be calculated counting the pulse signal.

b. Magnetic encoder

The magnetic field is generated by an eternal magnet. The propeller rotation varies the distribution of magnetic field and, thus, its detection allows to measure the rotation speed.

c. Electromagnetic induction type encoder

The variation of magnetic field between an exciting coil and a sensing coil is detected. The basic principle is the as same as a transformer using the electromagnetic induction.

Itagaki & Furuta (1983) measured the open-channel flow by using the 2 cm diameter propeller current meter, in which the sampling rate is 0.15 Hz. Although the response to turbulent current is comparatively low, they reported the measured data obeys well the –5/3 power law in the inertial subrange.

8.3.1.3 Electromagnetic velocimetry (EMV)

The electromagnetic velocimetry (EMV) has an intrusive probe, which is an application of Faraday's law of electromagnetic induction. The EMV has been developed over one century, as reviewed by Watral et al. (2015). It composes of an electromagnetic coil for generation of the magnetic field and the electro codes through the induced current flows. When an electro-conductive liquid flows within a pipe where the magnetic field is generated, the electromotive force is induced, which is proportional to the pipe radius, magnetic flux density and the flow velocity. Thus, the flow velocity can be evaluated by the value of the electromotive force.

This device is often used for discharge measurement within the pipe, as shown in FIG. 8.4(a). A pair of electrodes is placed across the pipe. The magnetic coils generate the magnetic field perpendicular to the pipe flow. The electro-conductive fluid such as water induces the voltage between the electrodes. Hence, we can measure the bulk-mean velocity in the pipe due to the correlation between the velocity and the induced voltage.

A similar setting that can measure the velocity in also an open channel. When the point measurement of the velocity is required, the tiny probe is used, as shown in FIG. 8.4(b). The magnetic field occurs around the probe and the fluid flow influences it. As a result, the velocity is measured near the probe (FIG. 8.4(c)).

The advantages in the EMV are that two or three components of velocities can be measured simultaneously and we can use it also in natural rivers. In contrast, the EMV may be susceptible to noise in the laboratory

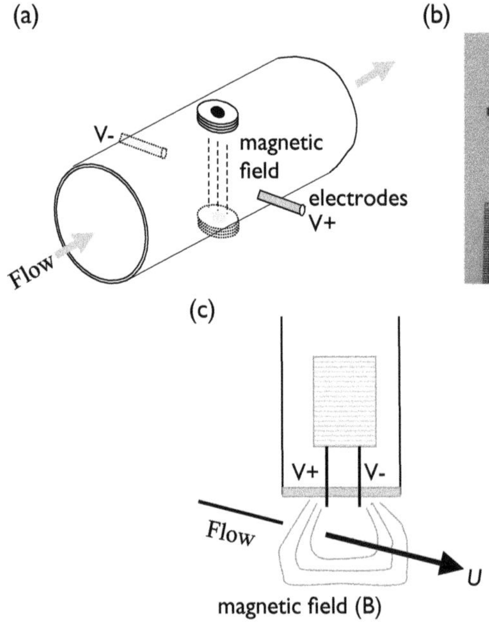

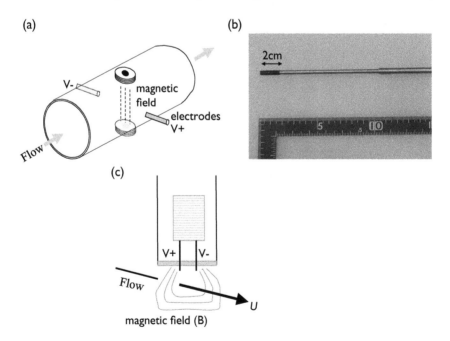

Figure 8.4 Electromagnetic velocimetry: (a) setup of discharge measurement in pipe flow; (b) KENEK-made EMV probe for the point measurement; (c) composition of the EMV probe.

room compared to the outdoor field. Hence, it is important to denoise properly for reliable evaluation of turbulence components or small currents such as secondary flows.

8.3.1.4 Ultrasonic velocity profiler (UVP)

The UVP evaluates the velocity using the Doppler shift frequency, transmitting the ultrasonic waves to the fluid. It is a non-intrusive velocimetry without disturbing the sampling fluid. Hence, this is one of popular velocimetry used in not only the laboratory flume but also the natural river. Note that small particles that follow fluid motion should be dissolved in the fluid, which reflects the ultrasonic waves transmitted by the UVP probe when we use it in the clear water. The receiver of the probe catches the back scatters and then the Doppler shift frequency is calculated.

Thus, the transmitter and receiver are placed inclined to the mainstream with an angle θ, as shown in FIG. 8.5(a). Here, we define f_0 as the frequency of an incident wave, and f of the frequency of the reflected wave, i.e., Doppler shift. It is generally given by a following form:

$$f = \frac{V - v_0}{V - v_s} f_0 \tag{8.3.1}$$

in which V is a sound speed in the fluid, v_0 is the velocity of the observer and v_s is the velocity of the sound source. The observer and the sound source are assumed to have a motion in one dimension.

Considering the wave propagation direction, the sound source and the observer come closer to each other, as shown in FIG. 8.5(b). Hence, relative velocities of the sound source and the observer to the unidircctional fluid with streamwise velocity, U, are given by $v_0 = -U \cos \theta$ and $v_s = U \cos \theta$, respectively. Eq. (8.3.1) leads to

$$f = \frac{V + U \cos \theta}{V - U \cos \theta} f_0 \simeq \left(1 + \frac{2U \cos \theta}{V}\right) f_0 \quad \because C \gg U \tag{8.3.2}$$

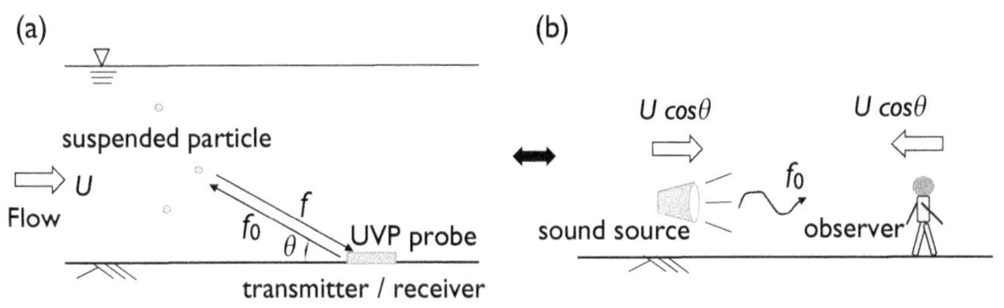

Figure 8.5 Ultrasonic velocity profiler: (a) setup of UVP measurement; (b) one-dimensional motion of the sound source and the observer.

Thus,

$$\Delta f = f - f_0 = \frac{2U \cos\theta}{V} f_0 \tag{8.3.3}$$

is obtained. The Doppler shift, Δf, if measured correctly, the fluid velocity can be evaluated. When another probe is added with a different incident angle, θ, two components of velocity can be measured in the two-dimensional flow. Furthermore, the velocity profile is obtained with the wave propagation axis under the condition that suspended seeding particles are distributed evenly. The principle of velocity profiler is explained in the section of ADCP.

The multi-bistatic UVP called acoustic Doppler velocity profiler (ADVP), e.g., Ubertone-made UB-Lab 2C, can measure accurately Reynolds stress, turbulence flux, etc. (Shen & Lemmin 1997, Mignot et al. 2009).

8.3.1.5 Acoustic Doppler velocimetry (ADV)

The basic principle is almost the same as that of the UVP, which uses the Doppler shift of the ultrasonic wave (see McLelland & Nicholas 2000). The devise for the point measurement is called ADV and that for the velocity profile is called ADCP, as mentioned later. This section focuses on the ADV.

The commercial ADV probe is composed of one transmitter and two or three receivers, as shown in FIG. 8.6. The SonTek-made micro ADV measured a point 5 cm apart from the transmitter. It is useful for the simultaneous measurement of two or three components of velocities.

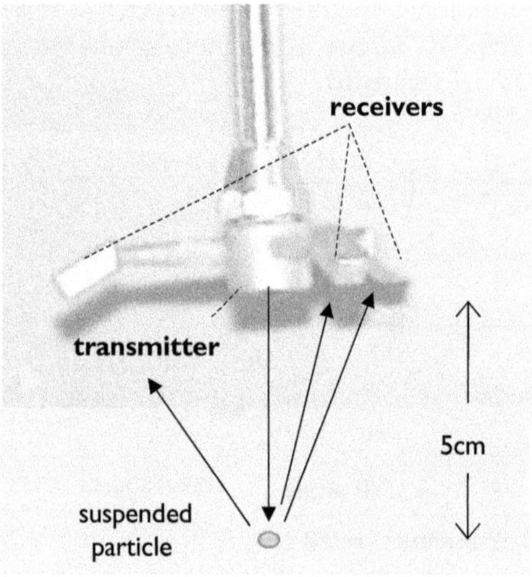

Figure 8.6 Down-looking 3-D ADV probe (SonTek made Micro ADV).

The probe is composed of several receivers and one transmitter. The ADV needs the seeding particles for the experiments in clear water as well as the UVP.

The ADV is expected to measure turbulent flow reasonably when the noise reduction is conducted. There following are causes of electrical noises and solutions.

a. Aliasing

ADV uses dual pulses with different pulse repetition rates, s τ_1 and τ_2. Two kinds of pulses are separated by a dwell time, τ_D. Hence, the time it takes to complete the three components of velocities is given by $T = 3 \times (\tau_1 + \tau_D + \tau_2 + \tau_D)$. The maximum sampling rate is $f = 1/T$, and the Nyquest frequency is given by $f/2$ (see section 8.2). This refers to fluctuation energy in the frequency range from $f/2$ to f is included in the lower frequency range less than $f/2$. Such an aliasing causes the measurement error. Garcia et al. (2005) pointed out that the fluctuation energy in the fluid with the convection velocity, U_c, is concentrated on the short wavelength range from $f/(2U_c)$ to f/U_c. They proposed a non-dimensional parameter, $F \equiv f_R L/U_c$, in which f_R is an ADV's user-set frequency and L is a length scale of a large eddy. They concluded that an influence of aliasing can be ignored in the data processing when $F > 20$.

b. Doppler noise

There is an incomplete understanding of what causes Doppler noise. Although Nikora & Goring (1998) and Hurther & Lemmin (2001) proposed some denoising methods, they are not a perfect way considering the detailed noise mechanism. When the Doppler noise is equal to the white noise, the time-averaged velocity is expected not to include the white noise. This is because white noise becomes zero after it is averaged in time. Voulgaris & Trowbridge (1998) suggested a following relation between the measured and true Reynolds stress.

$$\overline{uv} = \overline{\hat{u}\hat{v}} + K\sigma_t^2 \tag{8.3.4}$$

K is a constant used for coordinate transform, in which the velocity along the propagation path of the ultrasonic wave is transformed to the velocity in the Cartesian coordinates. An over hat refers to the true value. σ_t^2 is a velocity error variance along the transmit beam which is composed of sampling noise, Doppler noise and velocity shear noise. The constant K is generally small for the Reynolds stress, and thus it results in the noise is also small if σ_t^2 can be removed as much as possible (Voulgaris & Trowbridge 1998).

c. Wall reflection

ADV transmits pulse waves intermittently, as mentioned above, and uses two pulses for the velocity measurement. First, pulses are reflected in the seeding particle within the sampling volume, and then are received by the probe. However, some of them pass through the sampling volume and reach the wall. They also return to the receiver, and interfere with the second pulse. This induces a large measurement error depending on the sampling volume position and the dwell time of the pulse waves. Putting a plastic film on the wall surface can prevent the wave reflection in a certain degree.

8.3.1.6 Hot film velocimetry

Heat of a hot film sensor is transferred to the surrounding fluid. The heat flux depends on the fluid velocity. The hot film velocimetry uses this principle. The tiny wire with a high heat response as a sensor, and thus, it can obtain frequently a time-series of electrical signal. It allows the accurate turbulent measurement in the fluid.

The heat included in the hot film is dissipated, and temperature of the film reduces with a variation of the electrical resistance. Faster flow promotes the cooling of the film more efficiently. A relation between the fluid velocity and heat dissipation is known as King's law, and the following formula is used (Vieira & Mansur 2012):

$$E^2 = A + BU^{0.5} \tag{8.3.5}$$

in which E is an output bridge voltage, U is a fluid velocity and A and B are experimental constants.

A feedback circuit is designed in a constant-temperature-type velocimetry, in which a bridge and amplifier are incorporated, as shown in FIG. 8.7. The bridge circuit is controlled to be balanced. The electric resistance of the hot film sensor cooled by the surrounding fluid reduces and the bridge circuit becomes unbalanced. Then the amplifier controls the voltage in order to heat the hot film. It results in the recovery of the electric resistance in the hot film and the sensor temperature is kept constant. Instantaneous velocity can be evaluated by the measurement of the output voltage.

8.3.1.7 Laser Doppler anemometer (LDA)

A laser Doppler anemometer (LDA) projects two laser beams in the tracer particle followed by the fluid, and calculates the flow velocity analyzing a scattered light of the particle detected by the photo multiplier. A stripped pattern called fringe has a regular contrast of brightness and darkness at the

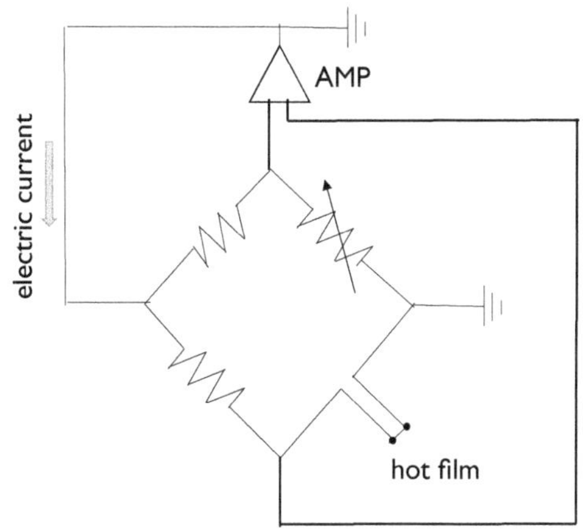

Figure 8.7 Bridge circuit used in the hot film velocimetry.

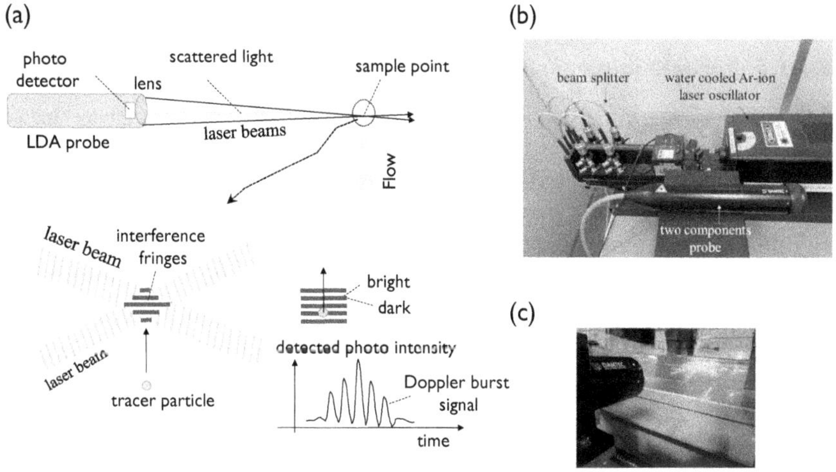

Figure 8.8 Setup of laser Doppler anemometer: (a) measurement principle; (b) optical system; (c) probe and laser beams.

cross point of two beams, as shown in FIG. 8.8. The spacing of fringe depends on the crossing angle of the beams and the beam wavelength. The tracer particle passing through the crossing point of the beams induces a scattered light having a contrast. The intensity has a time variation like a rainbow curve due to a Gaussian property of the laser beam. The scattered light is converted to the electric voltage by the photo multiplier, which is called a Doppler burst signal.

Counting the peak edge of the Doppler burst signal, the velocity can be calculated, the fast-speed fluid and the low-speed one make the gap of peak edges narrow and wide, respectively. Since the fridge spacing is constant, the passing tracer particle generates a signal with a regular frequency. Hence, the velocity is given by

$$V = \lambda \times f \tag{8.3.6}$$

in which V is the velocity of fluid or particle, λ is the fridge spacing and f is a frequency of the Doppler signal. Using the acoustic-optic modulator called a Bragg cell for the beam splitter, the frequency of one beam is shifted. That is to say, two beams with a different frequency and the same intensity are output through a Bragg cell. Further, such a frequency shift moves the interference fringe along the flow direction with a constant speed. Introducing the frequency shift, the LDA can measure countercurrent, including zero velocity.

The sampling volume is much smaller than other flow sensors, as shown in Table 8.1, and it is therefore possible to measure accurately a very small eddy comparable with the Kolmogorov scale.

8.3.1.8 Particle image velocimetry (PIV)

Two general methods can be used to obtain velocity vectors from the sequential digital images: individual particle tracking and statistical correlation of sequential images. The former is called particle tracking velocimetry (PTV). The latter is a particle image velocimetry (PIV). The spatial resolution and dynamic range depend on the number and the size of particles, respectively. Although various methods for PTV are used to obtain velocity distributions, it is difficult to identify particle pairs in high particle-density images. Hence, use of PTV is limited to low particle density conditions, and the number of velocity vectors obtained simultaneously is comparatively small. In contrast, PIV follows distribution patterns of particles by using a correlation function about a digital image pair.

FIG. 8.9(a) shows the basic digital PIV configuration, in which a high-sensitivity camera, glass flume and laser light source system are required. A laser beam is converted to a planar sheet of light, i.e., a laser-light sheet (LLS), via a cylindrical lens. The LLS should be as thin as possible to remove influences of out-of-plane velocities. The LLS is generally projected into the fluid body vertically or horizontally in laboratory measurements. Two components of instantaneous velocity are evaluated on the basis of the space and time distributions of tracer particles illuminated by LLS. Image resolution and data transfer rates are critical to the experimental method. High-performance CCD and CMOS cameras, which can rapidly transfer a large volume of data, are used. Also, to obtain sufficient dynamic range, higher sensitivity cameras that record images with a resolution greater than a megapixel are needed. The recording of consecutive images is spaced by a

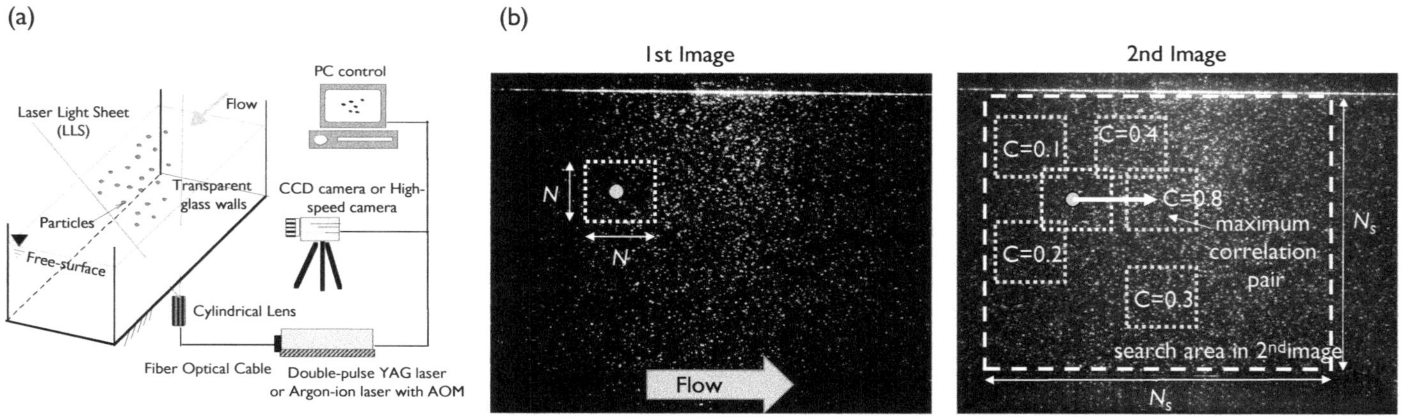

Figure 8.9 PIV measurement: (a) configuration for laboratory experiment; (b) correlation between consecutive two images.

sampling interval, Δt. Higher digital movie frame rates make for shorter interval times between images, which is desirable for accurately measuring high-speed fluids. However, a strong light intensity is needed, due to the short exposure time of a high-speed camera.

The direct cross-correlation method detects similarities in local patterns of brightness by cross-correlating the two images of an image pair. First, an interrogation region of $N \times N$ pixels is chosen within the first image, and then a search region of $N_s \times N_s$ pixels is considered in the second image, as shown in FIG. 8.9(b). The correlation value, C, between the first and second images is calculated by the following equation:

$$C(\Delta x, \Delta y) = \sum_{i=1}^{N} \sum_{j=1}^{N} f(x, y) g(x + \Delta x, y + \Delta y) \tag{8.3.7}$$

in which f and g are the brightness values in first and second images illuminated in the $x - y$ plane, respectively. The candidate region that has the largest correlation with the interrogation region of the first image is obtained mathematically. An FFT-based cross-correlation method proposed by Willert & Gharib (1990) is often used, which can calculate Eq. 8.3.7 in a short time.

The interrogation size significantly influences the result of PIV measurements. When a large interrogation is applied to turbulent flow where larger velocity shear and vorticity appear, such structures cannot be observed accurately. In contrast, small interrogations can induce many incorrect vectors. To provide the most appropriate interrogation size, recursive cross-correlation and hierarchical cross-correlation methods have been proposed.

Distributions of obtained velocity vectors are required at sub-pixel accuracy in PIV measurements. When a one-dimensional (1D) brightness pattern is considered, the peak correlation can appear close to a pixel that has a second peak correlation. If the distribution of the correlation function is assumed to be a Gaussian distribution, then the peak position of correlation function is evaluated at a sub-pixel location. Note that in sub-pixel analysis, the peak locking phenomena converges measured values to integer samples and measurement accuracy decreases as a result. To avoid this, each tracer particle should be taken as more than 2×2 to 3×3 pixels.

Post-processing is needed to examine the accuracy of raw velocity vectors and to evaluate instantaneous velocity data and statistical values precisely. This processing consists of: 1) the detection and removal of incorrect vectors and 2) vector interpolation.

When these processes are correctly completed, instantaneous velocity vectors in two-dimensional plains are obtained as shown in such FIGs. 3.31, 4.5 and 7.15. The basic principle and configuration are explained in detail by Nezu & Sanjou (2011).

8.3.2 Field Study

Field survey and velocity measurement in rivers have larger scales in space and time compared to the experiment in the laboratory flume. Hence, it is very difficult to use the previously mentioned devices for the fieldworks. Although measurement accuracy and reliability are required there as well as the indoor experiments, in addition, durability against use in severe natural condition and safety of operators are also of importance. Further, it is often the case that the sampling point is situated out of operator's reach. Therefore, further progress of remote surveying technique will be needed. Velocimetry used in the river observation is explained as bellow.

8.3.2.1 Floating method

This is the most popular method for the velocity and discharge measurements in natural rivers. The cross section is divided into several sub-cross-sections. Dropping the float from the bridge, a travel time for the certain distance is measured by eyes. Repeating this work changing the lateral position in the starting section, the free-surface velocities can be evaluated in each sub-area. The international standard such as ISO 478 states the no more than 5% of the total flow discharge should be included in each sub-cross-section for the reliable discharge measurement.

A surface float is used for shallow depth condition and, in contrast, a stick float is often used for deep condition. The advantage of the floating method is the simple principle compared to other velocimetries. That is why it is prevailing. Shin et al. (2016) proposed an advanced floating method called SFIV, which incorporates the imaging analysis.

However, there are some disadvantages. For example, the measurement of traveling time by eyes depends on the observer's skill. The place where the float is initially dropped is limited to a bridge across the river. When wind pushes the float, it cannot follow the streamline in the free surface. They are causes of reduction of the measurement accuracy. The author's group has been developing the drone-type float with a GPS receiver, as shown in FIG. 8.10. This float flies to the target point on a free surface according to the operator's control. After which, it runs freely downstream, detecting the time-series of self-position with a centimeter-order accuracy using the real-time kinetic GPS method. It also measures the local water depth using an ultrasonic sensor during the drifting downstream.

8.3.2.2 Price-type current meter

This is an intrusive velocimetry which has a vertical axis with several cups. The rotation speed of the current meter is proportion to the current velocity. Therefore, when the calibration curve between them is obtained in laboratory test, the velocity can be measured in the river and channel. The

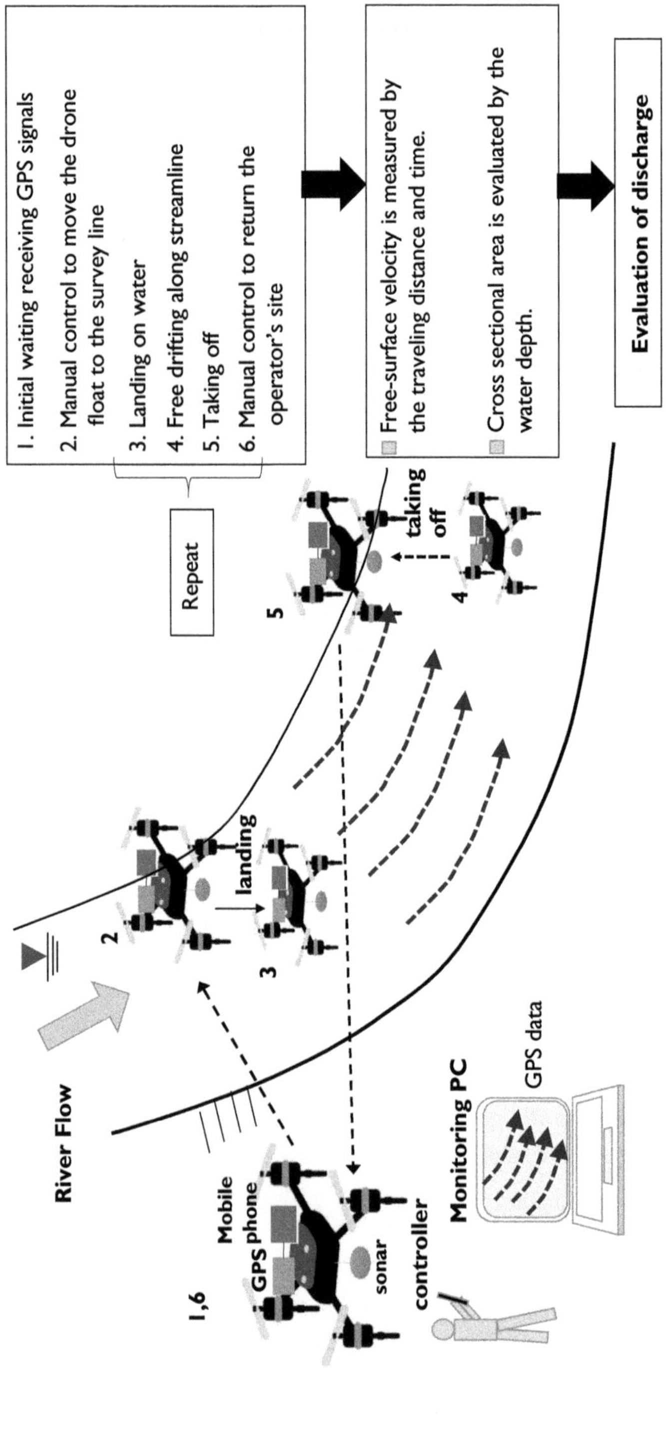
1. Initial waiting receiving GPS signals
2. Manual control to move the drone float to the survey line
3. Landing on water
4. Free drifting along streamline
5. Taking off
6. Manual control to return the operator's site
Repeat
Free-surface velocity is measured by the traveling distance and time.
Cross sectional area is evaluated by the water depth.
Evaluation of discharge
River Flow
1,6
Mobile phone
GPS
sonar
controller
Monitoring PC
GPS data
2
3
landing
5
4
taking off

Figure 8.10 Image of the drone-type float.

basic principle is the same as the propeller current meter. The mechanical design is so simple that it is easy to use. However, we cannot use deeper and fast flow situation in rivers. Further, drifting algae or flexible plant are coiled around the stay of the current meter.

8.3.2.3 Acoustic Doppler current profiler (ADCP)

ADCP conducts a remote sensing of the velocity profile under the free-surface analyzing the Doppler shift of ultrasonic wave transmitted by the transducers. ADCP has been originally developed for the oceanographic survey and could not be applied to a shallower river flow. However, the recent development of the broadband technology and high-speed pinging allows measurements in a river flow.

FIG. 8.11 shows an image of multi-layer measurement of the velocity profile. The several transducers transmit the ultrasonic waves and receive waves reflected at each layer. The velocity component with the wave propagation direction is calculated due to the Doppler shift. When at least three transducers are used, three-dimensional velocity vectors can be obtained at each layer. The four transducers are generally equipped for the commercial product for evaluation of error velocity. Note that the larger distance between the layer and the transducers increases horizontal distance of ultrasonic beams. This means that neighboring velocity components on each beam are not situated at the same point. Hence, the accuracy of three-dimensional velocity vectors converted to the Cartesian coordinates reduce farther from the transducers.

8.3.2.4 Microwave current meter

This device placed on the bridge transmits microwave to the free surface in the river flow with a depression, as shown in FIG. 8.12. The reflection wave

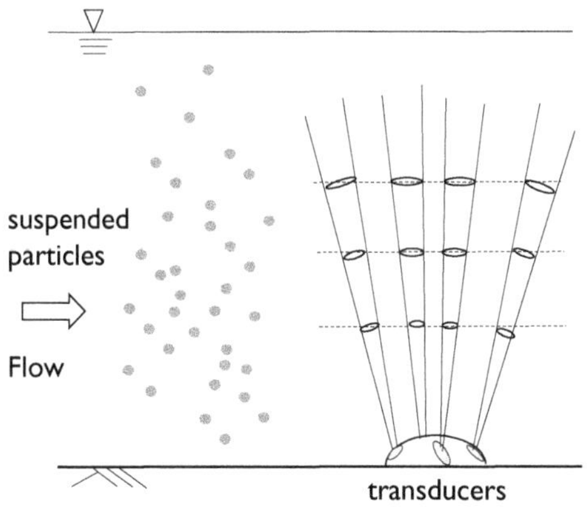

Figure 8.11 Multi-layer measurement of ADCP.

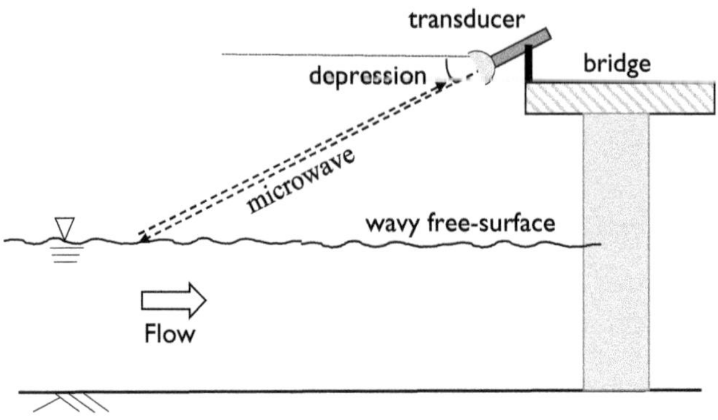

Figure 8.12 Field setup of microwave current meter.

includes the Doppler shift induced by the surface roughness, and then the free-surface velocity can be evaluated. It is possible to conduct a safe survey even in a flood because of perfect remote sensing. Therefore, the microwave current meter is expected to complement severe missions that the conventional float method cannot cover. Also, a combination with a drone may be able to measure even if there is no bridge. However, there are some disadvantages for practical use. A slow free-surface current under 50 cm/s cannot be measured (YDK technologies, model WJ7661-S3) because the small free-surface roughness does not allow the reflection of the microwave. Further, the reflection wave is influenced by the wind, rain and snow falls, and it reduces the measurement accuracy. The velocity under the free surface cannot be measured as well as the float method.

8.3.2.5 Large-scale PIV (LS-PIV)

Fujita et al. (1998) applied PIV to the measurement of free-surface current in rivers, which is called a large-scale PIV (LS-PIV), as shown in FIG. 8.13. Analyzing digital images taken by a video camera, the distribution of horizontal velocity vectors in the free surface is evaluated. A roughness pattern on the free surface is used for the correlation function between the consecutive images. Only the camera can take the free-surface images without specific light sources. In contrast, a geometric correction is required to correspond the coordinate in the image to that in the real field; thus, accurate calibration work including survey of reference points should be necessary.

LS-PIV can conduct a comprehensive non-intrusive measurement by placing the video camera near the riverbank. Therefore, it is possible without any bridge across a river flow. Further, we can measure during nighttime, introducing a IR camera.

However, the measurement accuracy and reliability reduce without visible patterns or drifting matters in the free surface. Also, wind-induced

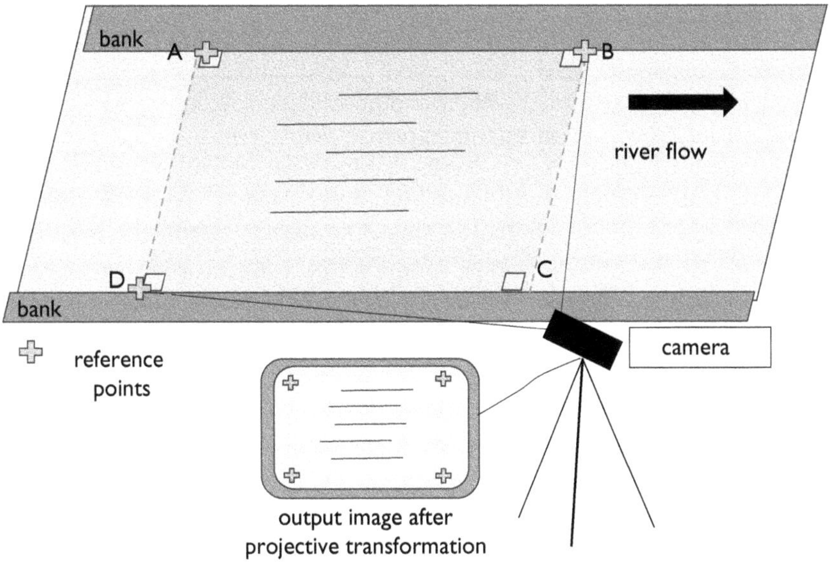

Figure 8.13 Example of LS-PIV setting for river survey.

waves may not follow the true free-surface stream. Although there are such disadvantages, it prevails rapidly as a useful practical method for the river monitoring, and further improvement is expected.

8.3.2.6 *Acoustic mapping velocimetry (AMV)*

AMV is a new technology applied to the river flow measurement, which has been originally developed for the purpose of ocean exploring as well as ADCP (Muste et al. 2016). This is a non-intrusive method that combines the ultrasonic technology with image sensing. Particularly, it specializes in the bottom flow measurement, including the bed road transport in rivers. The principle is composed of two steps. First, transmitting the ultrasonic wave toward the bottom in the river, the bottom configurations called acoustic map is made. Second, analyzing time-series of acoustic maps by the correlation method used for the PIV, the bottom velocities and sediment transport can be measured. Hence, AMV has following advantages, i.e., we can monitor sediment transport related to river environment, we can detect an ever-changing river current in a very short time and we can get both the velocity distribution and the bottom configuration instantaneously. Although suspended objects or boulders may reduce the measurement accuracy, this technology will develop further incorporating cutting-edge techniques of sonic prospecting, image sensing, scanning, and sounding and so on.

8.3.2.7 Unmanned surface vehicle (USV)

Sanjou & Nagasaka (2017) developed a small unmanned surface vehicle (USV) by using a one-board microcomputer called Arduino, which could automatically measure surface streamwise velocity in a straight creek. A combination of positioning camera-tracking systems and proportional/integral/derivative (PID) control methods were used to keep the robot in position against the main stream; then, the mean velocity was evaluated reasonably by the rotation speed of screw propeller.

The drag force, F_D, and the propulsion force, F_T, can be given in the following forms:

$$F_D = \frac{1}{2}\rho A U^2 C_D \tag{8.3.8}$$

$$F_T = \frac{1}{2}\rho m^2 D^4 C_T \tag{8.3.9}$$

where ρ is the water density, U is the streamwise mean velocity, A is the frontal area of the robot boat against the main stream, m is the rotation speed of the screw (rpm) and D is the diameter of screw. C_D and C_T are the coefficients of drag and propulsion, respectively. When they are balanced in the stationary stage, the attacking velocity is proportional to the screw speed, as follows:

$$U \propto Dm \tag{8.3.10}$$

When the relationship between the screw speed and the oncoming velocity was obtained by the preliminary experiment, the flow velocity can be evaluated by the screw speed required for the USV to remain in the water current. A propulsion motor controls the robot to keep it stationary at a measurement point. As the rotation speed of the propulsion screw varies with the oncoming velocity, the streamwise velocity can be measured in the river. A side thruster was also used to allow a planar two-dimensional motion. Their prototype remained in uniform flow, and oncoming mainstream velocity was measured using the propulsion force. Remarkably, the USV itself worked as a one-component velocimetry in which no external velocimetry was required. However, their system assumes that the mainstream flows parallel to the banks.

The advanced USV sensing both the oncoming velocity and the flow direction was developed, as shown in FIG. 8.14(a) (Sanjou et al. 2021). The automated measurement procedure composes of four control stages: 1) calculation of tentative propulsion force, 2) navigation to the target point, 3) velocity by staying at target and 4) detection of flow direction by

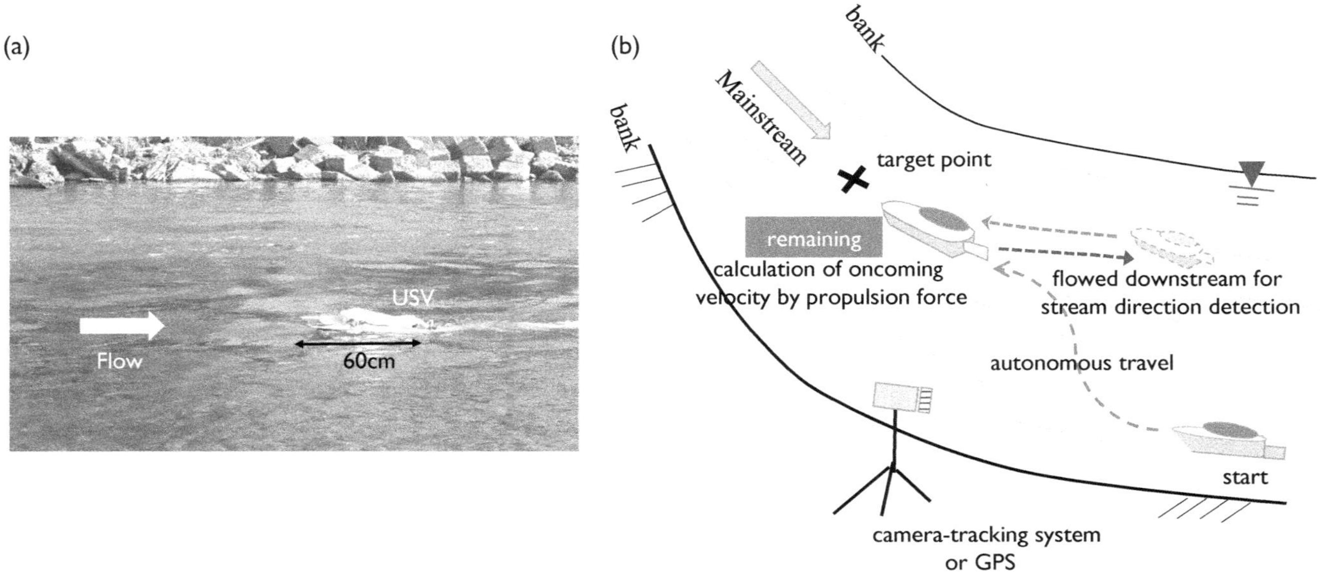

Figure 8.14 Automatic velocity measurement using autonomous USV: (a) prototype USV remaining in the river flow; (b) process of velocity and flow direction measurements.

following downstream (FIG. 8.14(b)). Field tests were conducted to examine the reliability and accuracy of the present automatic flow measurements in a river. Both local velocity and direction in the river flow were measured well by the USV. Error analysis was conducted by comparing with the existing velocimetry results, and the USV was found to possess a sufficient ability to meet practical performance for the flow measurement in a 1 m/s calm river. Further system improvement will make it possible to measure in faster rivers.

8.4 DETECTION TECHNIQUE OF COHERENT TURBULENCE

8.4.1 Quadrant Analysis

Many researchers have been challenged to detect organized or coherent turbulent motions in random-like velocity signals. Lu & Willmarth (1973) proposed the quadrant theory using time-series in a pair of two component velocity fluctuations (u, v). Nakagawa & Nezu (1977) conducted the quadrant analysis in a rough-bottom flow, and showed that sweeps are more significant than the ejections.

Contribution to Reynolds stress in quadrant i of (u, v) as shown in FIG. 3.28 except, for the hyperbolic hole region is expressed by

$$\langle uv \rangle_{i,H} = \lim_{T\to\infty} \frac{1}{T} \int_0^T u(t)v(t) I_{i,H}[u(t), v(t)]\, dt \tag{8.4.1}$$

in which

$$I_{i,H}[u(t), v(t)] = \begin{cases} 1, & \text{if}(u, v) \text{is in qudrant } i \text{ and if } |uv| \geq H|\overline{uv}| \\ 0, & \text{otherwise} \end{cases} \tag{8.4.2}$$

The contribution of the quadrant i to Reynolds stress is as follows:

$$RS_i(H) \equiv \frac{\langle uv \rangle_{i,H}}{\overline{uv}} \tag{8.4.3}$$

A time occupied rate of the quadrant i is calculated by

$$T_i(H) \equiv \lim_{T\to\infty} \frac{1}{T} \int_0^T I_{i,H}[u(t), v(t)]\, dt \tag{8.4.4}$$

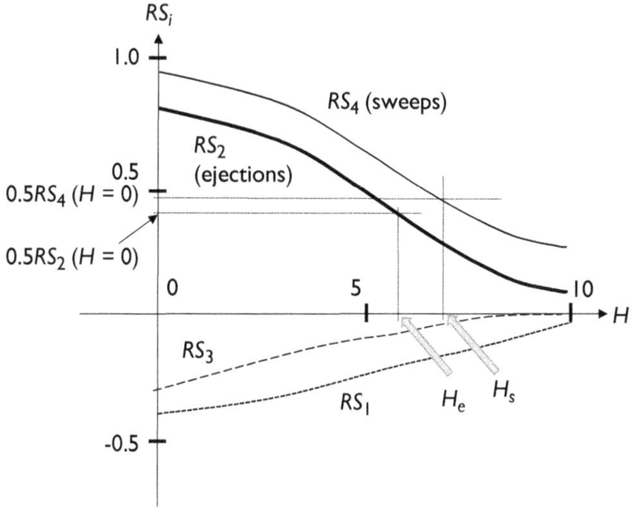

Figure 8.15 Example of variation of contributions to Reynolds stress against the hole value.

Nakagawa & Nezu (1977) proposed how to decide the hole value, H. First, they calculated the variation of contribution rate against the hole value, as shown in FIG. 8.15. Then, the hole value corresponding to the half contribution rate calculated for $H = 0$ is defined as the threshold values, H_e and H_s, for ejection and sweep, respectively.

8.4.2 Variable-Interval Time-Averaging (VITA)

Blackwelder & Kaplan (1976) proposed a variable-interval time-averaging (VITA) technique in order to concentrate on a localized region in space or time. The variable-time averaged value for the fluctuation component, Q, with the small time duration, T, is defined by

$$\hat{Q}(t, T) = \frac{1}{T} \int_{t-\frac{1}{2}T}^{t+\frac{1}{2}T} Q(s)ds \tag{8.4.5}$$

Referring to Eq. (8.4.5), a localized square of velocity fluctuation $\widehat{u^2}(t, T)$ and localized squared mean velocity fluctuation, $\{\hat{u}(t, T)\}^2$, are considered. The local variance is defined by

$$\widehat{\mathrm{var}}(t, T) = \widehat{u^2}(t, T) - \{\hat{u}(t, T)\}^2 \tag{8.4.6}$$

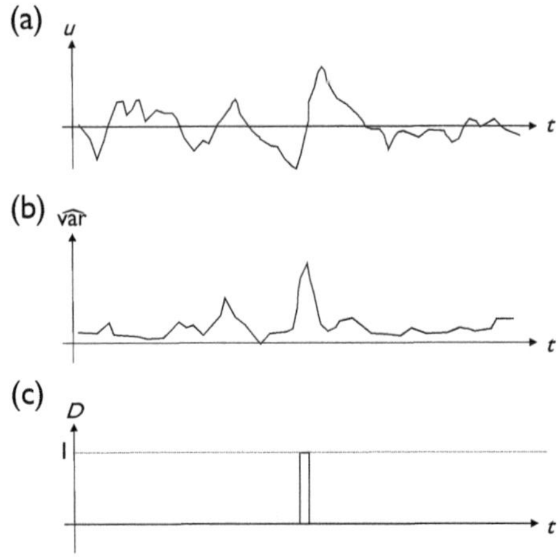

Figure 8.16 Example of time series in (a) velocity fluctuation; (b) variance; and (c) corresponding detection function drawn by referring to Blackwelder & Kaplan (1976).

The increase of the local variance implies the significant variation of velocity. To clear the occurrence of the bursting event, a following detection function is introduced:

$$D(t) = \begin{cases} 1 & if\ \widehat{\mathrm{var}} > ku'^2 \\ 0 & \text{otherwise} \end{cases} \tag{8.4.7}$$

in which $u'^2 = \lim_{T\to\infty} \widehat{\mathrm{var}}$. k is a threshold to decide how large the bursting is detected. The integral time scale, T, corresponds to mean time interval of bursting events. Blackwelder and Kaplan analyzed the streamwise velocity fluctuation using $k = 1.2$ and $T^+ \equiv TU_*/\nu = 10$, as shown in FIG. 8.16. Several peaks of the localized variation appear corresponding to the rapid change of the velocity fluctuation. One bursting point is detected, meeting the requirement expressed by Eq. (8.4.7).

8.4.3 Proper Orthogonal Decomposition (POD)

The conventional space-time correlation analysis is often conducted by the PIV data to reveal three-dimensional vortex structure. To decompose such a time-averaged vortex into velocity fluctuation modes, the present study examined the proper orthogonal decomposition technique (POD), which was developed by Lumley (1981). He applied the POD to turbulent flow and discussed the importance of POD. If there exists a sequence of orthogonal eigenfunctions and the associated eigenvalues that converge optimally and rapidly, the spatial structure of the turbulent field is assumed to be composed of such eigenfunctions. The convergence rate of the eigenvalues is used as a sensitive indicator of the presence of coherent structure. This method captures a function that has maximum correlation with instantaneous variables, such as $\tilde{u}(t)$.

Lumley (1981) introduced the following eigenvalue problem:

$$[R_u][\phi] = \lambda[\phi] \tag{8.4.8}$$

in which $[R_u]$ is $M \times M$ size space correlation matrix of velocity fluctuations of $\tilde{u}(t)$. $[\phi]$ is the eigenvector and λ is the eigenvalue. Since the eigenvectors are mutually orthogonal, the instantaneous velocity, $\tilde{u}$, can be decomposed in the following way:

$$\tilde{u}(x, y, z, t) = \sum_{m}^{M} a_m(t)\phi_m(x, y, z) \tag{8.4.9}$$

$$a_m(t) = \frac{\iiint \tilde{u}(x, y, z, t)\phi_m(x, y, z)dxdydz}{\iiint \phi_m{}^2(x, y, z)dxdydz} \tag{8.4.10}$$

in which a_m is the m-th order principal component and $\phi_m(x, y, z)$ is the m-th order eigenvector. If $\tilde{u}_k$ ($k = 1, 2, 3$) represents the velocity fluctuations at the k-th elevation of y, $\tilde{u}_k$ is given by

$$\tilde{u}_1(t) = \phi_{11}a_1(t) + \phi_{21}a_2(t) + \phi_{31}a_3(t) \tag{8.4.11a}$$

$$\tilde{u}_2(t) = \phi_{12}a_1(t) + \phi_{22}a_2(t) + \phi_{32}a_3(t) \tag{8.4.11b}$$

$$\tilde{u}_3(t) = \phi_{13}a_1(t) + \phi_{23}a_2(t) + \phi_{33}a_3(t) \tag{8.4.11c}$$

$\phi_{m,k}$ is the m-th order normalized component of eigenvector at k-th elevation. The contribution of m-th order fluctuations, C_m, is defined as

$$C_m = \frac{\lambda_m}{\lambda_1 + \lambda_2 + \lambda_3} \quad (m = 1, 2 \text{ and } 3) \tag{8.4.12}$$

8.4.4 Vortex Detection

Researchers have proposed several mathematical methods to detect a vortex core of coherent structure in shear layers. One of them is the delta method of Chong & Perry (1990). They considered the eigenvalue of the velocity shear tensor ($\partial\tilde{u}_i/\partial x_j$) in shear layers. The eigenvalue equation of 2-D shear flow is given by

$$\sigma^2 - P\sigma + Q = 0 \tag{8.4.13}$$

in which

$$P \equiv \frac{\partial \tilde{u}_i}{\partial x_i} = 0 \quad \text{and} \quad Q \equiv \frac{1}{2}\left(\left(\frac{\partial \tilde{u}_i}{\partial x_i}\right)^2 - \frac{\partial \tilde{u}_i}{\partial x_j}\frac{\partial \tilde{u}_j}{\partial x_i}\right) \tag{8.4.14}$$

$$\Delta \equiv P^2 - 4Q \tag{8.4.15}$$

where $\tilde{u}_i$ is the instantaneous velocity components with $i = 1$ and 2 herein. Complex eigenvalues occur if the discriminant, Δ, is negative. Chong & Perry (1990) assumed that $\Delta < 0$ corresponds to the existence of a vortex core, and verified the validity of such a vortex detection, referred to as the delta method ($\Delta < 0$) in the shear layers.

Please refer back to FIG. 3.31, which shows an example of the distribution of instantaneous velocity vectors subtracted by $U_c = 0.9U_s$ in open-channel flow, together with the contour of Δ value calculated by Eq. (8.4.15) in-channel flow case at intervals of 0.1 s. U_s is the time-averaged free-surface streamwise velocity, i.e., $U_s = U\,(y = H)$, which becomes the maximum velocity in open-channel flow. Large negative distributions of Δ, i.e., corresponding to the rotation motion, are observed in the hairpin heads. The swirling and ejection motions correspond well to the head and leg parts of the hairpin vortex model proposed by Adrian et al. (2000).

8.5 EXAMPLE OF MEASUREMENT SETUP

8.5.1 Recirculation Flume

FIG. 8.17(a) shows an example of components in a recirculation flume system. The side walls are likely to be made of transparent glass or acrylic board. This is because it is useful for flow observation. Also, using the transparent bottom wall, the light sheet can be projected from the flume bottom without being influenced by the free-surface fluctuation. The tail gate is needed in the channel outlet to adjust the water depth. Installing a contraction zone with honeycomb in the channel inlet is effective for flow straightening (FIG. 8.17(b)).

It is possible to adjust the channel slope, supporting the flume body by jack or chain block.

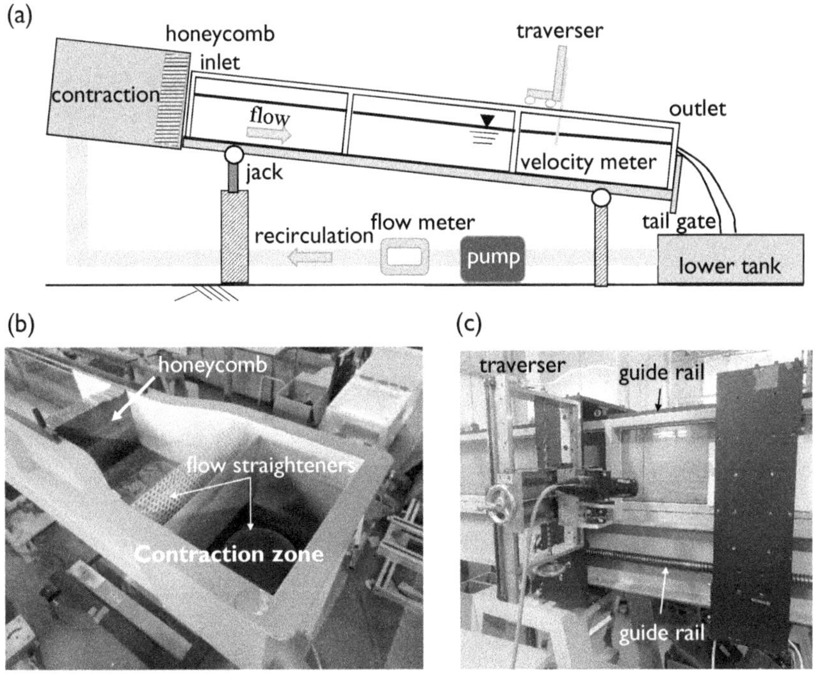

Figure 8.17 Example of typical experimental flume: (a) recirculation system; (b) contraction zone and flow straighteners; (c) traverser system.

The size of the lower tank should be chosen, considering the maximum discharge. Further, the electric pump and related piping are needed to convey water to a head tank. The inflow discharge can be adjusted by a valve in a drain pipe of the head tank. When the discharge is directly controlled by the inverter pump, the head tank is unnecessary. The discharge is usually monitored by the electromagnetic flow meter, the principle of which is explained in section 8.3. A feedback control of the inverter pump using the electromagnetic flow meter is able to generate any time-dependent inflow discharge curve. A bypass pipe is usually used to prevent the decrease of pump rotation speed for a low discharge condition.

8.5.2 Measurement Instrumentation

Instrumentation such as flow meter, water gauge and other sensing devices are fixed near the experimental flume. The traverser is needed to make the sensor meet with a sampling point accurately (FIG. 8.17(c)). The guide rail along the streamwise direction is useful to move the traverser in the wide range.

A data logger can record simultaneous output signals of several sensors, including velocimetry. Further, a function generator makes a trigger signal for synchronization of various kinds of instrumentations, such as a combination of PIV and capacitance wave gauge.

When we study the turbulence structure of in the straight open-channel flow, the distance from the inlet should be noticed. Kirlkgoz & Ardichoglu (1997) proposed the following formula about the developing distance, x_d, in the open-channel flows:

$$x_d/h = 76 - 0.0001\,\mathrm{Re}/\mathrm{Fr} \tag{8.5.1}$$

in which h is the water depth. Normalized developing distance, x_d/h, can be evaluated by a function of ratio of Re to Fr. It is generally known that a larger depth condition requires a longer developing distance of the open-channel turbulence or the boundary layer. For example, in the case of h = 20 cm with Re = 20,000 and Fr = 0.07, this formula results in the developing distance as x_d = 9.5 m.

8.5.3 Experimental Procedure

The example of experimental procedure for the case of measurement in the straight uniform flow is shown below:

a. generation of uniform flow

The uniform flow without spatial variation of the water depth is occurred by adjusting the tail gate and the flume gradient. Also, the inflow should not be included with a low frequency oscillation.

b. choice of measurement section

The developing distance should be considered as previously mentioned. Further, it must not be affected by the backwater due to the tail gate.

c. measurement of velocity

The sensing point of the velocimetry should be adjusted carefully.

d. measurement of water depth

A point gauge, an ultrasonic distance sensor and capacitance wave gauge are often used. The imaging analysis can measure instantaneous depth profile, as shown in FIG. 8.18 (Sanjou et al. 2010).

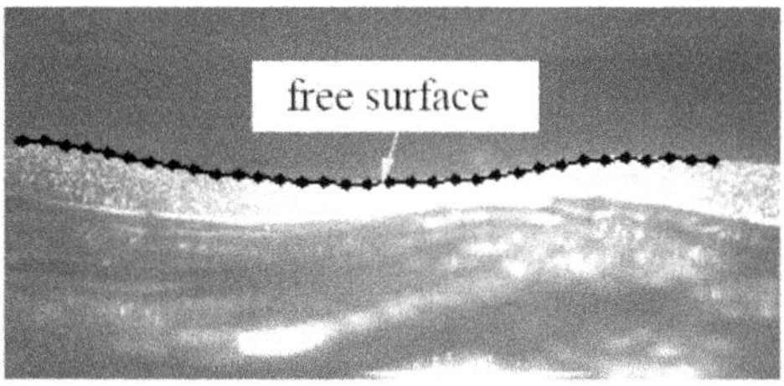

Figure 8.18 Instantaneous water surface profile measured by imaging technique.

e. measurement of water temperature

The water temperature should be measured, because the viscosity and density of the fluid depend on it.

8.5.4 Examples of Experimental Flume

Experimental studies have been conducted in the turbulent hydraulic laboratory in Kyoto University (FIG. 8.19(a)) in order to reveal open-channel turbulence. Characteristic flumes are used depending on respective purposes.

a. Steep-slope flume

The size has 10 m in length and 40 cm in width. The flame is made of iron. A specially designed jack realizes a maximum slope of 1/10 (FIG. 8.19(b)). Hence, it can generate a high-speed supercritical flow. The glass-made side and bottom walls are installed for the flow visualization measurements. Research products due to this flume were reported by Nezu & Onitsuka (2001), Nezu & Sanjou (2011), Sanjou et al. (2012), Okamoto & Nezu (2013), Sanjou & Nezu (2013), Sanjou et al. (2018a) and Okamoto et al. (2021).

b. Wide width flume and self-propelled carrier

The size is 9 m in length and 150 cm in width. Maximum discharge 120 l/s is generated by three inverter pumps (FIG. 8.19(c)). It is used for the study on mass and momentum transfer with the lateral direction like compound channel, side-cavity flow and inundation flow.

Lagrange tracking of horizontal vortex is possible using a high-speed camera place on a self-propelled carrier as shown in FIG. 8.19(d). Also, this flume is used for the development in autonomous USV introduced in section 8.3. Research products due to this flume system were reported by Sanjou et al. (2010), Sanjou & Nezu (2010), Sanjou & Nezu (2011b), Sanjou & Nezu (2017), Sanjou & Nagasaka (2017) and Sanjou et al. (2021).

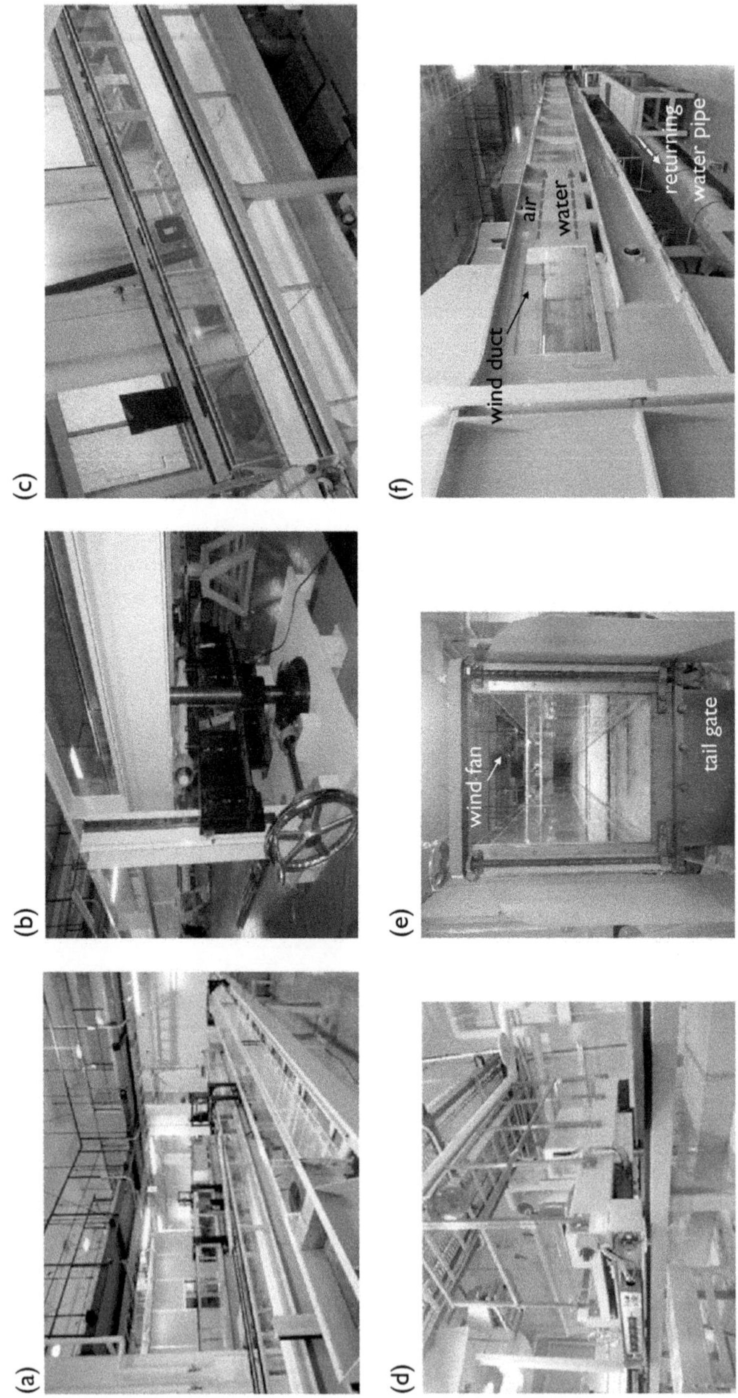

Figure 8.19 Examples of experimental flume: (a) turbulence hydraulics laboratory in Kyoto University; (b) steep-slope flume; (c) wide width flume; (d) self-propelled carrier; (e) multi-purpose flume; (f) wind tunnel tank.

c. Multi-purpose flume

The size is 10 m in length and 40 cm in width. A large fan and suction system is installed in order to generate air/water two-phase flow and seepage flow (FIG. 8.19(e)). Furthermore, it has been used for experimental researches on rough bed turbulence like a submerged vegetation flow and sediment-laden flow. A computer-controlled inverter pump can generate unsteady open-channel turbulent flow. Research products due to this flume system were reported by Nezu & Sanjou (2006), Nezu & Sanjou (2008), Okamoto & Nezu (2009), Okamoto & Nezu (2010), Okamoto et al. (2012), Okamoto et al. (2016) and Okamoto et al. (2018).

d. Recirculating tank with wind tunnel

The size is 16 m in length and 40 cm in width, as shown in FIG. 8.19(f). An electric fan generates air over the free surface of water current. Air/water interfacial dynamics, surface velocity divergence, Langmuir circulation and related gas transfer mechanism have been studied. Particularly, the gas transfer velocity through the free surface has been measured under various hydraulic condition. Research products due to this channel were reported by Sanjou et al. (2010b), Sanjou & Nezu (2011a), Sanjou & Nezu (2014), Sanjou et al. (2017), Sanjou et al. (2018b) and Sanjou (2020).

Appendix 1

Reynolds Decomposition

A.1.1 REYNOLDS AVERAGING

N-S equations hold for instantaneous velocity components and pressure. Decomposition of instantaneous value into time-mean and deviation is called Reynolds decomposition, as follows:

$$\begin{aligned}
\tilde{u}(x, y, z, t) &= U(x, y, z) + u(x, y, z, t) \\
\tilde{v}(x, y, z, t) &= V(x, y, z) + v(x, y, z, t) \\
\tilde{w}(x, y, z, t) &= W(x, y, z) + w(x, y, z, t) \\
\underline{\tilde{p}(x, y, z, t)} &= \underline{P(x, y, z)} + \underline{p(x, y, z, t)}
\end{aligned}$$

(instantaneous value) = (time mean value) + (deviation)

A time average of an instantaneous value $\tilde{f}$ is defined by $F = \overline{\tilde{f}} = \frac{1}{T}\int_0^T \tilde{f}dt$.

T is a time averaging period. A time average of deviation is zero, as indicated by (A.1.1). This fact implies that the instantaneous value varies with time around the time mean value.

$$\overline{u} = \overline{v} = \overline{w} = \overline{p} = 0 \tag{A.1.1}$$

The following formulae related to averaging operations are obtained.

(average of summation)

$$\overline{\tilde{f}_1 + \tilde{f}_2} = \frac{1}{T}\int_0^T \tilde{f}_1 + \tilde{f}_2 dt = \frac{1}{T}\int_0^T \tilde{f}_1 dt + \frac{1}{T}\int_0^T \tilde{f}_2 dt = \overline{\tilde{f}_1} + \overline{\tilde{f}_2} = F_1 + F_2 \tag{A.1.2}$$

(average of averaged value)

$$\overline{\overline{f}} = \frac{1}{T}\int_0^T \overline{f}\,dt = \frac{1}{T}\overline{f}\int_0^T dt = \overline{f} = F \tag{A.1.3}$$

(average of product of averaged value and instantaneous value)

$$\overline{\overline{f_1}\tilde{f_2}} = \frac{1}{T}\int_0^T \overline{f_1}\tilde{f_2}\,dt = \frac{1}{T}\overline{f_1}\int_0^T \tilde{f_2}\,dt = \overline{f_1}\,\overline{\tilde{f_2}} = F_1F_2 \tag{A.1.4}$$

(average of product of instantaneous values)

$$\begin{aligned}\overline{\tilde{f_1}\tilde{f_2}} &= \overline{(\overline{f_1} + f_1)(\overline{f_2} + f_2)} = \overline{(F_1 + f_1)(F_2 + f_2)} = \overline{F_1F_2 + F_1f_2 + f_1F_2 + f_1f_2} \\ &= \overline{F_1F_2} + \overline{F_1f_2} + \overline{f_1F_2} + \overline{f_1f_2} = F_1F_2 + \overline{f_1f_2}\end{aligned} \tag{A.1.5}$$

(average of differential)

$$\overline{\frac{d\tilde{f}}{dx}} = \frac{1}{T}\int_0^T \frac{d\tilde{f}}{dx}dt = \frac{d}{dx}\left(\frac{1}{T}\int_0^T \tilde{f}\,dt\right) = \frac{d\overline{\tilde{f}}}{dx} \tag{A.1.6}$$

(average of integral)

$$\overline{\int \tilde{f}\,dx} = \frac{1}{T}\int_0^T \int \tilde{f}\,dx\,dt = \int \frac{1}{T}\int_0^T \tilde{f}\,dt\,dx = \int \overline{\tilde{f}}\,dx \tag{A.1.7}$$

Averaging the governing equation according to the above formulae is called Reynolds averaging.

A.1.2 REYNOLDS AVERAGING OF CONTINUITY EQUATION

The instantaneous continuity equation is as follows:

$$\frac{\partial \tilde{u}}{\partial x} + \frac{\partial \tilde{v}}{\partial y} + \frac{\partial \tilde{w}}{\partial z} = 0 \tag{A.1.8}$$

The Reynolds decomposition of Eq. (A.1.8) leads to

$$\frac{\partial(U+u)}{\partial x}+\frac{\partial(V+v)}{\partial y}+\frac{\partial(W+w)}{\partial z}=0 \tag{A.1.9}$$

The time average of each term leads to

$$\overline{\frac{\partial(U+u)}{\partial x}}+\overline{\frac{\partial(V+v)}{\partial y}}+\overline{\frac{\partial(W+w)}{\partial z}}=0 \leftrightarrow \frac{\partial\overline{(U+u)}}{\partial x}+\frac{\partial\overline{(V+v)}}{\partial y}+\frac{\partial\overline{(W+w)}}{\partial z}=0$$

As the time average of each deviation component is zero, a following form is obtained.

$$\frac{\partial\overline{(U)}}{\partial x}+\frac{\partial\overline{(V)}}{\partial y}+\frac{\partial\overline{(W)}}{\partial z}=0 \leftrightarrow \frac{\partial U}{\partial x}+\frac{\partial V}{\partial y}+\frac{\partial W}{\partial z}=0 \tag{A.1.10}$$

Hence, the continuity equation holds for the time-averaged components. Furthermore, a combination of Eqs. (A.3.9) and (A.3.10) leads to

$$\frac{\partial u}{\partial x}+\frac{\partial v}{\partial y}+\frac{\partial w}{\partial z}=0 \tag{A.1.11}$$

Therefore, the continuity equation holds also for the deviation components.

A.1.3 REYNOLDS AVERAGING OF N-S EQUATIONS (DERIVATION OF RANS)

Although the external force term is ignored for simplicity here, only gravitational force is usually considered in open-channel flow. Since the gravity is constant, Reynolds averaging of the external force remains unchanged.

N-S equations are written by

$$\begin{aligned}
\frac{\partial\tilde{u}}{\partial t}+\tilde{u}\frac{\partial\tilde{u}}{\partial x}+\tilde{v}\frac{\partial\tilde{u}}{\partial y}+\tilde{w}\frac{\partial\tilde{u}}{\partial z}&=-\frac{1}{\rho}\frac{\partial\tilde{p}}{\partial x}+\nu\left(\frac{\partial^2\tilde{u}}{\partial x^2}+\frac{\partial^2\tilde{u}}{\partial y^2}+\frac{\partial^2\tilde{u}}{\partial z^2}\right)\\
\frac{\partial\tilde{v}}{\partial t}+\tilde{u}\frac{\partial\tilde{v}}{\partial x}+\tilde{v}\frac{\partial\tilde{v}}{\partial y}+\tilde{w}\frac{\partial\tilde{v}}{\partial z}&=-\frac{1}{\rho}\frac{\partial\tilde{p}}{\partial y}+\nu\left(\frac{\partial^2\tilde{v}}{\partial x^2}+\frac{\partial^2\tilde{v}}{\partial y^2}+\frac{\partial^2\tilde{v}}{\partial z^2}\right)\\
\frac{\partial\tilde{w}}{\partial t}+\tilde{u}\frac{\partial\tilde{w}}{\partial x}+\tilde{v}\frac{\partial\tilde{w}}{\partial y}+\tilde{w}\frac{\partial\tilde{w}}{\partial z}&=-\frac{1}{\rho}\frac{\partial\tilde{p}}{\partial z}+\nu\left(\frac{\partial^2\tilde{w}}{\partial x^2}+\frac{\partial^2\tilde{w}}{\partial y^2}+\frac{\partial^2\tilde{w}}{\partial z^2}\right)
\end{aligned} \tag{A.1.12}$$

We focus on each term in the equation of $\tilde{u}$.

$$\text{(acceleration term)} \quad \overline{\frac{\partial \tilde{u}}{\partial t}} = \frac{\partial \overline{\tilde{u}}}{\partial t} = \frac{\partial U}{\partial t}$$

This is zero because time variation of time averaged velocity. However, when U is considered to be averaged during a shorter period than time scale of the unsteady fluid phenomenon, it may not be zero.

(convection term)

$$\overline{\tilde{u}\frac{\partial \tilde{u}}{\partial x}} = \overline{(U+u)\frac{\partial (U+u)}{\partial x}} = \overline{U\frac{\partial U}{\partial x}} + \overline{U\frac{\partial u}{\partial x}} + \overline{u\frac{\partial U}{\partial x}} + \overline{u\frac{\partial u}{\partial x}} = \overline{U}\frac{\partial \overline{U}}{\partial x} + \overline{u\frac{\partial u}{\partial x}}$$
$$= U\frac{\partial U}{\partial x} + \overline{u\frac{\partial u}{\partial x}}$$

$$\overline{\tilde{v}\frac{\partial \tilde{u}}{\partial y}} = \overline{(V+v)\frac{\partial (U+u)}{\partial y}} = \overline{V\frac{\partial U}{\partial y}} + \overline{V\frac{\partial u}{\partial y}} + \overline{v\frac{\partial U}{\partial y}} + \overline{v\frac{\partial u}{\partial y}} = \overline{V}\frac{\partial \overline{U}}{\partial y} + \overline{v\frac{\partial u}{\partial y}}$$
$$= V\frac{\partial U}{\partial y} + \overline{v\frac{\partial u}{\partial y}}$$

$$\overline{\tilde{w}\frac{\partial \tilde{u}}{\partial z}} = \overline{(W+w)\frac{\partial (U+u)}{\partial z}} = \overline{W\frac{\partial U}{\partial z}} + \overline{W\frac{\partial u}{\partial z}} + \overline{w\frac{\partial U}{\partial z}} + \overline{w\frac{\partial u}{\partial z}} = \overline{W}\frac{\partial \overline{U}}{\partial z}$$
$$+ \overline{w\frac{\partial u}{\partial z}} = W\frac{\partial U}{\partial z} + \overline{w\frac{\partial u}{\partial z}}$$

(pressure gradient term) $\overline{\frac{\partial \tilde{p}}{\partial x}} = \frac{\partial \overline{\tilde{p}}}{\partial x} = \frac{\partial P}{\partial x}$

(diffusion term)

$$\overline{\nu\left(\frac{\partial^2 \tilde{u}}{\partial x^2} + \frac{\partial^2 \tilde{u}}{\partial y^2} + \frac{\partial^2 \tilde{u}}{\partial z^2}\right)} = \nu\left(\overline{\frac{\partial^2 \tilde{u}}{\partial x^2}} + \overline{\frac{\partial^2 \tilde{u}}{\partial y^2}} + \overline{\frac{\partial^2 \tilde{u}}{\partial z^2}}\right) = \nu\left(\frac{\partial^2 \overline{\tilde{u}}}{\partial x^2} + \frac{\partial^2 \overline{\tilde{u}}}{\partial y^2} + \frac{\partial^2 \overline{\tilde{u}}}{\partial z^2}\right)$$
$$= \nu\left(\frac{\partial^2 U}{\partial x^2} + \frac{\partial^2 U}{\partial y^2} + \frac{\partial^2 U}{\partial z^2}\right)$$

Therefore, a following form is obtained.

$$\frac{\partial U}{\partial t} + U\frac{\partial U}{\partial x} + V\frac{\partial U}{\partial y} + W\frac{\partial U}{\partial z} + \overline{u\frac{\partial u}{\partial x}} + \overline{v\frac{\partial u}{\partial y}} + \overline{w\frac{\partial u}{\partial z}} = -\frac{1}{\rho}\frac{\partial P}{\partial x} + \nu\left(\frac{\partial^2 U}{\partial x^2} + \frac{\partial^2 U}{\partial y^2} + \frac{\partial^2 U}{\partial z^2}\right) \tag{A.1.13}$$

Noting the fifth to seventh terms on the left-hand side, further expansion is as follows:

$$\frac{\partial U}{\partial t} + U\frac{\partial U}{\partial x} + V\frac{\partial U}{\partial y} + W\frac{\partial U}{\partial z} + \overline{\frac{\partial uu}{\partial x} - u\frac{\partial u}{\partial x}} + \overline{\frac{\partial uv}{\partial y} - u\frac{\partial v}{\partial y}} + \overline{\frac{\partial uw}{\partial z} - u\frac{\partial w}{\partial z}}$$
$$= -\frac{1}{\rho}\frac{\partial P}{\partial x} + \nu\left(\frac{\partial^2 U}{\partial x^2} + \frac{\partial^2 U}{\partial y^2} + \frac{\partial^2 U}{\partial z^2}\right)$$

$$\leftrightarrow \frac{\partial U}{\partial t} + U\frac{\partial U}{\partial x} + V\frac{\partial U}{\partial y} + W\frac{\partial U}{\partial z} + \frac{\partial \bar{u}u}{\partial x} + \frac{\partial \bar{u}v}{\partial y} + \frac{\partial \bar{u}w}{\partial z} - u\left(\frac{\partial u}{\partial x} + \frac{\partial \bar{v}}{\partial y} + \frac{\partial w}{\partial z}\right)$$
$$= -\frac{1}{\rho}\frac{\partial P}{\partial x} + \nu\left(\frac{\partial^2 U}{\partial x^2} + \frac{\partial^2 U}{\partial y^2} + \frac{\partial^2 U}{\partial z^2}\right)$$

An eighth term on the left-hand side is zero due to Eq. (A.1.11). Hence, a following $\tilde{u}$ component of RANS is derived.

$$\frac{\partial U}{\partial t} + U\frac{\partial U}{\partial x} + V\frac{\partial U}{\partial y} + W\frac{\partial U}{\partial z} = -\frac{1}{\rho}\frac{\partial P}{\partial x} + \nu\left(\frac{\partial^2 U}{\partial x^2} + \frac{\partial^2 U}{\partial y^2} + \frac{\partial^2 U}{\partial z^2}\right) - \frac{\partial \overline{uu}}{\partial x} - \frac{\partial \overline{uv}}{\partial y} - \frac{\partial \overline{uw}}{\partial z}$$

$$\frac{\partial V}{\partial t} + U\frac{\partial V}{\partial x} + V\frac{\partial V}{\partial y} + W\frac{\partial V}{\partial z} = -\frac{1}{\rho}\frac{\partial P}{\partial y} + \nu\left(\frac{\partial^2 V}{\partial x^2} + \frac{\partial^2 V}{\partial y^2} + \frac{\partial^2 V}{\partial z^2}\right) - \frac{\partial \overline{uv}}{\partial x} - \frac{\partial \overline{vv}}{\partial y} - \frac{\partial \overline{vw}}{\partial z}$$

$$\frac{\partial W}{\partial t} + U\frac{\partial W}{\partial x} + V\frac{\partial W}{\partial y} + W\frac{\partial W}{\partial z} = -\frac{1}{\rho}\frac{\partial P}{\partial z} + \nu\left(\frac{\partial^2 W}{\partial x^2} + \frac{\partial^2 W}{\partial y^2} + \frac{\partial^2 W}{\partial z^2}\right) - \frac{\partial \overline{uw}}{\partial x} - \frac{\partial \overline{vw}}{\partial y} - \frac{\partial \overline{ww}}{\partial z}$$

(A.1.14)

The third to fifth terms on the right-hand side of each equation are terms related to Reynolds stress. The above calculation means that Reynolds stress is produced by the Reynolds average of the convection term.

A.1.4 EINSTEIN NOTATION

Einstein notation makes complicated mathematical forms more simple. For example, N-S equations are described by

$$\frac{\partial \tilde{u}_i}{\partial t} + \tilde{u}_k \frac{\partial \tilde{u}_i}{\partial x_k} = -\frac{1}{\rho}\frac{\partial \tilde{p}}{\partial x_i} + \frac{\partial}{\partial x_k}(2\nu \tilde{d}_{ik}), \text{ in which}$$

$$\tilde{d}_{ik} = \frac{1}{2}\left(\frac{\partial \tilde{u}_i}{\partial x_k} + \frac{\partial \tilde{u}_k}{\partial x_i}\right) \text{ and } i = 1,2,3.$$

The convection term is transformed by the continuity equation. It results in

$$\frac{\partial \tilde{u}_i}{\partial t} + \frac{\partial \tilde{u}_i \tilde{u}_k}{\partial x_k} = -\frac{1}{\rho}\frac{\partial \tilde{p}}{\partial x_i} + \frac{\partial}{\partial x_k}(2\nu \tilde{d}_{ik}) \tag{A.1.15}$$

The Reynolds average of each term is $\overline{\frac{\partial \tilde{u}_i}{\partial t}} + \overline{\frac{\partial \tilde{u}_i \tilde{u}_k}{\partial x_k}} = \overline{-\frac{1}{\rho}\frac{\partial \tilde{p}}{\partial x_i}} + \overline{\frac{\partial}{\partial x_k}(2\nu \tilde{d}_{ik})}$.

Further, the previously mentioned formulae of Reynolds averaging are applied. Particularly, the convection term is expanded as follows:

$$\overline{\frac{\partial \tilde{u}_i \tilde{u}_k}{\partial x_k}} = \overline{\frac{\partial (U_i + u_i)(U_k + u_k)}{\partial x_k}} = \overline{\frac{\partial (U_i U_k + U_i u_k + u_i U_k + u_i u_k)}{\partial x_k}}$$

$$= \frac{\partial \overline{(U_i U_k)}}{\partial x_k} + \frac{\partial \overline{(U_i u_k)}}{\partial x_k} + \frac{\partial \overline{(u_i U_k)}}{\partial x_k} + \frac{\partial \overline{(u_i u_k)}}{\partial x_k}$$

$$= \frac{\partial (U_i U_k)}{\partial x_k} + \frac{\partial \overline{(u_i u_k)}}{\partial x_k}$$

The term related to Reynolds stress is produced. After similar calculations for other terms, RANS is derived as follows.

$$\frac{\partial U_i}{\partial t} + \frac{\partial U_i U_k}{\partial x_k} = -\frac{1}{\rho}\frac{\partial P}{\partial x_i} + \frac{\partial}{\partial x_k}(2\nu D_{ik} - R_{ik}) \tag{A.1.16}$$

in which $D_{ik} = \frac{1}{2}\left(\frac{\partial U_i}{\partial x_k} + \frac{\partial U_k}{\partial x_i}\right)$, $R_{ik} = \overline{u_i u_k}$. $\rho \overline{u_i u_k}$ is Reynolds stress.

Appendix 2

Calculation Procedure of Blasius Solution

The final goal is to solve the velocity profile in the laminar boundary layer as shown in FIG. A.2.1. Assuming the steady ($\partial/\partial t = 0$) and laminar ($\tilde{u} = U$) conditions, the boundary layer equation is simplified as follows.

$$U\frac{\partial U}{\partial x} + V\frac{\partial U}{\partial y} = \nu\frac{\partial^2 U}{\partial y^2} \tag{A.2.1}$$

Continuity equation is expressed by

$$\frac{\partial U}{\partial x} + \frac{\partial V}{\partial y} = 0 \tag{A.2.2}$$

Rayleigh problem provides a following form of the boundary layer thickness, $\delta \sim \sqrt{\frac{\nu x}{U_\infty}}$. Here, a following coordinate transform is introduced.

$$\xi = x, \quad \eta = \frac{y}{\delta} = y\sqrt{\frac{U_\infty}{\nu x}} \tag{A.2.3}$$

Further, a following stream function $\Psi(\xi, \eta)$ is introduced to reduce the number of variables. It is required to find the normalized function $f(\eta)$, which is applicable to the boundary layer flow.

$$\Psi(\xi, \eta) = \sqrt{\nu x U_\infty} f(\eta) \tag{A.2.4}$$

In the former part, the boundary layer equation (A.2.1) is expressed by $f(\eta)$, and the latter part solves the rewritten (A.2.1).

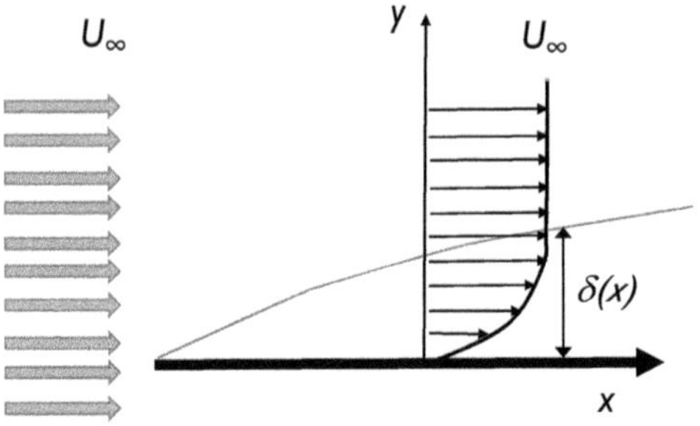

Figure A.2.1 Developing boundary layer.

A.2.1 FORMER PROCESS

Some partial differentials are calculated. First, the differential of η is as follows.

$$\frac{\partial \eta}{\partial x} = -\frac{y}{2}\sqrt{\frac{U_\infty}{\nu}}x^{-3/2} = -\frac{y}{2x}\sqrt{\frac{U_\infty}{\nu x}} = -\frac{\eta}{2x} \tag{A.2.5}$$

$$\frac{\partial \eta}{\partial y} = \sqrt{\frac{U_\infty}{\nu x}} \tag{A.2.6}$$

Second, the differential of $\Psi(\xi, \eta)$ is as follows.

$$\frac{\partial \Psi(\xi,\eta)}{\partial x} = \frac{\partial \Psi}{\partial \xi}\frac{\partial \xi}{\partial x} + \frac{\partial \Psi}{\partial \eta}\frac{\partial \eta}{\partial x} = \frac{\partial \Psi}{\partial \xi}\cdot 1 + \frac{\partial \Psi}{\partial \eta}\left(-\frac{\eta}{2x}\right) = \left(\frac{\partial}{\partial \xi} - \frac{\eta}{2x}\frac{\partial}{\partial \eta}\right)\Psi \tag{A.2.7}$$

$$\frac{\partial \Psi(\xi,\eta)}{\partial y} = \frac{\partial \Psi}{\partial \xi}\frac{\partial \xi}{\partial y} + \frac{\partial \Psi}{\partial \eta}\frac{\partial \eta}{\partial y} = 0 + \frac{\partial \Psi}{\partial \eta}\left(\sqrt{\frac{U_\infty}{\nu x}}\right) = \sqrt{\frac{U_\infty}{\nu x}}\frac{\partial \Psi}{\partial \eta} \tag{A.2.8}$$

$$\begin{aligned}\frac{\partial^2 \Psi(\xi,\eta)}{\partial y^2} &= \frac{\partial}{\partial y}\frac{\partial \Psi}{\partial y} = \frac{\partial}{\partial y}\left(\frac{1}{2}\sqrt{\frac{U_\infty}{\nu x}}\frac{\partial \Psi}{\partial \eta}\right) = \frac{1}{2}\sqrt{\frac{U_\infty}{\nu x}}\frac{\partial}{\partial y}\left(\frac{\partial \Psi}{\partial \eta}\right)\\ &= \frac{1}{2}\sqrt{\frac{U_\infty}{\nu x}}\frac{\partial}{\partial \eta}\frac{\partial \eta}{\partial y}\left(\frac{\partial \Psi}{\partial \eta}\right) = \frac{1}{2}\sqrt{\frac{U_\infty}{\nu x}}\frac{\partial}{\partial \eta}\left(\sqrt{\frac{U_\infty}{\nu x}}\right)\left(\frac{\partial \Psi}{\partial \eta}\right)\\ &= \frac{1}{2}\frac{U_\infty}{\nu x}\frac{\partial^2 \Psi}{\partial \eta^2}\end{aligned} \tag{A.2.9}$$

The velocities can be calculated by the stream function. Eq. (A.2.4) leads

$$\frac{\partial \Psi}{\partial \eta} = \sqrt{\nu x U_\infty}\frac{\partial f(\eta)}{\partial \eta} = \sqrt{\nu x U_\infty} f' \tag{A.2.10}$$

Further, a following differential is obtained.

$$\frac{\partial \Psi}{\partial \xi} = \frac{\partial \sqrt{\nu x U_\infty} f(\eta)}{\partial \xi} = \frac{\partial \sqrt{\nu \xi U_\infty} f(\eta)}{\partial \xi} = \frac{1}{2}\xi^{-1/2}\sqrt{\nu U_\infty} f = \frac{1}{2}x^{-1/2}\sqrt{\nu U_\infty} f$$
$$= \frac{1}{2}\sqrt{\frac{\nu U_\infty}{x}} f \tag{A.2.11}$$

$$U = \frac{\partial \Psi}{\partial y} = \sqrt{\frac{U_\infty}{\nu x}}\frac{\partial \Psi}{\partial \eta} = U_\infty f' \tag{A.2.12}$$

They lead to

$$V = -\frac{\partial \Psi}{\partial x} = -\left(\frac{\partial \Psi}{\partial \xi} - \frac{\eta}{2x}\frac{\partial \Psi}{\partial \eta}\right) = -\frac{1}{2}\sqrt{\frac{\nu U_\infty}{x}} f + \frac{\eta}{2x}\sqrt{\nu x U_\infty} f'$$
$$= -\frac{1}{2}\sqrt{\frac{\nu U_\infty}{x}}(f - \eta f') \tag{A.2.13}$$

Expansion of Ψ to an arbitrary function A gives $\frac{\partial}{\partial x}A = \left(\frac{\partial}{\partial \xi} - \frac{\eta}{2x}\frac{\partial}{\partial \eta}\right)A$ and $\frac{\partial}{\partial y}A = \sqrt{\frac{U_\infty}{\nu x}}\frac{\partial A}{\partial \eta}$. Using these expressions, the following equations are derived, respectively.

$$\frac{\partial U}{\partial x} = \frac{\partial}{\partial x}(U_\infty f') = \left(\frac{\partial}{\partial \xi} - \frac{\eta}{2x}\frac{\partial}{\partial \eta}\right)(U_\infty f') = -\frac{\eta U_\infty}{2x} f'' \tag{A.2.14}$$

$$\frac{\partial U}{\partial y} = \frac{\partial}{\partial y}(U_\infty f') = \sqrt{\frac{U_\infty}{\nu x}}\frac{\partial}{\partial \eta}(U_\infty f') = U_\infty\sqrt{\frac{U_\infty}{\nu x}} f'' \tag{A.2.15}$$

Furthermore, a following equation is obtained.

$$\frac{\partial^2 U}{\partial y^2} = \frac{\partial}{\partial y}\left(\frac{\partial U}{\partial y}\right) = \sqrt{\frac{U_\infty}{\nu x}}\frac{\partial}{\partial \eta}\left(\frac{\partial U}{\partial y}\right) = \sqrt{\frac{U_\infty}{\nu x}}\frac{\partial}{\partial \eta}\left(U_\infty\sqrt{\frac{U_\infty}{\nu x}} f''\right) = \frac{U_\infty^2}{\nu x} f''' \tag{A.2.16}$$

Substitution of Eqs. (A.2.12) to (A.2.16) into Eq. (A.2.1) leads to

$$\begin{aligned}
&U_\infty f' \cdot \left(-\frac{\eta}{2}\frac{U_\infty}{x}f''\right) - \frac{1}{2}\sqrt{\frac{\nu U_\infty}{x}}(f - \eta f') \cdot U_\infty \sqrt{\frac{U_\infty}{\nu x}} f'' = \nu \frac{U_\infty^2}{\nu x} f''' \\
&\leftrightarrow \; -\frac{\eta}{2}\frac{U_\infty^2}{x} f' f'' - \frac{U_\infty^2}{2x}(f - \eta f') f'' = \frac{U_\infty^2}{x} f''' \\
&\leftrightarrow \; -\frac{U_\infty^2}{2x} f f'' = \frac{U_\infty^2}{x} f''' \\
&\leftrightarrow \; ff'' + 2f''' = 0
\end{aligned} \tag{A.2.17}$$

This is the differential equation of f. Boundary conditions are as follows:

$$\begin{aligned}
&at \quad \eta = 0, \quad U = 0 \quad \Rightarrow \; f' = 0 \\
&at \quad \eta = 0, \quad V = 0 \quad \Rightarrow \; f = 0 \\
&at \quad \eta = \infty, \quad U = U_\infty \quad \Rightarrow \; f' = 1
\end{aligned}$$

A.2.2 LATTER PROCESS

It is difficult to solve Eq. (A.2.17) because of the boundary problem with two distant boundary points Here, we consider the connection of asymptotic solutions at each boundary. The calculation procedure is as follows:

1. Calculation of asymptotic solution near $\eta = 0$
2. Calculation of asymptotic solution near $\eta = \infty$
3. Connection of them

(Step-1) Solution near $\eta = 0$

Taylor expansion of the function $f(\eta)$ around $\eta = 0$ is

$$\begin{aligned}
f(\eta) &= F_0 + F_1\eta + \frac{F_2}{2!}\eta^2 + \frac{F_3}{3!}\eta^3 + \frac{F_4}{4!}\eta^4 + \cdots \\
f'(\eta) &= F_1 + F_2\eta + \frac{F_3}{2!}\eta^2 + \frac{F_4}{3!}\eta^3 + \frac{F_5}{4!}\eta^4 + \cdots \\
f''(\eta) &= F_2 + F_3\eta + \frac{F_4}{2!}\eta^2 + \frac{F_5}{3!}\eta^3 + \frac{F_6}{4!}\eta^4 + \cdots \\
f'''(\eta) &= F_3 + F_4\eta + \frac{F_5}{2!}\eta^2 + \frac{F_6}{3!}\eta^3 + \frac{F_7}{4!}\eta^4 + \cdots
\end{aligned}$$

The previously mentioned boundary conditions, $f(0) = 0, \;\; f'(0) = 0$ lead to $F_0 = F_1 = 0$. Substitution of an expression of the Taylor expansion into Eq. (A.2.17) leads to

$$2 \times \left(F_3 + F_4\eta + \tfrac{F_5}{2!}\eta^2 + \tfrac{F_6}{3!}\eta^3 + \tfrac{F_7}{4!}\eta^4 + \cdots\right)$$
$$= \left(F_0 + F_1\eta + \tfrac{F_2}{2!}\eta^2 + \tfrac{F_3}{3!}\eta^3 + \tfrac{F_4}{4!}\eta^4 +\right) \tag{A.2.18}$$
$$\times \left(F_2 + F_3\eta + \tfrac{F_4}{2!}\eta^2 + \tfrac{F_5}{3!}\eta^3 + \tfrac{F_6}{4!}\eta^4 + \cdots\right)$$

Calculation of lower-order terms on the right-hand side results in

$$\frac{F_2^2}{2!}\eta^2 + \frac{F_2F_3}{2!}\eta^3 + \frac{F_2F_4}{2!2!}\eta^4 + \cdots$$
$$+ \frac{F_2F_3}{3!}\eta^3 + \frac{F_3^2}{3!}\eta^4 + \frac{F_2F_4}{2!3!}\eta^5 + \cdots$$
$$+ \frac{F_2F_3}{4!}\eta^4 + \frac{F_3F_4}{4!}\eta^5 + \frac{F_4^2}{2!4!}\eta^6 + \cdots .$$

Therefore, Eq. (A.2.18) gives

$$2F_3 + 2F_4\eta + \frac{1}{2!}(F_2^2 + 2F_5)\eta^2 + \frac{1}{3!}(4F_2F + 2F_6)\eta^3 + \cdots = 0 \tag{A.2.19}$$

Zero coefficients in all terms guarantees that Eq. (A.2.19) is an identity of η. Hence, $2F_3 = 0$, $2F_4 = 0$, $F_2^2 + 2F_5$, $4F_2F + 2F_6 = 0$, $\cdots$ are obtained. It results in $F_3 = 0$, $F_4 = 0$, $F_5 = -F_2^2/2$, $F_6 = 0$, $F_7 = 0$, $F_8 = 11F_2^3/4$, $\cdots$.

This means that coefficients except for F_2 can be expressed by only F_2. That is to say, F_2 is a parameter controlling an integral curve as follows:

$$f(\eta) = \frac{F_2}{2!}\eta^2 - \frac{1}{2}\frac{F_2^2}{5!}\eta^5 + \frac{11}{4}\frac{F_2^3}{8!}\eta^8 - \frac{375}{8}\frac{F_2^3}{11!}\eta^{11}\cdots$$

Consequently, a following form is obtained.

$$f(\eta) = \sum_{n=0}^{\infty}\left(-\frac{1}{2}\right)^n \frac{F_2^{n+1}C_n}{(3n+2)!}\eta^{3n+2} \tag{A.2.20}$$

in which $C_0 = 1$, $C_1 = 1$, $C_2 = 11$, $C_3 = 375$, $C_4 = 27897\cdots$.

(Step-2) Solution near $\eta = \infty$

In this step, a perturbation method is use. $f(\eta)$ is expressed by known approximation solution f_0 and deviations. The deviation is assumed to be a summation of parameter, ε, and thus, $f(\eta)$ is as follows:

$$\begin{aligned}
f &= f_0 + \varepsilon f_1 + \varepsilon^2 f_2 + \varepsilon^3 f_3 + \cdots \\
f' &= f'_0 + \varepsilon f'_1 + \varepsilon^2 f'_2 + \varepsilon^3 f'_3 + \cdots \\
f'' &= f''_0 + \varepsilon f''_1 + \varepsilon^2 f''_2 + \varepsilon^3 f''_3 + \cdots \\
f''' &= f'''_0 + \varepsilon f'''_1 + \varepsilon^2 f'''_2 + \varepsilon^3 f'''_3 + \cdots
\end{aligned} \tag{A.2.21}$$

Although ε is generally minute, $\varepsilon \simeq 1$ is given here because of no clear scale. Substitution of Eq. (A.2.21) into Eq. (A.2.17) leads to

$$\begin{aligned}
&(2f'''_0 + f_0 f''_0)\varepsilon^0 + (2f'''_1 + f_0 f''_1 + f_1 f''_0)\varepsilon^1 \\
&\quad + (2f'''_2 + f_0 f''_2 + f_1 f''_1 + f_2 f''_0)\varepsilon^2 + \cdots = 0
\end{aligned}$$

Here, we assume coefficients of all terms are zero for this equation to hold. Removing more than third-order terms, the following relations are obtained.

$$\begin{aligned}
&2f'''_0 + f_0 f''_0 = 0 \quad \cdots (a) \\
&2f'''_1 + f_0 f''_1 + f_1 f''_0 = 0 \quad \cdots (b) \\
&2f'''_2 + f_0 f''_2 + f_1 f''_1 + f_2 f''_0 = 0 \quad \cdots (c)
\end{aligned}$$

First, approximate solution f_0 is considered. Considering the boundary condition *at* $\eta = \infty$, $U = U_\infty \Rightarrow f' = 1$, a linear function using a parameter β is assumed.

$$f_0 = \eta - \beta \tag{A.2.22}$$

Eq. (A.2.22) is found to satisfy Eq. (a).

Furthermore, substitution of Eq. (A.2.22) into Eq. (b) leads to

$$\begin{aligned}
&2f'''_1 + (\eta - \beta) f''_1 = 0 \\
&\quad \leftrightarrow \frac{f'''_1}{f''_1} = \frac{1}{2}(\beta - \eta).
\end{aligned}$$

Integral of this relation provides $\ln f''_1 = \frac{1}{2}\beta\eta - \frac{1}{4}\eta^2 + \chi$ (χ is a integral function). Further, χ is replaced by $\chi = -\beta^2/4 + \ln\gamma$, using a new parameter γ.

Hence, the following equation is obtained.

$$f''_1 = e^{\frac{1}{2}\beta\eta - \frac{1}{4}\eta^2 + \chi} = e^{\frac{1}{2}\beta\eta - \frac{1}{4}\eta^2 - \frac{1}{4}\beta^2 + \ln\gamma} = \gamma e^{-\frac{1}{4}(\eta - \beta)^2} \tag{A.2.23}$$

Consideration of $\varepsilon \to 1$ in the second line of Eq. (A.2.21) and the boundary condition $f'(\infty) = 1$ provides

$$f'(\infty) = f'_0(\infty) + f'_1(\infty) + f'_2(\infty) + f'_3(\infty) + \cdots = 1 + f'_1(\infty) + \cdots = 1 \tag{A.2.24}$$

Eq. (A.2.24) gives $f'_1(\infty) = 0$, and Integrating Eq. (A.2.23) twice, $f_1 = \int_\eta^\infty \int_\eta^\infty \gamma e^{-\frac{1}{4}(\eta-\beta)^2} d\eta d\eta$ is obtained.

Here, removing more than second-order terms in the first line of Eq. (A.2.21)

$$f(\eta) = f_0(\eta) + f_1(\eta) = \eta - \beta + \int_\eta^\infty \int_\eta^\infty \gamma e^{-\frac{1}{4}(\eta-\beta)^2} d\eta d\eta \tag{A.2.25}$$

This is an approximate solution at $\eta = \infty$.

(Step-3) Connection of asymptotic solutions

Two asymptotic solutions indicated in previous steps include parameters F_2, β, γ. After searching the combination of parameters, make one solution meet another, $F_2 = 0.332$, $\beta = 1.73$, $\gamma = 0.231$ is obtained. (Hino 1992).

Howarth (1938) provided an exact solution, i.e., numerical table of relation between η and $f(\eta)$.

Eq. (A.2.3) gives

$$y = \eta\sqrt{\frac{\nu x}{U_\infty}} \tag{A.2.26}$$

Eq. (A.2.12) gives $f' = \frac{U}{U_\infty}$. Here, we use the 99% thickness of the boundary layer, in which $\frac{U}{U_\infty} = 0.99$ at $y = \delta$. Howarth's numerical table tells $f' = \frac{U}{U_\infty} \simeq 0.99$ at $\eta = 5.0$. Hence, Eq. (A.2.26) is rewritten by

$$y = 5.0\sqrt{\frac{\nu x}{U_\infty}} \tag{A.2.27}$$

Appendix 3

Mathematical Derivation of Equation in the K-H Instability

The velocity potentials satisfy the following Laplace equations.

$$\nabla^2\phi_1 = 0 \tag{A.3.1}$$

$$\nabla^2\phi_2 = 0 \tag{A.3.2}$$

Assuming there is no velocity deviations in areas very far away from the boundary, A.3.3 and A.3.4 are given.

$$\nabla\phi_1 = (U_1, 0) \tag{A.3.3}$$

$$\nabla\phi_2 = (U_2, 0) \tag{A.3.4}$$

Applying the kinetic boundary condition to the boundary between two layers

$$\frac{\partial \eta}{\partial t} + \tilde{u}\frac{\partial \eta}{\partial x} = \tilde{v} \text{ at } y = \eta(x, t) \tag{A.3.5}$$

are obtained. Decomposing the instantaneous velocity into the mean and the deviation, Eq. (A.3.5) is replaced by

$$\frac{\partial \eta}{\partial t} + (U_1 + \breve{u}_1)\frac{\partial \eta}{\partial x} = \breve{v}_1 \text{ at } y = \eta(x, t) \tag{A.3.6}$$

Eqs. (A.3.7) and (A.3.8) of ϕ_1, ϕ_2 are obtained, referring to Eq. (A.3.6).

$$\frac{\partial \phi_1}{\partial y} = \frac{\partial \eta}{\partial t} + (U_1 + \breve{u}_1)\frac{\partial \eta}{\partial x} \text{ at } y = \eta(x, t) \tag{A.3.7}$$

$$\frac{\partial \phi_2}{\partial y} = \frac{\partial \eta}{\partial t} + (U_2 + \breve{u}_2)\frac{\partial \eta}{\partial x} \text{ at } y = \eta(x, t) \tag{A.3.8}$$

in which $\breve{v}_1 = \frac{\partial \phi_1}{\partial y}$ and $\breve{v}_2 = \frac{\partial \phi_2}{\partial y}$.

Applying the unsteady Bernoulli's equation to layers 1 and 2, Eqs. (A.3.9a) and (A.3.9b) are introduced.

$$\frac{\partial \phi_1}{\partial t} + \frac{P_1}{\rho_1} + \frac{(\nabla \phi_1)^2}{2} + gy = C_1 \tag{A.3.9a}$$

$$\frac{\partial \phi_2}{\partial t} + \frac{P_2}{\rho_2} + \frac{(\nabla \phi_2)^2}{2} + gy = C_2 \tag{A.3.9b}$$

Since the pressures of layers 1 and 2 are same ($P_1 = P_2$) at the boundary ($y = \eta$), Eq. (A.3.10) is obtained.

$$\rho_1\left(\frac{\partial \phi_1}{\partial t} + \frac{(\nabla \phi_1)^2}{2} - C_1\right) = \rho_2\left(\frac{\partial \phi_2}{\partial t} + \frac{(\nabla \phi_2)^2}{2} - C_2\right) \tag{A.3.10}$$

The Bernoulli's equation of mean flow introduces Eq. (A.3.11).

$$\rho_1\left(\frac{U_1^2}{2} - C_1\right) = \rho_2\left(\frac{U_2^2}{2} - C_2\right) \tag{A.3.11}$$

Here, the velocity potential is decomposed into the mean component and the deviation, as indicated in Eq. (A.3.12).

$$\phi_1 = U_1 x + \breve{\phi}_1, \quad \phi_2 = U_2 x + \breve{\phi}_2 \tag{A.3.12}$$

Substituting them into Eqs. (A.3.1) and (A.3.2), Eq. (A.3.13) is introduced.

$$\nabla^2 \breve{\phi}_1 = 0, \quad \nabla^2 \breve{\phi}_2 = 0 \tag{A.3.13}$$

There is no deviation in the area away from the boundary according to the assumption described above, and therefore

$$\nabla\breve{\phi}_1 = 0 \quad (y \to \infty), \qquad \nabla\breve{\phi}_2 = 0 \quad (y \to -\infty) \tag{A.3.14}$$

are obtained.

We replace $y = \eta$ by $y = 0$ in Eqs. (A.3.7) and (A.3.8). Removing second-order terms in these equations, linear equations are introduced, given by Eqs. (A.3.15a) and (A.3.15b).

$$\frac{\partial\breve{\phi}_1}{\partial t} = \frac{\partial\eta}{\partial t} + U_1\frac{\partial\eta}{\partial x} \quad \text{at } y = 0 \tag{A.3.15a}$$

$$\frac{\partial\breve{\phi}_2}{\partial t} = \frac{\partial\eta}{\partial t} + U_2\frac{\partial\eta}{\partial x} \quad \text{at } y = 0 \tag{A.3.15b}$$

Substituting Eq. (A.3.12) to Eq. (A.3.9a), Eq. (A.3.9a) is transformed in the following way.

$$\frac{\partial(U_1x + \breve{\phi}_1)}{\partial t} + \frac{P_1}{\rho_1} + \frac{1}{2}\,|\nabla(U_1x + \breve{\phi}_1)|^2 + g\eta = C_1$$
$$\leftrightarrow \frac{\partial\breve{\phi}_1}{\partial t} + \frac{P_1}{\rho_1} + \frac{1}{2}\left[\frac{\partial}{\partial x}(U_1x + \breve{\phi}_1)^2 + \frac{\partial}{\partial y}(U_1x + \breve{\phi}_1)^2\right] + g\eta = C_1$$
$$\leftrightarrow \frac{\partial\breve{\phi}_1}{\partial t} + \frac{P_1}{\rho_1} + \frac{1}{2}\left[\frac{\partial}{\partial x}(U_1x + \breve{\phi}_1)^2 + \frac{\partial}{\partial y}(\breve{\phi}_1)^2\right] + g\eta = C_1$$

Here, neglecting a product of deviation variables, the pressure in layer 1 is expressed by

$$-P_1 = \rho_1\left(\frac{\partial\breve{\phi}_1}{\partial t} + \frac{1}{2}U_1^2 + U_1\frac{\partial\breve{\phi}_1}{\partial x} + g\eta - C_1\right).$$

The pressure in layer 2 is obtained in the same way.

$$-P_2 = \rho_2\left(\frac{\partial\breve{\phi}_2}{\partial t} + \frac{1}{2}U_2^2 + U_2\frac{\partial\breve{\phi}_2}{\partial x} + g\eta - C_2\right)$$

Eq. (A.3.16) is derived from Eq. (A.3.11) and the pressure condition at the boundary, i.e., $P_1 = P_2$ at $(y = \eta)$.

$$\rho_1\left(\frac{\partial\breve{\phi}_1}{\partial t}+U_1\frac{\partial\breve{\phi}_1}{\partial x}+g\eta\right)=\rho_2\left(\frac{\partial\breve{\phi}_2}{\partial t}+U_2\frac{\partial\breve{\phi}_2}{\partial x}+g\eta\right) \quad \text{(A.3.16)}$$

The governing equations are Eqs. (A.3.15a), (A.3.15b) and (A.3.16). The final goal is to solve η, $\breve{\phi}_1$, $\breve{\phi}_2$ by using these equations.

Let us express η, $\breve{\phi}_1$, $\breve{\phi}_2$ like Eqs. (A.3.17), (A.3.18a) and (A.3.18b), respectively.

$$\eta=\hat{\eta}e^{ik(x-ct)} \quad \text{(A.3.17)}$$

$$\breve{\phi}_1=\hat{\phi}_1(y)e^{ik(x-ct)} \quad \text{(A.3.18a)}$$

$$\breve{\phi}_2=\hat{\phi}_2(y)e^{ik(x-ct)} \quad \text{(A.3.18b)}$$

Substituting Eq. (A.3.18a) into Eq. (A.3.13), the differential equation $\frac{d^2\hat{\phi}_1}{dy^2}=k^2\hat{\phi}_1$, is obtained. The solution of this is $\hat{\phi}_1=Ae^{-ky}+Ce^{ky}$. Similarly, $\hat{\phi}_2=De^{-ky}+Be^{ky}$ is obtained. These forms are substituted into Eq. (A.3.18). Then, $C=D=0$ is required to satisfy Eq. (A.3.14). Hence, the deviation of the velocity potential is given by the following forms.

$$\hat{\phi}_1=Ae^{-ky} \quad \text{(A.3.19a)}$$

$$\hat{\phi}_2=Be^{ky} \quad \text{(A.3.19b)}$$

Therefore, Eqs. (A.3.17), (A.3.18a) and (A.3.18b) are rewritten by Eqs. (A.3.20), (A.3.21a) and (A.3.21b), respectively.

$$\eta=\hat{\eta}e^{ik(x-ct)} \quad \text{(A.3.20)}$$

$$\breve{\phi}_1=Ae^{-ky}(y)e^{ik(x-ct)} \quad \text{(A.3.21a)}$$

$$\breve{\phi}_2=Be^{ky}e^{ik(x-ct)} \quad \text{(A.3.21b)}$$

Substituting Eqs. (A.3.20) and (A.3.21a) into Eq. (A.3.15a)

$$\frac{\partial}{\partial y}(Ae^{-ky}e^{ik(x-ct)}) = \frac{\partial}{\partial t}(\hat{\eta}e^{ik(x-ct)}) + U_1\frac{\partial}{\partial x}(\hat{\eta}e^{ik(x-ct)})$$
$$\leftrightarrow\ -kAe^{-ky}e^{ik(x-ct)} = -ikc\hat{\eta}e^{ik(x-ct)}$$
$$+ ikU_1\hat{\eta}e^{ik(x-ct)}$$
$$\leftrightarrow\ Ae^{-ky} = -i\hat{\eta}(U_1 - c)$$

is introduced. Because this holds at $y = 0$, Eq. (A.3.22a) is derived.

$$A = -i(U_1 - c)\hat{\eta} \tag{A.3.22a}$$

Similarly, Eq. (A.3.22b) is obtained.

$$B = i(U_2 - c)\hat{\eta} \tag{A.3.22b}$$

Substituting Eqs. (A.3.20) and (A.3.21) into Eq. (A.3.16), a following form is introduced.

$$\rho_1\left(\frac{\partial\breve{\phi}_1}{\partial t} + U_1\frac{\partial\breve{\phi}_1}{\partial x} + g\eta\right) = \rho_2\left(\frac{\partial\breve{\phi}_2}{\partial t} + U_2\frac{\partial\breve{\phi}_2}{\partial x} + g\eta\right)$$
$$\leftrightarrow \rho_1 e^{ik(x-ct)}(-ikcAe^{-ky} + ikU_1Ae^{-ky} + g\hat{\eta})$$
$$= \rho_2 e^{ik(x-ct)}(-ikcBe^{ky} + ikU_2Be^{ky} + g\hat{\eta})$$

Since this holds at $y = 0$, $\rho_1[ikA(U_1 - c) + g\hat{\eta}] = \rho_2[ikB(U_2 - c) + g\hat{\eta}]$ is obtained.

Eqs. (A.3.22a) and (A.3.22b) introduce Eq. (A.3.23).

$$\rho_1[k(U_1 - c)^2\hat{\eta} + g\hat{\eta}] = \rho_2[-k(U_2 - c)^2\hat{\eta} + g\hat{\eta}]$$
$$\leftrightarrow \rho_1 k(U_1 - c)^2 + \rho_1 k(U_2 - c)^2 = g(\rho_2 - \rho_1) \tag{A.3.23}$$

Solving Eq. (A.3.23)

$$c = c_r + ic_i = \frac{\rho_1 U_1 + \rho_2 U_2}{\rho_1 + \rho_2} \pm \left[\frac{g}{k}\frac{\rho_2 - \rho_1}{\rho_1 + \rho_2} - \rho_1\rho_2\left(\frac{U_1 - U_2}{\rho_1 + \rho_2}\right)^2\right]^{1/2} \tag{A.3.24}$$

is obtained. This is Eq. (4.2.6) in Chapter 4.

References

Abe, K., Nagano, Y. and Kondo, T. (1992) Numerical prediction of separating and reattaching flows with a modified low-Reynolds-number k–ε model, *J. Wind Eng.*, 52, 213–218.

Abhari, M. N., Ghodsian, M., Vaghefi, M. and Panahpura, N. (2010) Experimental and numerical simulation of flow in a 90, bend, *Flow Meas. Instrum.*, 21, 292298.

Adrian, R. J. and Marusic, I. (2012) Coherent structures in flow over hydraulic engineering surfaces, *J. Hydraulic Res.*, 50(5), 451–464.

Adrian, R. J., Meinhart, C. D. and Tomkins, C. D. (2000) Vortex organization in the outer region of the turbulent boundary layer, *J Fluid Mech.*, 422, 1–54.

Albayrak I. and Lemmin, U. (2011) Secondary currents and corresponding surface velocity patterns in a turbulent open-channel flow over a rough bed, *J. Hydraulic Eng.*, 137(11), 1318–1334.

Ali, H. S. M. (1991) Effect of roughened-bed stilling basin on length of rectangular hydraulic jump, *J. Hydraulic Eng.*, 117(1), 83–93.

Ali, S. Z. and Dey, S. (2019) Bed particle saltation in turbulent wall-shear flow: a review , *Proc. R. Soc. A*, 475(2223), 20180824.

Andersson, J., Oliveira, D. R., Yeginbayeva, I., Leer-Andersen, M. and Bensow, R. E. (2020) Review and comparison of methods to model ship hull roughness, *Appl. Ocean Res.*, 99, 102119.

Arattano, M. and Marchi, L. (2005) Measurements of debris flow velocity through cross-correlation of instrumentation data, *Nat. Hazards Earth Syst. Sci.*, 5, 137–142.

Asai, M. and Nishioka, M. (1989) Origin of the peak-valley wave structure leading to wall turbulence, *J. Fluid Mech.*, 208, 1–23.

Auel, C., Albayrak, I., Sumi, T. and Boes, R. M. (2017) Sediment transport in high-speed flows over a fixed bed: 1. Particle dynamics, *Earth Surf. Proc. Landf.*, 42, 1365–1383.

Barenblatt, G. I. (1996) Scaling, self-similarity, and intermediate asymptotics, Cambridge University Press, Cambridge, U.K.

Batchelor, G. K. (1970) An introduction to fluid dynamics, Cambridge University Press, Cambridge, U.K.

Bennett, S. J., Bridge, J. S. and Best, J. L. (1998) Fluid and sediment dynamics of upper plane beds, *J. Geophys. Res.*, 103, 1239–1274.

Best, J., Bennett, S., Bridge, J. and Leeder, M. (1997) Turbulence modulation and particle velocities over flat sand beds at low transport rates, *J. Hydraulic Eng.*, 123(12), 1118–1129.

Blackwelder, R. F. and Kaplan, R. E. (1976) On the wall structure of the turbulent boundary layer, *J. Fluid Mech.*, 76, 89–112.

Becker, H. A. and Brown, A. P. G. (1974) Response of Pitot probes in turbulent streams. *J. Fluid Mech.*, 62, 85–114.

Blanckaert, K. and Graf, W. H. (2004) Momentum transport in sharp open-channel bends, *J. Hydraulic Eng.*, 130(3), 186–198.

Booij, R. (2004) Laboratory experiments of shallow free surface flows. In: G. H. Jirka and W. S. J. Uijttewaal, eds. *Shallow flows*. vol. 2. Rotterdam: Balkema Publishers, 317–323.

Bousmar, D. and Zech, Y. (2001) Periodic turbulent structures modeling in a symmetric compound channel, *Proc. of 29th IAHR Congress*, Beijing, 244–249.

Bradshaw, P. (1987) Turbulent secondary flows, *Annu. Rev. Fluid Mech.*, 19, 53–74.

Brereton, G. J. and Hwang, J. L. (1994) The spacing of streaks in unsteady turbulent wall-bounded flow, *Phys. Fluids*, 6, 2446.

Brereton, G. J. and Mankbadi, R. R. (1995) Review of recent advances in the study of unsteady turbulent internal flows, *Appl. Mech. Rev.*, ASME, 48, 189–212.

Brunet, Y., Finnigan, J. J. and Raupach, M. R. (1994) A wind tunnel study of air flow in waving wheat:single-point velocity statistics, *Bound. Layer Meteorol.*, 70, 95–132.

Bridge, J. S. (2003) Rivers and floodplains: Forms, Processes, and Sedimentary Record, Wiley – Blackwell.

Cameron, S. M., Nikora, V. I. and Stewart, M. T. (2017) Very-large-scale motions in rough-bed open-channel flow. *J. Fluid Mech.*, 814, 416–429.

Carollo, F. G., Ferro, V. and Termini, D. (2005) Flow resistance law in channel with flexible submerged vegetation, *J. Hydraulic Eng.*, 131, 554–564.

Carollo, F. G., Ferro, V. and Pampalone, V. (2007) Hydraulic jumps on rough beds, *J. Hydraulic Eng.*, 133(9), 989–999.

Cartigny, M. J. B., Ventra, D., Postma, G. and van Denberg, J. H. (2014) Morphodynamics and sedimentary structures of bedforms under supercritical-flow conditions: New insights from flume experiments, *Sedimentology*, 61, 712–748.

Chen, H., Hu, K., Cui, P. and Chen, X. (2017) Investigation of vertical velocity distribution in debris flows by PIV measurement, *Geomatics, Nat. Hazards Risk*, 8(2), 1631–1642.

Chong, M. S. and Perry, A. E. (1990) A general classification of three-dimensional flow fields. *Phys. Fluids A*, 2, 765–777.

Cioffi, F. and Gallerano, F. (1991) Velocity and concentration profiles of solid particles in a channel with movable and erodible bed, *J. Hydraulics Res.*, 29, 387–401.

Clauser, F. H. (1954) Turbulent boundary layers in adverse pressure gradients, *J. Aeronaut. Sci.*, 21, 91–108.

Di Cicca, G. M., Morvan, P. and Onorato, M. (2018) Turbulence investigation in the roughness sub-layer of a near wall flow. *Environ. Fluid Mech.*, 18, 1413–1433.

Duan, J. G. (2009) Mean flow and turbulence around a laboratory spur dike, *J. Hydraulic Eng.*, 135(10), 803–811.

de Vriend, H. J. (1981) Velocity redistribution in curved rectangular channels, *J. Fluid Mech.*, 107, 423–439.

Ead, S. A. and Rajaratnam, N. (2002) Hydraulic jumps on corrugated beds, *J. Hydraulic Eng.*, 128(7), 656–663.

Egashira, S., Ashida, K., Yajima, H. and Takahama, J. (1989) Constitutive equations of debris flow, (in Japanese) *Disaster Prevention Research Institute Annuals, Kyoto University*, 32, B-2, 487–501.

Egiazaroff, I. V. (1965) Calculation of nonuniform sediment concentrations, *J. Hydraulics Div.*, 91(4), 225–247.

Einstein, H. A. and Li, H. (1956) The viscous sublayer along a smooth boundary, *J. Eng. Mech. Div.*, 82, 1–27.

Ercan, A. and Younis B. A. (2009) Prediction of bank erosion in a reach of the Sacramento river and its mitigation with groynes. *Water Resour. Manag.*, 23(15), 3121–3147.

Escudier, M. P., Abdel-Hameed, A., Johnson, M. W. and Sutcliffe, C. J. (1998) Laminarization and re-transition of a turbulent boundary layer subjected to favorable pressure gradient, *Exp. Fluids*, 25, 491–502.

Ettema, R. and Muste, M. (2004) Scale effects in flume experiments on flow around a spur dyke in flatbed channel, *J. Hydraulic Eng.*, 130(7), 635–646.

Ferraro, D. and Dey, S. (2015) Principles of mechanics of bed forms, *Rivers – Physical, Fluvial and Environmental Processes*, Springer, 79–98.

Finnigan, J. (2000) Turbulence in plant canopies, *Annu. Rev. Fluid Mech.*, 32, 519–572.

Flack, K. A., and Schultz, M. P. (2014) Roughness effects on wall-bounded turbulent flows, *Phys. Fluids.*, 26, 101305, doi: 10.1063/1.4896280

Fujita, I., Muste, M. and Kruger, A. (1998) Large-scale particle image velocimetry for flow analysis in hydraulic engineering applications, *J. Hydraulic Res.*, 36, 397–414.

Garcia, C. M., Cantero, M. I., Nino, Y. and Garcia, M. H. (2005) Turbulence measurements using acoustic Doppler velocimeters, *J. Hydraulic Eng.*, 131, 1062–1073.

Gavrilakis, S. (1992) Numerical simulation of low Reynolds-number turbulent flow through a straight square duct, *J. Fluid Mech.*, 244, 101–129.

Gessner, F. B. (1973) The origin of secondary flow in turbulent flow along a corner, *J. Fluid Mech.*, 58, 1–25.

Ghisalberti, M. and Nepf, H. (2006) The structure of the shear layer in flows over rigid and flexible canopies, *Environ. Fluid Mech.*, 6, 277–301.

Ghisalberti, M. and Nepf, H. (2009) Shallow flows over a permeable medium: the hydrodynamics of submerged aquatic canopies, *Transport Porous Med.*, 78, 385–402.

Gholami, A., Akhtari, A. A., Minatour, Y., Bonakdari, H. and Javadi, A. A. (2014) Experimental and numerical study on velocity fields and water surface profile in a strongly-curved 90° open channel bend, *Eng. Appl. Comput. Fluid Mech.*, 8(3), 447–461.

Grass, A. J., Stuartr, J. and Mansour- Tehramn, I. (1991) Vortical structures and coherent motion in turbulent flow over smooth and rough boundaries. *Phil. Trans. R. Soc. Lond. A*, 336, 35–65.

Grinvald, D. I. and Nikora, V. I. (1988) River turbulence, Hydrometeoizdat, Leningrad, (in Russian).

Guo, J. (2020) Empirical model for shields diagram and its applications, *J. Hydraulic Eng.*, 146(6), 04020038
Guo, X., and Shen, L. (2010) Interaction of a deformable free surface with statistically steady homogeneous turbulence, *J. Fluid Mech.*, 658, 33–62.
Gupta, S. K., Mehta, R. C. and Dwivedi, V. K. (2013) Modeling of relative length and relative energy loss of free hydraulic jump in horizontal prismatic channel, *Procedia Eng.*, 51, 529–537.
Hama, F. R. (1954) Boundary layer characteristics for smooth and rough surfaces, *Trans. Soc. Nav. Arch. Mar. Engrs.*, 62, 333–358.
Han, S. S., Biron, P. M. and Ramamurthy, A. S. (2011) Three-dimensional modelling of flow in sharp open-channel bends with vanes, *J. Hydraulic Res.*, 49, 64–72.
Handler, R. A., Saylor, J. R., Leighton, R. I. and Rovelstad, A. L. (1999) Transport of a passive scalar at a shear-free boundary in fully developed turbulent open channel flow, *Phys. Fluids*, 11, 2607–2625.
Hashimoto, H. and Hirano, M. (1996) Gravity flow of sediment-water mixtures in a steep open channel, *Proc. JSCE*, 545(ii-36), 33–42. (in Japanese)
Hill, D. F. (2014) Simple model for the recirculation velocity of open-channel embayments, *J. Hydraulic Eng.*, 140(4), 06014004.
Hino, M. (1992) *Fluid mechanics*, Tokyo: Asakura Shoten, (in Japanese).
Hjulstrom, F. (1935) Studies of the morphological activity of rivers as illustrated by the River Fyris, Bulletin, *Geological Institute Upsalsa*, 25, 221–527.
Ho, C. M. and Huerre, P. (1884) Perturbed free shear layers, *Annu. Rev. Fluid Mech.*, 16, 365–424.
Howarth, L. (1938) On the society of the boundary layer equations, *Proc. Roy. Soc. A*, 164(919), 547–579.
Hurther, D. and Lemmin, U. (2001) A Correction method for turbulence measurements with a 3-D acoustic Doppler velocity profiler. *J. Atoms. Ocean. Technol.*, 18, 446–458.
Hurther, D., Lemmin, U. and Terray, E. A. (2007) Turbulent transport in the outer region of rough-wall open-channel flows: the contribution of large coherent shear stress structures (LC3S). *J. Fluid Mech.*, 574, 465–493.
Hutchins, N. and Marusic, I. (2007) Evidence of very long meandering features in the logarithmic region of turbulent boundary layers, *J. Fluid Mech.*, 579, 1–28.
Itagaki, H. and Furuta, T. (1983) Characteristics of current meter and turbulence phenomena in open-channel flow, *Research Bulletin of the Faculty College of Agriculture Gifu University*, 48, 159–171. (in Japanese)
Iverson, R. M. (2012) Elementary theory of bed-sediment entrainment by debris flows and avalanches. *J. Geophys. Res. Earth Surf.*, 117, 3006.
Iwasa, Y. (1955) The criterion for instability of steady uniform flow in open channels, *J. Japan Soc. Civil Eng.*, 40-6, 297–303. (in Japanese)
Jackson, R. G. (1976) Sedimentological and fluid-dynamic implications of the turbulent bursting phenomenon in geophysical flows, *J. Fluid Mech.*, 77(3), 531–560.
Jarrett, E. L. and Sweeney, T. L. (1967) Mass transfer in rectangular cavities, *AIChE J.*, 13(4), 797–800.
Jeffreys, H. J. (1925) The flow of water in an inclined channel of rectangular section. *Phil. Mag.*, 6, 793–807.

Jensen, B., Sumer, B. M. and Fredsoe, J. (1989) Turbulent oscillatory boundary layers at high Reynolds numbers, *J. Fluid Mech.*, 206, 265–297.

Jiang, X. Y., Lee, C. B., Smith, C. R., Chen, J. W. and Linden, P. F. (2020) Experimental study on low-speed streaks in a turbulent boundary layer at low Reynolds number, *J. Fluid Mech.*, 903, A6.

Jimenez, J. (2004) Turbulent flows over rough walls, *Annu. Rev. Fluid Mech.*, 36, 173–196.

Jones, W. P. and Launder, B. E. (1972) The prediction of laminarization with a two-equation model of turbulence, *Int. J. Heat Mass Transfer*, 15, 301–314.

Kadivar, M., Tormey, D. and McGranaghan, G. (2021) A review on turbulent flow over rough surfaces: Fundamentals and theories, *Int. J. Thermofluids*, 10, 100077.

Kaftori, D., Hetsroni, G. and Banerjee, S. (1995a) Particle behaviour in the turbulent boundary layer I. Motion and entrainment, *Phys. Fluids*, 7, 1095–1106.

Kaftori, D., Hetsroni, G. and Banerjee, S. (1995b) Particle behaviour in the turbulent boundary layer II. Velocity and distribution profiles, *Phys. Fluids*, 7, 1107–1121.

Kim, K. C. and Adrian, R. J. (1999) Very large-scale motion in the outer layer, *Phys. Fluids*, 11(2), 417–422.

Kennedy, J. F. (1963) The mechanics of dunes and antidunes in erodible-bed channels, *J. Fluid Mech.* 16(4), 521–544.

Kimura, I. and Hosoda, T. (1997) Fundamental properties of flows in open channels with dead zone, *J. Hydraulic Eng.*, 123(2), 98–107.

Kimura, I., Hosoda, T. and Tomochika, H. (1995) Characteristics of spatially growing disturbances in a mixing shear layer of open channel flows, *J. JSCE*, 509(II-30), 99–109. (in Japanese).

Kirlkgoz, M. S. and Ardichoglu, M. (1997) Velocity profiles of developing and developed open channel flow, *J. Hydraulic Eng.*, 123(12), 1099–1105.

Kline, S. J., Reynolds, W. C., Schraub, F. A. and Runstadler, P. W. (1967) The structure of turbulent boundary layers, *J. Fluid Mech.*, 30(4), 741–773.

Kolmogorov, A. (1941). The local structure of turbulence in incompressible viscous fluid for very large Reynolds' numbers. *Dokl. Akad. Nauk SSSR*, 39, 301–305.

Komori, S., Ueda, H., Ogino, F. and Mizushina, T. (1982) Turbulence structure and transport mechanism at the free surface in an open channel flow, *Int. J. Heat Mass Transfer*, 25(4), 513–521.

Kothyari, U., Hayashi, K. and Hashimoto, H. (2009) Drag coefficient of unsubmerged rigid in open channel flows, *J. Hydraulic Res.*, 47(6), 691–699.

Lee J. H. and Sung, H. J. (2011) Direct numerical simulation of a turbulent boundary layer up to Reh = 2500. *Int. J. Heat Fluid Flow*, 32, 1–10.

Lien, H. C., Hsieh, T. Y., Yang, J. C. and Yeh, K. C. (1999) Bend-flow simulation using 2D depth-averaged model, *J. Hydraulic Eng.*, 125, 1097–1108.

Lu, S. S. and Willmarth, W. W. (1973) Measurements of the structure of the Reynolds stress in a turbulent boundary layer, *J. Fluid Mech.*, 60, 481–511.

Lumley, J. L. (1981) Coherent structures in turbulence, *In Transition and Turbulence*, Academic press, New York, 215–241.

Lyn, D. A. (1991) Resistance in flat-bed sediment-laden flows, *J. Hydraulic Eng.*, 117(1), 94–114.

Lyn, D. A. (1992) Turbulence characteristics of sediment-laden flows in open-channels, *J. Hydraulic Eng.*, 118(7), 971–988.

McGregor, O. W. and White, R. A. (1970) Drag of rectangular cavities in supersonic and transonic flow including the effects of cavity resonance. *AIAA J.*, 8(11), 1959–1964.

McLelland, S. and Nicholas, A. (2000) A new method for evaluating errors in high-frequency ADV measurements, *Hydrol. Process.*, 14, 351–366.

Meile, T., Boillat, J. L. and Schleiss, A. J. (2011) Water-surface osci- llations in channels with axi-symmetric cavities. *J. Hydraulic Res.*, 49(1), 73–81.

Michel, R., Cousteix, J. and Houdeville, R. (eds.) (1981) Unsteady turbulent shear flows, Springer-Verlag, Berlin.

Mignot, E., Hurther, D. and Barthelemy, E. (2009) On the structure of shear stress and turbulent kinetic energy flux across the roughness layer of a gravel-bed channel flow, *J. Fluid Mech.*, 32, 423–452.

Mizumura, K. and Yamasaka, M. (2002) Flow in open-channel embayments, *J. Hydraulic Eng.*, 128(12), 1098–1101.

Moin, P. and Kim, J. (1982) Numerical investigation of turbulent channel flow, *J. Fluid Mech.*, 118, 341–377.

Moin, P. and Kim, J. (1985) The structure of the vorticity field in turbulent channel flow. Part 1. Analysis of instantaneous fields and statistical correlations, *J. Fluid Mech.*, 155, 441–464.

Monty, J. P., Hutchins, N., Ng, H. C. H., Marusic, I. and Chong, M. S. (2009) A comparison of turbulent pipe, channel and boundary layer flows, *J. Fluid Mech.*, 632, 431–442.

Muste, M. and Patel, V. C. (1997) Velocity profiles for particles and liquid in open-channel flow with suspended sediment, *J. Hydraulic Eng.*, 123, 742–571.

Muste, M., Baranya, S., Tsubaki, R., Kim, D., Ho, H., Tsai, H. and Law, D. (2016) Acoustic mapping velocimetry, *Water Resour. Res.*, 52, 4132–4150.

Nadaoka, K. and Yagi, H. (1998) Shallow-water turbulence modeling and horizontal large eddy computation of river flow, *J. Hydraulic Eng.*, 124, 493–500.

Nagaosa, R. and Handler, R. A. (2003) Statistical analysis of coherent vortices near a free surface in a fully developed turbulence, *Phys. Fluids*, 15, 375–394.

Nakagawa, H. and Nezu, I. (1977) Prediction of the contributions to the Reynolds stress from the bursting events in open channel flows, *J. Fluid Mech.*, 80, 99–128.

Naot, D., Nezu, I. and Nakagawa, H. (1993) Hydrodynamic behavior of compound rectangular open-channel flows, *J. Hydraulic Eng.*, 119, 390–408.

Narayanan, M. A. B. and Ramjee, V. (1969) On the criteria for reverse transition in a two-dimensional boundary layer flow, *J. Fluid Mech.*, 35, 225–241.

Neary, V. S. and Odgaard, A. J. (1993) Three-dimensional flow structure at open channel diversions, *J. Hydraulic Eng.*, 119(11), 1224–1230.

Nepf, H. M. (1999) Drag, turbulence, and diffusion in flow through emergent vegetation, *Water Resour. Res.*, 35(2), 479–489.

Nepf, H. M. (2012) Flow and transport in regions with aquatic vegetation, *Annu. Rev. Fluid Mech.*, 44, 123–142.

Nepf, H. M. and Vivoni, E. R. (2000) Flow structure in depth-limited, vegetated flow, *J. Geophys. Res.*, 105, 28547–28557.

Nepf, H. M., Sullivan, J. A. and Zavistoski, R. A. (1997) A model for diffusion within emergent vegetation, *Linnol. Oceanogr.*, 42(8), 1735–1745.
Nepf, H. M., Ghisalberti, M., White, B. and Murphy, E. (2007) Retention time and dispersion associated with submerged aquatic canopies, *Water Resour. Res.*, 43, W04422.
Nezu, I. (2005) Open-channel flow turbulence and its research prospect in the 21st century, *J. Hydraulic Eng.*, 131, 229–246.
Nezu, I. and Nakagawa, H. (1993) Turbulence in open channel flows. *IAHR Monograph*, A. A. Balkema, Rotterdam.
Nezu, I. and Azuma, R. (2004) Turbulence characteristics and interaction between particles and fluid in particle-laden open channel flows, *J. Hydraulic Eng.*, 130(10), 988–1001.
Nezu, I. and Onitsuka, K. (2001) Turbulent structures in open-channel flows with strong unsteadiness, *Proc., 2nd Int. Symp. on Turbulence and Shear Flow Phenomena*, Stockholm, 1, 341–346.
Nezu, I. and Onitsuka, K. (2002) PIV Measurements of side-cavity open-channel flows -Wando model in rivers-, *J. Visualization*, 5, 77–84. 10.1007/BF03182606
Nezu, I. and Rodi, W. (1986) Open-channel flow measurements with a laser Doppler anemometer, *J. Hydraulic Eng.*, 112(5), 335–355.
Nezu, I., Sanjou, M. and Goto, K. (2003) Transition process of coherent vortices in depth-varying unsteady compound open-channel flows, *Proc. Int. Symp. on Shallow Flows*, Delft, 2, 215–222.
Nezu, I. and Sanjou, M. (2004) 3-D numerical calculation of shallow free-surface flow in timed ependent stages from rectangular to compound channels, *Shallow Flows* (eds. G. H. Jirka and W. S. J. Uijttewaal), Balkema, pp. 339–346.
Nezu, I. and Sanjou, M. (2006) Numerical calculation of turbulence structure in depth-varying unsteady open-channel flows, *J. Hydraulic Eng.*, 132(7), 681–695.
Nezu, I. and Sanjou, M. (2008) Turbulence structure and coherent motion in vegetated open-channel flows, *J. Hydro-Environ. Res.*, 2, 62–90.
Nezu, I. and Sanjou, M. (2011) PIV and PTV measurements in hydro-sciences with focus on turbulent open-channel flows, *J. Hydro-Environ. Res.*, 5, 215–230.
Nezu, I., Kadota, A. and Nakagawa, H. (1997) Turbulent structure in unsteady depth-varying open-channel flows, *J. Hydraulic Eng.*, 123(9), 752–762.
Nezu, I., Kadota, A. and Nakagawa, H. (1997) Turbulent structure in unsteady depth-varying open-channel flows, *J. Hydraulic Eng.*, 123, 752–763.
Nezu, I., Onitsuka, K. and Iketani, K. (1999) Coherent horizontal vortices in compound open-channel flows, *Hydraulic Modeling* (ed. Singh et al.) Water Resources Pub., Colorado, 17–32.
Nichols, A., Tait, S. J., Horoshenkov, K. V. and Shepherd, S. J. (2016) A model of the free surface dynamics of shallow turbulent flows, *J. Hydraulic Res.*, 54(5), 516–526.
Nikora, V. (2010) Hydrodynamics of aquatic ecosystems, An interface between ecology, biomechanics and environmental fluid mechanics, *River Research and Applications*, 26, 367–384.
Nikora, V. and Goring, D. (2002) Fluctuations of suspended sediment concentration and turbulent sediment fluxes in an open-channel flow, *J. Hydraulic Eng.*, 128(2), 214–224.

Nikora, V., Koll, K., Mclean, S. R., Ditrich, A. and Aberie, J. (2002) Zero-plane displacement for rough-bed open-channel flows, *Proc. Riverflow* 2002, Louvain-La-Neuve, 2002, 83–92.

Nikora, V., McEwan, I., McLean, S., Coleman, S., Pokrajac, D. and Walters, R. (2007) Double-averaging concept for rough-bed open-channel and overland flows: Theoretical background, *J. Hydraulic Eng.*, 133(8), 83–883.

Nikora, V., Nkes, R., Davidson, M. and Jirka, G. H. (2007) Large-scale turbulent structure of uniform shallow free-surface flows, *Environ. Fluid Mech.*, 7, 59–172.

Nikora, V. I. and Goring, D. G. (1998) ADV measurements of turbulence: Can we improve their interpretation? *J. Hydraulic Eng.*, 124, 630–634.

Nikuradse, J. (1933) Stromungsgesetze in Rauhen Rohren, Technical Report No. 361.

Nino, Y. and Garcia, M. (1996) Experiments on particle—turbulence interactions in the near–wall region of an open channel flow: Implications for sediment transport, *J. Fluid Mech.*, 326, 285–319.

Nomicos, G. N. (1956) Effects of sediment load on the velocity field and friction factor of turbulent flow in an open channel(Dissertation (Ph.D)), California Institute of Technology, DOI:10.7907/QZ0J-A652

Okamoto, T., and Nezu, I. (2009) Turbulence structure and "Monami" phenomena in flexible vegetated open-channel flows, *J. Hydraulic Res.*, 47, 798–810.

Okamoto, T. and Nezu, I. (2010) Large eddy simulation of 3-d flow structure and mass transport in open-channel flows with submerged vegetations, *J. Hydro-Environ. Res.*, Elsevier, 4(3), 185–197.

Okamoto, T. and Nezu, I. (2013) Spatial evolution of coherent motions in finite-length vegetation patch flow, *Environmental Fluid Mechanics*, Springer, 13(5), 417–434.

Okamoto, T., Nezu, I. and Ikeda, H. (2012) Vertical mass and momentum transport in open-channel flows with submerged vegetations, *J. Hydro-Environ. Res.*, 6(4), 287–297.

Okamoto, T., Nezu, I. and Sanjou, M. (2016) Flow–vegetation interactions: length-scale of the "monami" phenomenon, *J. Hydraulic Res.*, 54, 251–262.

Okamoto, T., Takebayashi, H., Sanjou, M., Suzuki, R. and Toda, K. (2018) Log jam formation at bridges and the effect on floodplain flow – a flume experiment, *J. Flood Risk Manag.*, 13(S1), e12562.

Okamoto T., Tanaka K., Matsumoto K. and Someya, T. (2021) Influence of velocity field on driftwood accumulation at a bridge with a single pier, *Environ. Fluid Mech.*, 21(3), 693–711.

Onitsuka, K. and Nezu, I. (2000) Prediction of the shear stress distributions in accelerated open-channel flows, *J. JSCE*, 642(II-50), 67–75. (in Japanese)

Ouillon, S. and Dartus, D. (1997) Three-dimensional computation of flow around groyne, *J. Hydraulic Eng.*, 123(11), 962–970.

Park, T. S. and Sung, H. J. (1997) A new low-Reynolds-number k–ε–f_μ model for predictions involving multiple surfaces. *Fluid Dyn. Res.*, 20, 97–113.

Parsons, J. D., Whipple, K. X., Simoni, A. (2001) Experimental study of the grain-flow, fluid-mud transition in debris flows. *J.Geology*, 109(4), 427–447.

Patel, V. C., Rodi, W. and Scheuerer, G. (1985) Turbulence models for near-wall and low-Reynolds-number flows. *AIAA J.*, 23, 1308–1319.

Pedinotti, S., Mariottigi, G. and Banerjees, S. (1992) Direct numerical simulation of particle behavior in the wall region of turbulent flows in horizontal channels, *lnt. J. Multiphase Flow*, 18, 927–941.

Perkins, H. J. (1970) The formation of streamwise vorticity in turbulent flow, *J. Fluid Mech.*, 44, 721–740.

Poggi, D., Porpotato, A. and Ridolfi, L. (2004) The effect of vegetation density on canopy sub-layer turbulence, *Bound. Layer Meteorol.*, 111, 565–587, 2004.

Qian, J. (1996) Experimental values of Kolmogorov constant of turbulence, *J. Phys. Soc. Japan*, 65(8), 2502–2505.

Ranjan, R. and Narasimha, R. (2017) An assessment of the two-layer quasi-laminar theory of relaminarization through recent high-Re accelerated TBL experiments, *J. Fluids Eng.*, 139, 111205.

Rashidi, M. and Banerjee, S. (1988) Turbulence structure in free-surface channel flows, *Phys. Fluids*, 31, 2491.

Rashidi, M., Hetsroni, G. and Banerjee, S. (1990) Particle-turbulence interaction in a boundary layer, *Int. J. Multiphase Flow*, 16(6), 935–949.

Rasmussen, H. O. (1999) A new proof of Kolmogorov's 4/5-law, *Phys. fluids*, 11, 3495.

Raupach, M. R., Antonia, R. A. and Rajagopalan, S. (1991) Rough-wall turbulent boundary layers, *Appl. Mech. Rev.* 44, 1–25.

Raupach, M. R., Finnigan, J. J. and Brunet, Y. (1996) Coherent Eddies and Turbulence in Vegetation Canopies: The Mixing-Layer Analogy, *Bound. Layer Meteorol.*, 78, 351–382.

Raupachi, M. R. (1981) Conditional statistics of Reynolds stress in rough-wall and smooth wall turbulent boundary layers, *J. Fluid Mech.*, 108, 363–382.

Richardson, L. P. (1926) Atmospheric diffusion shown on a distance-neighbor graph, *Proc. Roy. Soc. London*, A110, 709–727.

Robinson, S. K. (1991) Coherent motions in the turbulent boundary layer, *Annu. Rev. Fluid Mech.*, 23, 601–639.

Rodriguez, J. F. and Garcia, M. H. (2008) Laboratory measurements of 3-D flow patterns and turbulence in straight open channel with rough bed, *J. Hydraulic Res.*, 46, 454–465.

Rouse, H. (1937) Modern conceptions of the mechanics of turbulence, *Trans. ASCE*, 102, 463–543.

Sana, A. and Tanaka, H. (1995) The Performance of low Reynolds number model to analyse an oscillatory boundary layer, *Proc. Mathematical Modeling of Turbulent Flows*, pp. 291–296.

Sanjou, M. (2020) Local gas transfer rate through the free surface in spatially accelerated open-channel turbulence, *Phys. Fluids*, 32(10), 105103

Sanjou, M. and Nagasaka, T. (2017) Development of autonomous boat-type robot for automated velocity measurement in straight natural river, *Water Resour. Res.*, 53(11), 9089–9105.

Sanjou, M. and Nezu, I. (2009) Turbulence structure and coherent motion in meandering compound open-channel flows, *J. Hydraulic Res.*, 47(5), 598–610.

Sanjou, M. and Nezu, I. (2010) Large eddy simulation of compound open-channel flows with emergent vegetation near the floodplain edge, *J. Hydrodynamics, Ser.B*, 22(5, Sup.1), 582–586.

Sanjou, M. and Nezu, I. (2011a) Turbulent structure and coherent vortices in open-channel flows with wind-induced water waves, *Environ. Fluid Mech.*, 11(2), 113–131.

Sanjou, M. and Nezu, I. (2011b) Turbulence structure and concentration exchange property in compound open-channel flows with emergent trees on the floodplain edge, *Int. J. River Basin Manag.*, 9, 181–193.

Sanjou, M. and Nezu, I. (2013) Hydrodynamic characteristics and related mass transfer properties in open-channel flows within a rectangular shaped embayment zone, *Environ. Fluid Mech.*, 13, 527–555.

Sanjou, M. and Nezu, I. (2014) Secondary current properties generated by wind-induced water waves in experimental conditions, *Advances in Oceanography and Limnology,* 5(1), 1–17.

Sanjou, M. and Nezu, I. (2017) Fundamental study on mixing layer and horizontal circulation in open channel flows with rectangular embayment zone, *J. Hydrodynamics Ser.B*, 29(1), 840–853.

Sanjou, M. and Nezu, I. (2003) Numerical study on depth-varying time-dependent compound open-channel flows, Inland Waters (ed. I. Nezu et al.), *Proc. 30th IAHR Congress,* Thessaloniki, Greece, 1, 25–32.

Sanjou, M., Nezu, I. and Itai, K. (2010) Space-time correlation and momentum exchanges in compound open-channel flow by simultaneous measurements of two-sets of ADVs, *Proc. Riverflow 2010*, Braunshweig, pp. 495–502.

Sanjou, M., Nezu, I., Suzuki, S. and Itai, K. (2010a) Turbulence structure of compound open-channel flows with one-line emergent vegetation, *J. Hydrodynamics, Ser.B*, 22(5, Sup. 1), 577–581.

Sanjou, M., Nezu, I. and Toda, A. (2010b) PIV studies on turbulence structure in air/water interface with wind-induced water waves, *J. Hydrodynamics, Ser.B*, Elsevier, 22(5, Sup.1), 349–353.

Sanjou, M., Akimoto, T. and Okamoto, T. (2012) Three-dimensional turbulence structure of rectangular side-cavity zone in open-channel stream, *Int. J. River Basin Manag.*, 10(4), 293–305.

Sanjou, M., Nezu, I. and Okamoto, T. (2017) Surface velocity divergence model of air/water interfacial gas transfer in open-channel flows, *Phys. Fluids*, 29(4), 045107.

Sanjou, M., Okamoto, T. and Nezu, I. (2018a) Experimental study on fluid energy reduction through a flood protection forest, *J. Flood Risk Manag.*, 11(4), e12339.

Sanjou, M., Okamoto, T. and Nezu, I. (2018b) Dissolved oxygen transfer into a square embayment connected to an open-channel flow, *Int. J. Heat Mass Transfer*, 125, 1169–1180.

Sanjou, M., Shigeta, A., Kato, K. and Aizawa, W. (2021) Portable unmanned surface vehicle that automatically measures flow velocity and direction in rivers, *Flow Meas. Instrum.*, 80, 101964.

Savelsberg, R. and van deWater, W. (2009) Experiments on free surface turbulence, *J. Fluid Mech.*, 619, 95–125.

Schlichting, H. (1979) *Boundary Layer Theory*, 7th Ed., McGraw-Hill, New York.

Schroder, M., Stein, C. J. and Rouve, G. (1991) Application of the 3D-LDV technique on physical model of meandering channel with vegetated floodplain, *Proc. 4th Int. Conf. on Laser Anemometry-Advances and Applications*, Ohio, pp. 1–9.

Sellin R. H. J. (1964) A laboratory investigation into the interaction between the flow in the channel of a river and that over its flood plain, *La Houille Blanche*, 50(7), 793–802.

Sellin, R. H. J., Ervine, D. A. and Willetts, B. B. (1993) Behavior of meandering two-stage channels, *Proc. Insti. of Civil Eng.*, 101, 99–101.

Sharma, K. and Mohapatra, P. K. (2012) Separation zone in flow past a spur dike on rigid bed meandering channel, *J. Hydraul. Eng.*, 138(10), 897–901.

Shen, C. and Lemmin, U. (1997) A two-dimensional acoustic sediment flux profiler, *Measurement Science and Technology*, 8(8), 880–884.

Shields, A. (1936) *Application of similarity principles and turbulence research to bed-load movement*, California Institute of Technology, Pasadena, CA. (Unpublished)

Shin, S. S., Park, S. D. and Lee, S. K. (2016) Measurement of Flow Velocity Using Video Image of Spherical Float, *Procedia Engineering*, 154, 885–889.

Shiono, K. and Knight, D. W. (1991) Turbulent open-channel flow with variable depth across the channel, *J. of Fluid Mech.*, 222, 617–646.

Shiono, K. and Muto, Y. (1998) Complex flow mechanisms in compound meandering channels with overbank Flow, *J. of Fluid Mech.*, 376, 221–261.

Shiono, K., Ishigaki, T., Kawanaka, R. and Heatlie, F. (2009) Influence of one line vegetation on stage-discharge rating curves in compound channel, *Proc. 33rd IAHR Congress*, Vancouver, 8pages on CD-ROM.

Silva, A. M. F., El-Tahawy, T. and Tape, W. D. (2006) Variation of flow pattern with sinuosity in sine-generated meandering streams, *J. of Hydraulic Eng.*, 132, 1003–1014.

Siniscalchi, F. and Nikora, V. (2012) Flow-plant interactions in open-channel flows: A comparative analysis of five freshwater plant species, *Water Resour. Res.*, 48, W05503-1–13.

Siniscalchi, F. and Nikora, V. (2013) Dynamic reconfiguration of aquatic plants and its interrelations with upstream turbulence and drag forces, *J. Hydraulic Res.*, 51, 46–55.

Smalley, R. J., Leonardi, S., Antonia, R. A., Djenidi, L. and Orlandi, P. (2002) Reynolds stress anisotropy of turbulent rough wall layers, *Exp. Fluid*, 33(1), 31–37.

Smith, C. R. (1984) A synthesized model of the near-wall behavior in turbulent boundary layers, *Proc. 8th Symposium on Turbulence*, University of Missouri-Rolla, 299–325.

Smith, C. R. and Schwartz, P. (1983) Observation of streamwise rotation in the near-wall region of a turbulent boundary layer, *Phys. Fluids*, 26, 641–652.

Song, T. and Graf, W. H. (1994) Nonuniform open-channel flow over a rough bed, *J. Hydrosci. Hydr. Eng.*, 12, 1–25.

Song, T. and Graf, W. H. (1994) Non-uniform open-channel flow over a rough bed, *JHHE*, 12(1), 1–25.

Spalart, P. R. (1988) Direct numerical simulation of a turbulent boundary layer up to Re = 1410, *J. Fluid Mech.*, 187, 61–98.

Squire, D. T., Baars, W. J., Hutchins, N. and Marusic, I. (2016) Inner–outer interactions in rough-wall turbulence, *J. Turbulence*, 17, 1159–1178.

Stone, B. M. and Shen, H. T. (2002) Hydraulic Resistance of Flow in Channels with Cylindrical Roughness, *J. Hydraulic Eng.*, 128(5), 500–506.

Sukhodolov, A., Engelhard, C., Kruger, A. Bungartz, H. (2004) Case study: Turbulent flow and sediment distributions in a groyne field, *J. Hydraulic Eng.*, 130(1), 1–9.

Sukhodolov, A. N. and Sukhodolova, T. A. (2010) Case study, Effect of submerged aquatic plants on turbulence structure in a lowland river, *J. Hydraulic Eng.*, 136, 434–446.

Sumer, B. M. and Deigaard, R. (1981) Particle motions near the bottom in turbulent flow in an open channel. *Part 2, J. Fluid Mech.*, 109, 311–337.

Sumer, B. M., Lloyd, H. C., Chua, N. S. and Chen, J. (2003) Fredsoe, Influence of turbulence on bed load sediment transport, *J. Hydraulic Eng.*, 129(8), 585–596.

Sun, X. and Shiono, K. (2009) Flow resistance of one-line emergent vegetation along the floodplain edge of a compound channel, *Adv. Water Resour.*, 32, 430–438.

Takahashi, T. (1982) Study on the deposition of debris flows (3) – erosion of debris fan -, *Disaster Prevention Research Institute Annuals, Kyoto University*, 25 (B-2), 327–348. (in Japanese).

Takahashi, T. (2019) *Debris Flow: Mechanics, Prediction and Countermeasures*, 2nd edition, CRC press.

Tamburrino, A. and Gulliver, J. S. (1999) Large flow structures in a turbulent open channel flow, *J. Hydraulic Res.*, 37(3), 363–380.

Tamburrino, A. and Gulliver, J. S. (2007) Free-surface visualization of streamwise vortices in a channel flow, *Water Resour. Res.*, 43, wi1410.

Tardu, S. F., Binder, G. and Blackwelder, R. F. (1994) Turbulent channel flow with large-amplitude velocity oscillations, *J. Fluid Mech.*, 267, 109–151.

Tavoularis, S. (2005) *Measurement in fluid mechanics*, Cambridge University press.

Taylor, G. I. (1922) Diffusion by continuous movements, *Proc. London Math. Soc., Ser.*, 2(20), 196–211.

Tennekes, H. and Lumley, J. L. (1972) *First course in turbulence*, MIT Press.

Theodorsen, T. (1952) Mechanism of turbulence. Proc. 2nd Midwestern Conf. on Fluid Mech. 1–19.

Tingsanchali, T. and Maheswaran, S. (1990) 2D depth-averaged flow computation near Groyne. *J. Hydraulic Eng.*, 116(1), 71–86.

Tominaga, A. and Nezu, I. (1991) Turbulent structures in compound open-channel flow, *J. Hydraulic Eng.*, 117, 21–41.

Tominaga, A., Nezu, I., Ezaki, K. and Nakagawa, H. (1989) Three dimensional turbulent structure in straight open-channel flows, *J. Hydraulic Res.*, 27, 149–173.

Tomkins, C. D. and Adrian, R. J. (2003) Spanwise structure and scale growth in turbulent boundary layers, *J. Fluid Mech.*, 490, 37–74.

Tsubaki, T., Hashimoto, H. and Suetsugu, T. (1982) Grain stresses and flow properties of debris flow, *Proc. JSCE*, 317, 79–90. (in Japanese)

Tsuji, Y. (2009) High-Reynolds-number experiments: the challenge of understanding universality in turbulence, *Fluid Dyn. Res.*, 41, 064003.

Tu, S. W. and Ramaprian, B. R. (1983) Fully developed periodic turbulent pipe flow. Part 1. Main experimental results and comparison with predictions, *J. Fluid Mech.*, 137, 31–58.

Tuna, B. A., Tinar, E. and Rockwell, D. (2013) Shallow flow past a cavity: globally coupled oscillations as a function of depth, *Exp. Fluids*, 54(8), 1586–1606.
Turney, D. E. and Banerjee, S. (2013) Air-water gas transfer and near-surface motions, *J. Fluid Mech.*, 33, 588–624.
Uijttewaal, W. S. J., Lehmann, D. and Van Mazijk, A. (2001) Exchange processes between a river and its groyne fields: Model experiments, *J. Hydraulic Eng.*, 127(11), 928–936.
Van Proojen, B. C. and Uijttewaal, W. S. J. (2002) A linear approach for the evolution of coherent structures in shallow mixing layers, *Phys. Fluids*, 14(12), 4105–4114.
Vedernikov, V. V. (1945) Conditions at the front of a translation wave disturbing a steady motion of real fluid, *Dokl. Acad. Sci. USSR*, 48, 239–242.
Vieira, E. D. R. and Mansur, S. S. (2012) Hot film calibration system using free jet in water, *Proc. 14th Brazilian Congress of Thermal Sciences and Engineering*, Rio de Janeiro, RJ, Brazil. https://www.researchgate.net/publication/235919835
Voulgaris, G. and Trowbridge J. (1998) Evaluation of the acoustic Doppler velocimeter ADV for turbulence measurements, *J. Atmos. Ocean. Technol.*, 151, 272–288.
Waleffe, F. (1997) On a self-sustaining process in shear flows. *Phys. Fluids*, 9(4), 883–900.
Wang, Z. Q. and Cheng, N. S. (2006) Time-mean structure of secondary flows in open channel with longitudinal bed forms. *Adv. Water Resour.*, 29(11), 1634–1649.
Warnack, D. and Fernholz, H. H. (1998) The effects of a favourable pressure gradient and of the Reynolds number on an incompressible axisymmetric turbulent boundary layer. Part 2. *J. Fluid Mech.*, 359, 357–381.
Watral, Z., Jakubowski, J and Michalski, A. (2015) Electromagnetic flowmeters for open channels: Current state and development prospects, *Flow Meas. Instrum.*, 42, 16–25.
Webel, G. and Schatzmann, M. (1984) Transverse mixing in open channel flow, *J. Hydraulic Eng.*, 110(4), 423–435.
Weitbrecht, V., Socolofsky, S. A. and Jirka, G. (2008) Experiments on mass exchange between groin fields and main stream in rivers. *J. Hydraulic Eng.*, 134(2), 173–183.
Weitbrecht, V., Seol, D. G., Negretti, E. et al. (2011) PIV measurements in environmental flows: Recent experiences at the institute for hydromechanics in Karlsruhe, *J. Hydro-Environ. Res.*, 5(4), 231–245.
Wells, M. R. and Stock, D. E. (1983) The effects of crossing trajectories on the dispersion of particles in a turbulent flow. *J. Fluid Mech.*, 136, 31–62.
White, F. M. (1991) *Viscous Fluid Flow*, 3rd Ed., McGraw-Hill, New York.
Whiting, P. J. and Dietrich, W. E. (1993) Experimental studies of bed topography and flow patterns in large-amplitude meanders, 1. observations, *Water Resour. Res.*, 29, 3605–3614.
Willert, C. E. and Gharib, M. (1990) Digital particle image velocimetry, *Fluid Measurement and Instrumentation Forum*, Toronto, ASME FED, 95, 39–44.
Williamson, C. H. K. (1992) The natural and forced formation of spot-like "Vortex Dislocations" in the transition of a wake, *J. Fluid Mech.*, 243, 393–441.
Wilson, C. A. M. E., Stoesser, T., Bates, P. D. and Batemann Pinzen, A. (2003) Open channel flow through different forms of submerged flexible vegetation, *J. Hydraulic Eng.*, 129, 847–853.

Wolfinger, M., Ozen, C. A. and Rockwell, D. (2012) Shallow flow past a cavity: Coupling with a standing gravity wave, *Phys. Fluids*, 24(10), 104103.

Wormleaton, P. R. and Marriot, M. J. (2007) An experimental and numerical study of the effects of width of meandering main channel on overbank flows, *Proc. 32th IAHR Congress*, Venice, Paper 090 (CD-ROM).

Wormleaton, P. R., Hey, R. D., Sellin, R. H. J., Bryant, T., Loveless, J. and Catmur, S. E. (2005) Behavior of meandering overbank channels with graded sand beds, *J. Hydraulic Eng.*, 131, 665–681.

Yalin, M. S. (1964) Geometrical properties of sand waves, *Proc. ASCE*, 90(HY5), 105–119.

Yang, S.-Q., Tan, S. K. and Wang, X.-K. (2012) Mechanism of secondary currents in open channel flows, *J. Geophys. Res.*, 117, F04014.

Yang, S. Q. and Lim, S. Y. (1997) Mechanism of energy transportation and turbulent flow in a 3D channel, *J. Hydraulic Eng.*, 123(8), 684–692.

Yang, S. Q. and McCorquadale, A. J. (2004) Determination of boundary shear stress and Reynolds shear stress in smooth rectangular channel flows, *J. Hydraul. Eng.*, 130(5), 458–462.

Yang, Z. and Shih, T. H. (1993) New time scale based k–ε model for near-wall turbulence. *AIAA J.*, 31, 1191–1198.

Yossef, M. F. M. and De Vriend, H. (2010) Sediment exchange be- tween a river and its groyne fields: mobile-bed experiment. *J. Hydraulic Eng.*, 136(9), 610–625.

Yossef, M. F. M. and De Vriend, H. (2011) Flow details near river groynes: Experimental investigation, *J. Hydraulic Eng.*, 137(5), 504–516.

Yung, B. P. K., Merry, H. and Bott, T. R. (1988) The role of turbulent bursts in particle re-entrainment in aqueous systems, *Chem. Eng. Sci.*, 44, 873–882.

Zhou, J., Adrian, R. J., Balachandar, S. and Kendall, T. M. (1999) Vortex organization in the outer region of the turbulent boundary layer, *J. Fluid Mech.*, 387, 353–396.

Zhuang, Y., Wilson, J. D. and Lozowski, E. P. (1989) A trajectory-simulation model for heavy particle motion in turbulent flow. *J. Fluids Eng.*, 111, 492–494.

Index

accelerated flow, 209–214
acoustic Doppler current profiler (ADCP), 239
acoustic Doppler velocimetry (ADV), 142, 167, 180, 225, 230–232
acoustic mapping velocimetry (AMV), 241
aliasing, 224, 231
anti-dune, 178–179
aspect ratio, 117
averaging procedures, 11–13

bed load, 179
Blasius's solution, 39–40, 261–267
boil, 82–83
bottom roughness effects, 88–92
bottom shear generation, 144
bottom shear stress, 202
boundary layer, 3, 29, 33, 50–51
boundary layer equations, 36–40
boundary layer theory, 29–42
boundary layer thickness, 29, 33–36
Bragg cell, 234
bridge peer wake, 99
buffer layer, 59
bursting, 78–80, 86
bursting period, 80

cavity, 100, 164
cascade process, 7, 74–75
channel transition, 126–127
coherent structures, 77–83
compound channel, 99, 126, 152
compound meandering flow, 125–127
compound open-channel flow, 152–164, 205–209
confluence zone, 99
consecutive groynes, 166–168
continuity equation, 14, 36, 108
convection term, 17, 25, 74, 258, 260
correlation coefficient, 66–67
critical bed shear stress, 175
critical grain Reynolds stress, 175
cross correlation, 67
cross-correlation function, 69
cross-correlation method, 236
crossing-trajectories effect, 185
curved channel, 121–122

deadwater zone, 164–174
debris flow, 186–199
depth averaged continuity equation, 108
depth averaged momentum equations, 106
developing distance, 250
diffusibility, 6
diffusion coefficient, 129
diffusion in turbulence, 8–10
diffusion terms, 63, 65
diffusion time, 134, 136
dimensionless critical bed shear stress, 175
dispersion, 136–138
dispersion coefficient, 138
displacement thickness, 34–35, 145
dissipation, 6
Doppler burst signal, 233
Doppler noise, 231
Doppler shift, 229
drag coefficient, 144
drag force, 174
drone-type float, 237–238
dual-layer PIV, 158
dune, 178

eddy-dependent trajectories, 136
eddy diffusion coefficient, 9
eddy viscosity, 56

eddy viscosity model, 55–57
Einstein notation, 14, 260
ejection, 78–80, 141
electromagnetic velocimetry (EMV), 225, 227–229
embayment, 164, 166, 168
emergent vegetation, 142–144
energy dissipation, 214–222
energy equilibrium, 64–66
energy in the mean current, 64
ensemble average, 13
entrainment, 78, 182
ergodicity, 13
erosion, 168, 175
Euler equations, 16–19
Euler specification, 16–17
equivalent sand roughness, 88
experimental flume, 251–253
experimental procedure, 250–251

Fick's law, 129
field survey, 237
flexible vegetation, 149–151
floating method, 237
floodplain, 152
flood protection forest, 217–218
free-stream layer, 38, 50
free surface, 3
free-surface effects, 83–88
free-surface deformation, 87
frequency spectrum, 73–74
friction force, 174
friction velocity, 57, 200
Froude number, 96, 124, 178–179, 216–217, 222

generation of turbulence, 7–8
growth of disturbance, 46
groyne, 100

hairpin vortex, 77, 80–82
Hjulstrom curve, 175–176
hole size, 79
hole value, 245
horizontal shear instability, 93–96
horizontal vortex, 96–100, 152, 158–164
hot film velocimetry, 232
hydraulic jump, 216–217

inertial force, 2–3, 5
inertial subrange, 75–76
inflection instability, 152
integral scale, 69
interrogation region, 236
inner layer, 58–59
integral constant, 59
inundation on a floodplain, 207–209
inverse cascade, 76–77
irregularity, 5
isolated groyne, 166–167

$k - \varepsilon$ model, 56
Karman's integral equation of momentum, 40–42
Kelvin Helmholtz (K-H) instability, 96–98, 269–273
K-H waves, 142
kinematic boundary condition, 105–106
kinetic viscosity, 10, 25–26, 175
King's law, 232
Kolmogorov number, 180
Kolmogorov scales, 10–11
Kolmogorov's -5/3 power law, 74–76

Lagrange correlation function, 132
Lagrangian specification, 16, 131–132
Lagrangian tracking of horizontal vortex, 164
laminar boundary layer (LBL), 33, 42–44
laminar flow, 3, 5
large eddy simulation (LES), 153
large-scale motion (LSM), 118–121
large-scale PIV (LS-PIV), 118, 240–241
laser Doppler anemometer (LDA), 11, 58, 147, 153, 225, 232–234
laser-light sheet (LLS), 161, 234
law of the wall, 57–58
Leibniz rule, 104–105
lift force, 174
local isotropy, 75–76
log law, 59–60, 203
log law layer, 59
logarithmic zone, 145
loop property, 199, 202
low-speed streak, 182–183

main-channel, 152
mass and momentum exchanges, 152
mean velocity profile, 89–90, 151
meandering compound channel, 125
meandering flow, 124
microwave current meter, 239–240
minimum vortex, 10–11
mixing layer zone, 145

mixing length, 54–55, 58, 195
mixing length model, 54–55, 196
molecular diffusion, 8–9, 131
momentum equation, 19, 26
momentum thickness, 35–36
monami, 150
multi-scales, 7

Navier Stokes (N-S) equations, 19–25, 29
Newton's viscosity law, 20, 25, 55
Nezu-Rodi curve, 59
Nyquist frequency, 224

one-particle analysis, 131–134
open-channel flow, 1, 3
Orr-Sommerfeld equation, 46, 48–49

packet structure, 81
particle image velocimetry (PIV), 66, 100, 153, 167, 173, 225, 234–236
particle tracking analysis, 131–136
particle tracking velocimetry (PTV), 234
Peclet number, 138–139
Pitot tube, 226
potential core, 219–220
1/7 power law, 44
Prandtl number, 10
Prandtl's first kind of secondary currents, 111–112
Prandtl's second kind of secondary currents, 112–113
pressure gradient parameter, 210–212
pressure strain term, 62
price-type current meter, 237, 239
primary gyre, 169, 172
propeller current meter, 225–227
proper orthogonal decomposition (POD), 246–247

quadrant analysis, 79, 244–245

Rayleigh problem, 30–33
Rayleigh's instability theory, 49–50
reaction force, 174
recirculation flume, 248–249
redistribution, 62, 64
relaminarization, 212–213
Reynolds-averaged Navier-Stokes (RANS) equations, 26–27, 156, 259
Reynolds averaging, 255–256
Reynolds decomposition, 5, 26, 40, 130, 255
Reynolds experiment, 3
Reynolds number, 2, 5, 10, 37, 103, 138, 185
Reynolds stress, 27, 41, 51–55, 61–62, 113, 151, 170, 173–174, 181, 195, 210, 244
Richardson's 4/3 power law, 135
ripple, 178
river flow, 1
river management, 2
rolling/sliding, 176
rough flow regime, 90–91
roughness height, 145, 193
Rouse equation, 180–182

saltation, 177
sampling theorem, 223–224
sand dune, 82
sand ribbon, 83
search region, 236
secondary cells, 117–118
secondary currents, 111–117, 124, 126, 205, 208
secondary gyre, 169, 172
sediment entrainment, 182
sediment transport, 174–186
seiche, 173
shear length scale, 149
shear instability, 43, 125, 141, 145, 150, 151, 152, 166, 173
shear stress, 52
Shields diagram, 175
Shiono and Knight method (SKM), 158
side cavity, 164, 166
side walls, 100
single-layer debris flow, 189–193
skewness, 91–92
space average, 13, 147
spatial averaged velocity, 13
spectrum, 69–73
spillway, 215
spread width, 131
still basin, 215
Stokes number, 186
streaks, 80–81
Strouhal number, 173
structure function, 76–77
submergent vegetation, 145–149
surface renewal, 3, 86
surface velocity divergence, 100–104
suspended sediment, 179

suspension, 177
sweep, 78–80, 141

Taylor's frozen turbulence hypothesis, 74, 119
thermal diffusivity, 9
99% thickness, 34
three dimensionality, 6–7
time average, 11–12
time-averaged energy, 72
time-averaged velocity, 11, 201
time-mean velocity profile, 57–60
Tollmien-Schlichting (TS)waves, 42, 80–81
transport equation of vorticity, 113
turbulence, 2, 5
turbulence generation, 42–50
turbulence generation term, 62
turbulence intensity, 60–62
turbulence transport, 129–139
turbulent boundary layer (TBL), 42–46
turbulent diffusion, 9–10, 131
turbulent diffusion coefficient, 131–132, 137
turbulent diffusion equation, 129–131
turbulent energy dissipation rate, 3, 10, 56, 62, 144, 180
turbulent flow, 3, 5
turbulent kinetic energy, 56, 61–62, 63, 144, 204, 213
turbulent Schmidt number, 180
two-dimensional parallel jet, 220
two-layer debris flow, 194–198
two-particle analysis, 134–136
two-stage flow, 152

ultrasonic velocity profiler (UVP), 225, 229
unidirectional bed forms, 178–179
unmanned surface vehicle (USV), 242–244
unsteadiness parameter, 200
unsteady turbulence flow, 12
unsteady turbulent open channel, 199–209

van der Waals force, 174
variable-interval time-averaging (VITA), 245–246
vegetation, 100, 141–152
vegetation allocation density, 144
vegetation density, 147
velocimetry, 224–244
velocity dip, 117, 155
velocity shear, 8, 65, 86
very large-scale motion (VLSM), 118–121
virtual zero level, 89
viscosity, 10, 25–26
viscous fluid, 29
viscous force, 2–3, 5
viscous friction, 30
viscous normal stress, 21–23
viscous shear stress, 19–21, 51
viscous sublayer, 58, 200
von Karman constant, 58, 145, 186, 193, 199–200
Λ-vortex, 80
vortex detection, 247–248
vorticity, 113, 161

wake function, 59
wake zone, 145
wall reflection, 232
water waves, 86
wave number spectrum, 73–74
Wiener Khinchin theorem, 73

For Product Safety Concerns and Information please contact our EU representative GPSR@taylorandfrancis.com
Taylor & Francis Verlag GmbH, Kaufingerstraße 24, 80331 München, Germany

www.ingramcontent.com/pod-product-compliance
Ingram Content Group UK Ltd.
Pitfield, Milton Keynes, MK11 3LW, UK
UKHW021117180425
457613UK00005B/127

* 9 7 8 0 3 6 7 6 3 0 9 6 6 *